The Acadian Orogeny:
Recent Studies in New England, Maritime Canada, and the Autochthonous Foreland

Edited by

David C. Roy
and
James W. Skehan

Department of Geology and Geophysics
Boston College
Chestnut Hill, Massachusetts 02167

1993

Published by The Geological Society of America, Inc.
3300 Penrose Place, P.O. Box 9140, Boulder, Colorado 80301

Printed in U.S.A.

GSA Books Science Editor Richard A. Hoppin

Library of Congress Cataloging-in-Publication Data
The Acadian Orogeny : recent studies in New England, maritime Canada,
    and the autochthonous foreland / edited by David C. Roy and James W.
    Skehan.
        p.   cm. — (Special paper / Geological Society of America ;
    275)
    Based on papers from a symposium held at the annual meeting of the
    Northeastern Section of the Geological Society of America.
    Includes bibliographical references and index.
    ISBN 0-8137-2275-6
    1. Geology, Stratigraphic—Paleozoic—Congresses.   2. Geology,
Structural—Appalachian Mountains—Congresses.   3. Geology,
Structural—New England—Congresses.   4. Geology, Structural-
-Maritime Provinces—Congresses.   I. Roy, David C.   II. Skehan,
James William, 1923–   .   III. Series: Special papers (Geological
Society of America) ; 275.
QE654.A28   1993
551.7'2'0974—dc20   92-43325
                                            CIP

Cover photo: Acadian anticline in phyllite and calcareous
siltstone of the Silurian Allsbury Formation in an outcrop at mile
262 on northbound Interstate 95 south of Island Falls, Maine.
This regional F2 anticline plunges northward (into the picture).
The face of the outcrop is controlled by the axial planar S2
foliation. A weakly developed S1 foliation is subparallel to bedding and the face displaying the anticlinal hinge is parallel to S3
foliation associated with steeply plunging S3 folds. Photograph
by D. C. Roy.

10   9   8   7   6   5   4   3   2   1

Contents

Preface v

1. *Mid-Paleozoic Orogenesis in the North Atlantic: The Acadian Orogeny* 1
   N. Rast and J. W. Skehan

2. *Paleogeography, Accretionary History, and Tectonic Scenario:*
   A Working Hypothesis for the Ordovician and Silurian Evolution
   of the Northern Appalachians 27
   B. A. van der Pluijm, R.J.E. Johnson, and R. Van der Voo

3. *Comments on Cambrian-to-Carboniferous Biogeography and Its Implications*
   for the Acadian Orogeny 41
   A. J. Boucot

4. *The Sequence of Acadian Deformations in Central New Hampshire* 51
   J. D. Eusden, Jr., and J. B. Lyons

5. *Nature of the Acadian Orogeny in Eastern Maine* 67
   A. Ludman, J. T. Hopeck, and P. C. Brock

6. *Acadian Deformations in the Southwestern Quebec Appalachians* 85
   P. A. Cousineau and A. Tremblay

7. *Timing of the Deformation Events from Late Ordovician to Mid-Devonian*
   in the Gaspé Peninsula 101
   M. Malo and P.-A. Bourque

8. *Acadian Orogeny in Newfoundland* 123
   H. Williams

9. *Acadian Orogeny in West Newfoundland: Definition, Character,*
   and Significance 135
   P. A. Cawood

10. *Stratigraphic Effects of the Acadian Orogeny in the Autochthonous*
    Appalachian Basin 153
    T. Hamilton-Smith

Index 165

Preface

Over the past several decades, extensively studied mountain ranges such as the Appalachians have undergone systematic restudy, stimulated by the paradigm of plate tectonics. The new efforts have been varied, from large-scale remapping to detailed studies in petrology, structural geology, geochemistry, geochronology, paleomagnetism, and many other specialized fields. These studies have led to much-revised interpretations of the tectonic histories of these ranges; these new interpretations show the importance of plate tectonics on the broad scale but cause much debate about the details. The principal difficulty in interpreting the tectonic histories of these mountain belts is they have undergone multiple deformations that have varied considerably in timing, location within the belt, and intensity. Unraveling the effects of these deformations and relating them to the movements of plate tectonic elements (blocks, terranes, etc.) is the present challenge for those working in these complex mountain systems.

For much of the northern Appalachians in New England and adjacent Canada, the Acadian orogeny is the most widespread and appears to be the youngest orogenic event. It is usually found to be responsible for most of the folds, foliations, faults, and plutons within the region. In places, older and younger orogeneses obscure the effects of the Acadian. Along the western margin of the range, for example, the effects of the older Taconian orogeny are prominent and must be separated eastward from those of the Acadian in the more metamorphic interior of the range. The effects of the younger Alleghanian orogeny overprint the Acadian principally along the eastern margin of the range.

The earliest efforts to use a plate tectonic analysis on the northern Appalachians were by Bird and Dewey (1970). They interpreted the Taconian orogeny along the western margin of the range to be the result of a collision between a volcanic arc and the Ordovician margin of North America. Their model has since been refined and extended by others, notably by St. Julien and Hubert (1975) for the Quebec Appalachians, by Stanley and Ratcliffe (1985) and Bradley and Kusky (1986) for western New England and eastern New York, and by Coleman-Sadd (1982) and others for Newfoundland. It is generally accepted that the Taconian orogeny was an ocean-closing event because there is a well-defined belt of ophiolitic rocks that locates an oceanic suture; there is also a well-developed foreland thrust belt.

The Acadian orogeny, on the other hand, is considerably more intractable to model because there are no recognized ophiolitic sutures to mark the presence of closed ocean basins, and the western foreland has been difficult to differentiate from that of the Taconian. A comprehensive overview of the entire Acadian orogen is provided by Osberg and others (1989). Thoughtful regional summaries and plate tectonic models of the Acadian in New England and nearby Canada have been provided by Bradley (1983) and Berry and Osberg (1989). These models presume an Acadian closure of a post-Taconian ocean basin between the Bronson Hill–Boundary Mountain terrane (the accreted Taconian "arc") and the complex coastal terranes to the east. In these models, ophiolites indicating the existence and location of the ocean basin have been obscured by burial beneath thrusts during late Acadian compression. The absence of ophiolitic sutures that can be associated with the Acadian indicates that detailed stratigraphic, structural, metamorphic, plutonic, and geochronologic studies are keys to unraveling its deformational history. The critical "boundaries" within the deformed zone at the surface may be difficult to find and may be represented by very late structural features in "cover rocks." The search for these critical boundaries within the range has been narrowed by transverse geophysical profiles and by paleomagnetic, stratigraphic, and paleontologic studies that estimate distances of separation of rock units as a function of time.

In this volume are summaries of recent structural and stratigraphic studies in the "heart" of the Acadian orogen in central New Hampshire (Eusden and Lyons), eastern Maine (Ludman, Hopeck, and Brock) and Newfoundland (Williams). Also presented is recent work on large portions of the western structural "foreland" of southeastern Quebec (Cousineau and Tremblay), the Gaspé Peninsula (Malo and Bourque), western Newfoundland

(Cawood), and the autochthonous foreland of the Appalachian Basin well west of the range (Hamilton-Smith). The volume begins with a summary of the Acadian and its relations to orogenic events in Europe (Rast and Skehan); an examination of the paleomagnetic constraints on the timing of terrane accretions (van der Pluijm, Johnson, and Van der Voo); and comments on the limits placed on terrane separations by the biogeographic provincialism of fauna in the New England and Canadian Appalachians (Boucot). These studies show both the complexities of the problems faced in interpreting the history of a deeply eroded mid-Paleozoic mountain range, and the new interpretations that are emerging.

This volume grew out of a symposium held at an annual meeting of the Northeastern Section of the Society. The editors wish to thank the authors for their substantial efforts and for their patience during the "evolution" of the volume. In addition, numerous individuals assisted the authors with careful and detailed reviews. Reviewers included: Dwight C. Bradley, Wallace A. Bothner, Rodger T. Faill, Leslie Fyffe, James Hibbard, Mark Loiselle, J. Brendan Murphy, Richard S. Naylor, Brian H. O'Brien, A. R. Palmer, Rolfe S. Stanley, Peter J. Thompson, Walter E. Trzcienski Jr., and four anonymous reviewers.

David C. Roy
James W. Skehan, S. J.

REFERENCES CITED

Berry, H. N., IV, and Osberg, P. H., 1989, Stratigraphy of eastern Maine and western New Brunswick: Maine Geological Survey, Studies in Maine Geology, v. 2, p. 1–32.

Bird, J. M., and Dewey, J. F., 1970, Lithosphere plate-continental margin tectonics and the evolution of the Appalachian orogen: Geological Society of America Bulletin, v. 81, p. 1031–1059.

Bradley, D. C., 1983, Tectonics of the Acadian Orogeny in New England and adjacent Canada: Journal of Geology, v. 91, p. 381–400.

Bradley, D. C., and Kusky, T. M., 1986, Geologic evidence for rate of plate convergence during the Taconic arc-continent collision: Journal of Geology, v. 94, p. 667–681.

Coleman-Sadd, S. P., 1982, Two stage continental collision and plate driving forces: Tectonophysics, v. 90, p. 263–282.

Osberg, P. H., and others, 1989, The Acadian orogen, *in* Hatcher, R. D., Jr., Thomas, W. A., and Viele, G. W., eds., The Appalachian-Ouachita Orogen in the United States: Geological Society of America, The Geology of North America, v. F-2, p. 179–232.

St. Julien, P., and Hubert, C., 1975, Evolution of the Taconian orogen in the Quebec Appalachians: American Journal of Science, v. 275A, p. 337–362.

Stanley, R. S., and Ratcliffe, N. M., 1985, Tectonic synthesis of the Taconian orogeny in western New England: Geological Society of America Bulletin, v. 96, p. 1227–1250.

Geological Society of America
Special Paper 275
1993

Mid-Paleozoic orogenesis in the North Atlantic:
The Acadian orogeny

Nicholas Rast
Department of Geological Sciences, University of Kentucky, Lexington, Kentucky 40506-0059
James W. Skehan
Weston Observatory, Department of Geology and Geophysics, Boston College, Weston, Massachusetts 02193

ABSTRACT

The Acadian orogeny in the North Atlantic region is assessed in this chapter in the light of mid-Paleozoic tectonics; throughout, plate tectonic nomenclature is used, and cycles are avoided. In North America nine regions bearing the imprint of the Acadian orogeny are recognized.

In Newfoundland, in the Maritime Provinces of Canada, and in Vermont and New Hampshire a continuous sequence of lithotectonic belts correlates along the orogen. The Bronson Hill belt, although a continuous structure in southern New England, is not recognized as such but splits into two structures northeast of the Maine–New Hampshire border: the Boundary Mountain anticlinorium and the Lobster Mountain anticlinorium. Other lithotectonic belts are partly continuous from Canada into the United States; they include: (1) North-Central Maine belt, (2) Aroostook-Matapedia belt, (3) Miramichi belt, (4) Fredericton–Central Maine belt, (5) Richmond belt, (6) Casco Bay belt, (7) Benner Hill belt, (8) St. Croix–Ellsworth belt, (9) Mascarene belt, and (10) Avalon belt. The decision as to whether each of these belts represents a separate terrane is at present reserved. In the coastal Maine zone the situation is particularly complex, and belts 6 through 10 can be recognized there.

In Massachusetts, we interpret the Merrimack Trough belt as in fault contact with both the Kearsarge–Central Maine and Bronson Hill belts to the northwest, and in Connecticut, with the Bronson Hill belt alone. Additionally, the Merrimack Trough belt is in fault contact with the Putnam-Nashoba belt to the southeast. The latter shows mainly a Taconian metamorphism and extensive intrusion of granites; clear evidence for Acadian orogenic effects in the Putnam-Nashoba belt is lacking.

In Newfoundland the main orogeny appears to be Silurian in age, and the same is true of New Brunswick, whereas in the Meguma of Nova Scotia the Devonian deformation and intrusive activity continue from the Devonian to the Carboniferous.

Correlations with the south-central Appalachians indicate a possibility of significant Acadian transpressional effects. The most recent evidence of a new microfossil find, however, implies that considerable Acadian deformation occurred in the Southern Appalachians, although it may have been directly continuous with earlier Taconian events.

The Acadian metamorphism in the Northern Appalachians is associated with numerous granites, in general ranging in age from the Silurian to the Carboniferous. The

Rast, N., and Skehan, J. W., 1993, Mid-Paleozoic orogenesis in the North Atlantic: The Acadian orogeny, *in* Roy, D. C., and Skehan, J. W., eds., The Acadian Orogeny: Recent Studies in New England, Maritime Canada, and the Autochthonous Foreland: Boulder, Colorado, Geological Society of America Special Paper 275.

earlier Silurian granites may have originated along the Iapetus suture or may be associated with transcurrent faults.

The plate tectonic interpretation of the orogenic system is based on a model of successive blocks (terranes) approaching and colliding with North America and squeezing intervening sediments and volcanics. This took place over a fairly prolonged period of time.

INTRODUCTION

In this chapter we examine current research on the Acadian orogeny in the light of mid-Paleozoic tectonics of the North Atlantic region. We propose changes in terminology required for discussions of related orogenic events, and we reinterpret the regional relationships of deformation and metamorphism. In the contemporary literature the term *orogeny* is understood as a severe structural deformational event followed or accompanied by metamorphism and the intrusion of syntectonic to posttectonic granites. In plate tectonic theory an orogeny is usually explained by the collision of two continents or blocks or of an island arc with a continent. It is possible that the collision of a continent and midocean ridge may also result in an orogenic episode. Because a major plate-tectonic scenario may be of considerable complexity involving several types of deformational events, the word *cycle* sometimes has been applied to encompass the whole sequence of such events. *Cycle,* however, can be and often is confused with the so-called orogenic cycle employed in the geosynclinal theory, and we therefore avoid its use, preferring the term *tectonic regime* for a sequence of interrelated tectonic events in a comparatively short period of time.

The term *Acadian* was introduced to the geological literature by H. S. Williams as the "Acadian revolution" in 1895 (p. 42). Elsewhere, in Scandinavia and the British Isles, for instance, Suess (1908) introduced the term *Caledonian orogeny.* The formal relationship between the two was accepted by Schuchert (1930), despite the apparent difference in stratigraphic timing. The Acadian orogeny was then considered Devonian or even post-Devonian in age, whereas the Caledonian orogeny has been considered late Silurian or even end-Silurian. Later, when the concepts of orogenic phases and cyclicity became fashionable, it was recognized that the period of Acadian deformation was prolonged (see Naylor, 1971, for a contrary view) and the term *revolution* was superseded by *orogeny.* In any case, in the North Atlantic region many areas of Silurian-Devonian (mid-Paleozoic) tectonism were detected (Fig. 1) and given different names.

In the northern Appalachians, Boucot (1968), on the basis of a mid-Devonian unconformity, suggested that the Acadian orogeny has a mid-Devonian, post-Eifelian age. On the other hand, Donohoe and Pajari (1974), on the basis of radiometric data, proposed a time-transgressive Acadian orogeny across the northern Appalachian orogenic belt of Maritime Canada. They referred the Acadian orogeny to the Geddinian in southern New Brunswick and Maine and the Emsian in northern New Brunswick and Quebec. In the central and southern Appalachians, an

undifferentiated Devonian age has been adduced to the Acadian orogeny, although some extend it into the Lower Carboniferous (Rodgers, 1970, p. 32). This suggestion has found its way into most of the official glossaries (cf. Dennis, 1967; Bates and Jackson, 1987).

Until the advent of plate tectonic theory, virtually all workers in the Appalachians divided the Late Proterozoic to end-Paleozoic tectonism into a series of episodes or pulses (cf. Lyons and Faul, 1968, p. 313) and assumed that Acadian orogenic movements were distinct and widespread (Table 1). Accordingly, in addition to the traditional Taconian and Acadian, other episodes such as the Penobscottian, or Penobscotian (Neuman, 1967; Skehan, 1988), of late Cambrian to early Ordovician age became widely recognized, and preceding the Acadian a late Silurian (Ludlovian) Salinic episode, based on the detection of a widespread unconformity, was defined by Boucot (1962). The Penobscottian and Salinic events were referred to as disturbances because they were not accompanied by a strong metamorphism and plutonic activity. The Salinic, in particular, was said by Hatch (1982) to be unaccompanied by penetrative cleavage, although a cleavage is recognized in association with the Penobscottian.

In Europe, until recently, the Acadian orogeny had been considered as an event later than the Caledonian, although de Sitter (1964, p. 336) expressed the view that it could be alternatively regarded as either a Caledonian or Variscan (late Paleozoic) pulse. More recently, it has been suggested that the Acadian is late Caledonian (Osberg, 1988) or indeed the Caledonian (McKerrow, 1988). In a different sense the term *late Caledonian* was applied to large-scale transcurrent faults affecting the Caledonian of Svalbard (Harland, 1985).

In Scandinavia the main mid-Paleozoic orogeny (Scandian) is dated as middle to late Silurian and a somewhat later (Devonian) period of tectonic rejuvenation as late Scandian (cf. Roberts, 1988; Robinson and others, 1988). In France (Brittany and Massif Central, Fig. 1) the main mid-Devonian episode is named Ligerian (Autran and Cogné, 1980), and its inferred continuation in the northeastern Iberian peninsula (Choukroune and others, 1990) has been referred to as Eohercynian (early Variscan). The continuation of the Ligerian belt into the mainland of Europe has been called the Auverno-Vosgian (Autran and Guillot, 1975). In Morocco, Piqué and Michard (1989) refer to a late Devonian deformation belt as the Eovariscides (cf. Skehan and Piqué, 1989). The development of the plate tectonics hypothesis and the enhanced precision of isotopic dating suggest that orogenic episodes are time transgressive. Thus long-

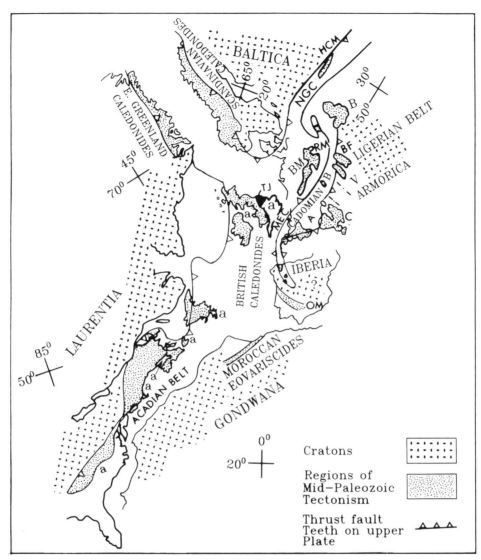

Figure 1. Regions of mid-Paleozoic tectonism in the North Atlantic. Source: Ziegler (1982), Williams (1984), Pickering and others (1988), Keppie and Dallmeyer (1989). Orogenic belts: MEC—Mid-European Caledonides; NGC—North German Caledonides. Massifs: A—Armorican; a—Avalonian; B—Bohemian; BF—Black Forest; BM—Brabant; C—Massif Central, of France; HCM—Holy Cross Mountains; OM—Ossa/Morena; RM—Rhenish; V—Vosges. Other abbreviations: TJ—Triple junction. Post-Carboniferous, pre-Triassic fit is used; oblique Mercator projection with pole at 18°S, 49°E.

distance correlations aimed at producing synchroneity of such episodes are unrealistic, and many of the aforementioned qualifying terms can be discarded.

The diversity of the suggested long-distance correlations, in our view, arises from attempts to fit all orogenic episodes into a straitjacket of pulsating orogenic cycles. As has been pointed out by Hsu (1989), this is not consistent with plate tectonic theory and should be avoided. We propose much more local designations of events even within a relatively small area such as the northern Appalachians, and a restriction of transatlantic correlations to regions that were demonstrably proximate during middle

to late Paleozoic times. Moreover, it is always necessary to state the basic evidence for such correlations: stratigraphy, deformation, metamorphism, and plutonic intrusions. Each of these criteria, sometimes independently of each other, has been used to designate orogenic episodes. When several criteria are used, the resulting timing is rarely concordant, because isotopic time scales, isotopic systems, and methods of analysis vary. Thus in this reappraisal of the Acadian and other mid-Paleozoic orogenies, all the specific quoted dates (Table 2) have been tabulated according to the radiometric methods employed.

Because the term *Acadian* principally refers to the deforma-

**TABLE 1. TIME AND OROGENIC EVENTS
IN THE NORTHERN APALACHIANS***

Time		Orogenic Events in Northern Apalachians
Permian	L	
	E	
		Alleghanian
Carboniferous	L	
	E	
Devonian	L	
	M-E	
Silurian	L	Acadian Caledonian-Ligerian
	E	Salinic Scandian
Ordovician	L	Taconian
	M	
	E	Penobscottian

*Adopted partly from Lyons and Faul (1968).

tion, metamorphism, and plutonism of rocks in the northern Appalachians, we begin in this area. For the purpose of this chapter we divide this region, on the basis of lithotectonic characteristics, into the following areas (Fig. 2): 1—northern New Brunswick, the Gaspé of Quebec, and parts of Maine; 2—Newfoundland; 3—southern New Brunswick, northern Nova Scotia, and coastal Maine; 4—southern Nova Scotia; 5—western New England and parts of Quebec; and 6—Boston block. In the central and southern Appalachians, three areas can be isolated: 7—New York, New Jersey, and Pennsylvania; 8—south-central Appalachians; and 9—southern Appalachians. In this chapter we mainly confine our discussion of the Acadian to the northern Appalachians since the relevant data are abundant. We will treat the Acadian in the southern Appalachians only in passing, as the published evidence at present does not allow a comprehensive treatment.

THE ACADIAN OF GREATER ACADIA

Schuchert (1930) associated the Acadian orogeny with Greater Acadia, which according to him included not only the

**TABLE 2. STRATIGRAPHIC AND CRITICAL DATA ON RADIOMETRIC DATING
OF THE ACADIAN AND RELATED OROGENIES IN THE NORTH ATLANTIC REGION**

	Stratigraphic Time	Corresponding Isotopic Time (Ma)	Dates and Sources*
Carb.	Tournasian		
		335	
	Fammenian		361 ± 6 (K-Ar), deformed edge of pluton (doming event), Ashwal and others (1979) Belchertown pluton in New Hampshire
	Frasnian		
		375	
	Givetian		ca. 380, Pin and Peucat (1986), Ligerian anatexis of North France 382.2–380.0, Pb$^{207/206}$ zircon, undeformed core of pluton
Devonian	Eifelian		Boucot (1962, S) - New York State, S
	Emsian		McKerrow (1988, S), Devon to Newfoundland, some I data
		390	
	Pragian		392 ± 3, monazite in granulite gneiss, Bristol, NH, Chamberlain and Rumble (1988) Zircon overgrowth in quartz-graphite veins yields U-Pb of 409 ± 6 Ma (I)
	Lochkovian		Boucot (1962, S), Nova Scotia
		410	
	Pridoli		415 ± 2, abraded zircon (I, U/Pb)
	Ludlow		
		424	Dunning and others (1990), Burgeo intrusive suite, south Newfoundland, ≈ Salinic? (I)
Silurian	Wenlock		
		428	
	Llandovery		429 +5/-3, abraded zircon (I, U/Pb) ca. 430, Pin and Peucat (1986), U/Pb on zircon in the Armorican and Central massifs of France

*S = Stratigraphic; I = Isotopic.

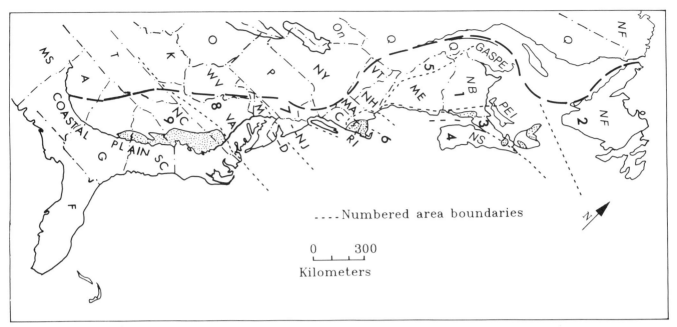

Figure 2. Distribution of areas of Acadian tectonism in the Appalachian Orogenic Belt. Area boundaries indicated by dashed lines: 1—northern New Brunswick, the Gaspé of Quebec, and parts of Maine; 2—Newfoundland; 3—southern New Brunswick, northern Nova Scotia, and southern Maine; 4—southern Nova Scotia (Meguma); 5—western New England and parts of Quebec; 6—Boston block (southeastern New England); 7—New York, New Jersey, and Pennsylvania (north-central Appalachians); 8—Virginia, south-central Appalachians; 9—southern Appalachians. Canadian provinces: NB—New Brunswick; NF—Newfoundland; NS—Nova Scotia; ON—Ontario; PEI—Prince Edward Island; Q—Quebec. American states: A—Alabama; C—Connecticut; D—Delaware; F—Florida; G—Georgia; K—Kentucky; M—Maryland; MA—Massachusetts; ME—Maine; MS—Mississippi; NC—North Carolina; NH—New Hampshire; NJ—New Jersey; NY—New York; O—Ohio; P—Pennsylvania; P—Pennsylvania; RI—Rhode Island; SC—South Carolina; T—Tennessee; VA—Virginia; VT—Vermont; WV—West Virginia. Outcrops of Avalon blocks are stippled. Broken line is Appalachian orogenic front, which is of several different ages along its extent, ranging from Ordovician to Permian-Carboniferous. To the northwest of the front lies Laurentian craton (mainly in Canada) and the North American platform (mainly in the United States).

Gaspé, New Brunswick, and Nova Scotia but also New England and parts of New York, or areas 1, 3, 4, and much of 5 of our subdivisions (Fig. 2). The northern Appalachians in this region are characterized by thick Silurian-Devonian sequences of strata that commonly rest unconformably on the older basement of Proterozoic and lower Paleozoic rocks.

In area 1 the Silurian-Devonian succession in general is folded into upright folds, in places with axial planes transecting steep cleavage (Stringer, 1974, 1975). These deformed rocks are intruded by mainly posttectonic granites. In many parts of the area an unconformity exists between Ordovician and Silurian strata (Rast and Stringer, 1974). In northwest New Brunswick (Fig. 3), Devonian strata generally rest paraconformably on upper Silurian sequences. A similar situation exists in the Gaspé peninsula (Beland, 1974). Yet in between, at the border of Quebec and New Brunswick (Fig. 3), the situation is more complex. A significant paleontological break appears to exist within the so-called Honorat Formation (Riva and Malo, 1988). Here three tectonostratigraphic units consisting of Ordovician to Devonian

strata are separated by large strike-slip east-west trending faults with ≈150-km dextral sense of displacement. The faults are situated within strongly deformed shear zones (Malo and Beland, 1989). The regional cleavage referred to by these authors as S2 transects the earlier oblique folds, and trends northeast-southwest. This cleavage underwent some rotation during the transcurrent deformational event, according to Malo and Beland (1989).

In Quebec to the north of the New Brunswick–Quebec transcurrent fault zone (Fig. 3) lie the rocks of the Gaspé Peninsula, considered as the extension of Connecticut Valley–Gaspé synclinorium of Cady (1960), or of the Connecticut Valley trough of Hatch (1988). To avoid genetic connotations we propose to call this package of rocks the Connecticut Valley–Gaspé belt. This continuous, slightly curved belt lies parallel to and east of the Taconian orogen (Figs. 3 and 4) that also contains inliers of Grenville basement.

In Newfoundland (area 2, Fig. 2; Fig. 4) both in the Humber and particularly the western part of the Dunnage Zone (Notre Dame subzone), the Silurian-Devonian strata form a more or less

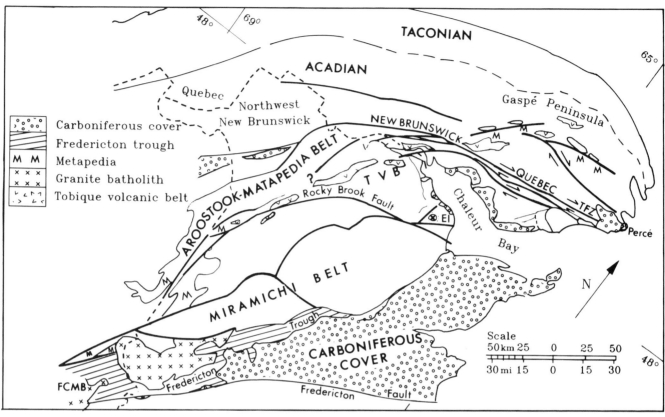

Figure 3. General geological outlines of northern New Brunswick and southern Gaspé, showing divisions (belts) of the Acadian orogen and approximate boundary of the Acadian and Taconian orogens. Sources: Fyffe (1982), Ludman (1978), Malo and Beland (1989), Keppie (1982), Potter and others (1968). Abbreviations: EI—Elmtree Inlier (xmarks Elmtree Granite); FCMB—Fredericton—central Maine belt; M—rocks of Aroostook-Matapedia belts; TFZ—Transcurrent fault zone; TVB—Tobique volcanic belt.

continuous belt that is equivalent to the Connecticut Valley–Gaspé belt (Fig. 4). The belt in the United States is characterized by mainly marine sediments, but in the upper (Devonian) part by subaerial sediments, the latter containing, both in Vermont and Quebec, plant remains (Hueber and others, 1990). However, it must be noted here that in Vermont Spear and Harrison (1990), on the basis of $^{40}Ar/^{39}Ar$ cooling ages, consider these rocks as Ordovician.

The curvilinear nature of the Connecticut Valley–Gaspé–Notre Dame belt (Fig. 4) reflects the former edge of the Laurentian continent and the Acadian deformation in this belt as marginal to the former continent (cf. Thomas, 1977). The Acadian deformation in this belt is complex, and its kinematics vary along the trend. In Newfoundland Cawood and Williams (1988) recorded both west-northwest– and south-southeast–directed thrusts and claimed an overlap in space between the Taconian (Middle Ordovician) and Acadian structures. In Vermont and New Hampshire the Connecticut Valley belt is bordered on the west and east, respectively, by the Dog River fault zone (Westerman, 1987) and the Monroe line (Hatch, 1987), both in all probability being thrusts. Hatch (1988) proposed that the rocks of

the belt were deposited in a trough that stretched along the western front of the Bronson Hill island arc of Taconian age.

The Bronson Hill belt stretches as a slightly curved structure (Fig. 4) from southern Connecticut to the border of New Hampshire and Maine, where it terminates as a single structure just south of the Chain Lakes Massif. The Bronson Hill anticlinorium in western Maine splits into the Boundary Mountain anticlinorium and the Lobster Mountain anticlinorium (Osberg and others, 1985). Bird and Dewey (1970) referred to the Bronson Hill anticlinorium as the Oliverian volcanic arc and continued it, with an inflection, into Newfoundland, although they recorded an interruption in Maine (p. 1049–1050). Thus the local parallelism of the Bronson Hill belt and the Connecticut Valley–Gaspé belt is in part coincidental. Moreover, Boone and Boudette (1989) have suggested that Ordovician inliers of northern Maine belong to the Boundary Mountain terrane—a block that was accreted to the North American (Laurentian) continent in late Cambrian time, causing the Penobscottian orogeny. The Boundary Mountain terrane incorporates the Chain Lakes Massif (Fig. 4). The Connecticut Valley–Gaspé belt has the Boundary Mountain terrane as its basement in New Hampshire, Maine, and the Gaspé Peninsula.

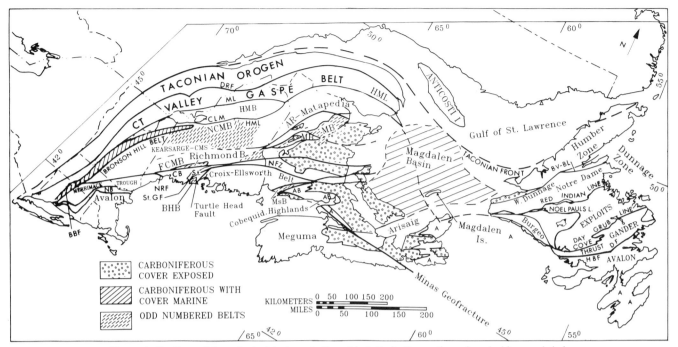

Figure 4. Outlines of the principal tectonostratigraphic belts in the northern Appalachians and their boundaries. Modified after Williams (1978). Silurian-Devonian belts: (1) NCMB—North-Central Maine belt; (2) AR-Matapedia—Aroostook-Matapedia belt; (3) MB—Miramichi belt; (4) FCMB—Fredericton–Central Maine belt; (5) Richmond B—Richmond belt; (6) CB—Casco Bay belt; (7) BHB—Benner Hill belt; (8) St. Croix–Ellsworth belt; (9) MsB—Mascarene belt; (10) AB—Avalon belt of southern New Brunswick. Other features identified by abbreviation: A—Avalon blocks (and in southern Newfoundland, Avalon and Mascarene equivalents); BBF—Bigelow Brook fault; BV-BL—Baie Verte–Brompton Line; CLM—Chain Lakes massif; CMS—Central Maine synclinorium; DF—Dover fault; DRF—Dog River fault; HBF—Hare Bay fault; HMB—Hurricane Mountain belt; HML—Hurricane Mountain line; ML—Monroe line; NB—Putnam-Nashoba belt; NFZ—Norumbega fault zone; NRF—Nonesuch River fault; St.GF—St. George fault. Dunnage Zone and Gander collectively form the Central Mobile belt.

Whether the basement is necessarily Cambrian is at present uncertain, since the ages of metamorphism recently claimed for it are around 470 Ma (Dunning and Cousineau, 1990), and parts of the terrane are clearly Ordovician (Boone and Boudette, 1989, p. 20). Throughout this belt Silurian-Devonian rocks are unconformable on the Ordovician. The same applies to Newfoundland (Williams and others, 1989a), where west of the Red Indian Line (Fig. 4) of Williams and others (1988) Silurian and/or Devonian strata are unconformable on the Ordovician. In the mainland of North America the equivalent of this line is the southeast border of the Hurricane Mountain belt and can be referred to as the Hurricane Mountain line (Fig. 4, HML).

Southeast of the Hurricane Mountain line (Fig. 4), which continues into northern New Brunswick as a series of aforementioned transcurrent faults (Fig. 3), lies a complex stretch that is subdivided in New Brunswick by Fyffe and Fricker (1987) into a series of terranes with distinctions based mainly on the cover rocks. There have been attempts to correlate these terranes into Maine, for instance by Ludman (1986) and Berry and Osberg (1989). For the time being we avoid the terrane concept, since at

present there are disagreements about the position and extent of the actual terranes, and we substitute the terminology based on belts, which in general is more objective. However, we will return to an interpretation of terranes in our final discussion. In Maine to New Brunswick (Fig. 4), between the exposed Avalon belt and the Connecticut Valley–Gaspé belt, we recognize a series of Silurian-Devonian belts, mainly adopted from Berry and Osberg (1989) and partly from Fyffe and Fricker (1987). Of these belts, 1 to 5 inclusive constitute parts of area 1, and belts 6 to 10 inclusive constitute parts of area 3 of Figure 2.

The North-Central Maine belt (Fig. 4, NCMB), which lies to the southeast of the Hurricane Mountain belt (HMB) of Lower Paleozoic, Cambrian-Ordovician inliers, consists of a thick sequence of Silurian-Devonian strata. The most up-to-date map available, in the western part of this area, is by Moench and Pankiwskyj (1988a). The eastern part has been extensively treated by Roy (1980). The Silurian-Devonian strata of this belt from place to place include the extensive volcanic formations, well exposed in parts of New Brunswick (Dostal and others, 1989) and central Maine (Rankin, 1968) where they are known in the Pisquataquis volcanic belt consisting of two horizons of calcalkaline volcanics, the lower at the Silurian-Devonian boundary and the higher in the mid-Devonian.

In western Maine (area 5, Fig. 2) as well as New Hampshire, Silurian strata can be divided into a thin sequence to the northeast and a thick sequence to the southwest. The line separating the two is known as the Silurian tectonic hinge (Hatch and others, 1983; Moench, 1989). The hinge is partly supported by the Bronson Hill belt, which has long been interpreted as an anticlinorium. The Silurian-Devonian rocks of the belt to the southeast of the hinge zone have been interpreted to have been deposited in an elongated basin into which turbidites arrived from the northwest and southeast. The basin was almost filled by the end of Silurian time and was covered by a much more extensive Devonian Seboomook Formation that consists of pelitic schists, metagraywackes, and the aforementioned metavolcanic intercalations, interpreted as products of an Andean-type volcanic arc (Osberg, 1988).

Deformation in this area began during the deposition of Silurian rocks, although sedimentation probably began in Caradocian time, resulting ultimately in complex premetamorphic faults (fold-folds) first described by Moench (1970) and reinterpreted by Moench and Pankiwskyj (1988b). The southwestern part of the North-Central Maine belt constitutes the northwestern flank of the so-called Kearsarge–Central Maine synclinorium (Fig. 4, CMS), which had suffered a very complex sequence of deformational events involving early recumbent folds (F_1 according to Osberg, 1988), that on later refolding yielded downward-facing structures. The F_2 folds refold the earlier structures and produce major upward-facing folds with steep axial plane cleavage and some asymmetry. Osberg (1988) suggested that the first episode involved folding and thrusting with a northwestward transport and the second, southeastward vergence.

The Aroostook-Matapedia belt (Figs. 3 and 4) forms an inflected, curved anticlinal belt from Houlton, Maine (Fig. 5), to Percé on the Gaspé peninsula. The belt is characterized by banded carbonate-pelite-psammite sequences straddling the Upper Ordovician–Silurian boundary and the Canada–United States border. The rocks within it are strongly deformed into unusually symmetrical folds, but these in places can be shown to refold earlier recumbent structures (Rast and others, 1980). Along the southeast margin of this belt are the Silurian-Devonian strata of Chaleur Bay, south of which is the Miramichi belt, and finally the Fredericton–Central Maine belt (Fig. 3, FCMB). These relationships indicate that the Aroostook-Matapedia belt wraps around a block of which the Miramichi belt (Fig. 3, MB) forms the core and the Chaleur Bay strata form the cover. Thus here we consider the Chaleur Bay sequence as a part of the Miramichi belt.

The Chaleur Bay rocks consist of a sequence of Silurian and Devonian shallow water clastic and thin carbonate sediments interbedded with volcanics. Dostal and others (1989) point out that the Chaleur Bay succession is associated with what they call the Tobique volcanic belt (Fig. 3, TVB). Its orientation, oblique to the general strike, they interpret as a result of rifting in a sinistral shear regime in association with a continental graben. The belt was later telescoped by dextral shear during the Acadian orogeny. The belt of volcanics lies parallel to the margin of the exposed Miramichi belt, which consists mainly of Ordovician Tetagouche Group intruded by posttectonic Silurian granitoids and gabbros (Bevier, 1988; Fyffe and others, 1988). The outline of the Miramichi belt is shown in Figures 3 and 4 (MB). The basement to this block, according to geochemical determinations of Bevier and Whalen (1988), is continental and of possibly Proterozoic Grenville age. We suggest that this block has had influence both on the sedimentation of the Aroostook-Percé belt and on its resultant curvilinear shape after deformation. In other words, the Miramichi belt together with the adjacent Tobique belt (Fig. 3) formed a separate Paleozoic tectonic unit. There is some evidence that the Tobique cover of the Miramichi lower Paleozoic basement is in fact overthrust upon it in a southeasterly direction (Ludman and others, 1990).

In area 3 (Fig. 2), the Fredericton–Central Maine belt (Fig. 4, FCMB) of principally Silurian graywackes and pelites in New Brunswick lies immediately to the southeast of the Miramichi belt (Fig. 3). Our limited observation on its load casts suggests that the turbidity currents within the belt were flowing from northeast to southwest. The belt in Maine had been continued by Berry and Osberg (1989) to as far as the Deblois granite pluton and then stopped. On lithological grounds we maintain that the central Maine flysch belt of Ludman (1986) is the continuation of the Fredericton graywacke belt, including a substantial part of the south-central Maine belt of Berry and Osberg and in particular the Smalls Falls Formation and possibly the so-called Vassalboro Formation. The Fredericton–Central Maine belt (Fig. 4, FCMB) is conventionally included in the Kearsarge–Central Maine synclinorium. The synclinorium is bounded on the southeast by the Richmond belt (Fig. 4), and both are truncated by the Norumbega fault zone. On the south side of the Norumbega fault the Merrimack trough extends as far northeast as the Nonesuch River fault that separates it from the Casco Bay belt; both lie south of the Norumbega fault (Fig. 4, NRF, CB, and NFZ respectively).

To the southeast of the Norumbega fault zone and to the northwest of the Lubec–Belle Isle fault (Fig. 5, N and LBI respectively) lies a very complex composite coastal Maine zone that is principally exposed in Maine but in New Brunswick is partly covered by Carboniferous strata. This zone was comprehensively treated by Berry and Osberg (1989): They recognized the Ellsworth belt (Figs. 4 and 6), the Isleboro-Rockport belt (Fig. 6), and the Benner Hill and the St. Croix belts (Fig. 4). The Avalon belt lies on the southeast side of the Belle Isle fault in New Brunswick (Fig. 6). In addition, parts of the Fredericton belt lie to the southeast of the Norumbega fault zone. The belts within the coastal Maine zone are separated by faults. Berry and Osberg (1989) recognized separate Ellsworth and St. Croix belts, although in places they are gradational. Therefore in this paper, in common with Fyffe and Fricker (1987), we refer to the two subunits collectively as the St. Croix–Ellsworth belt (Fig. 6). We

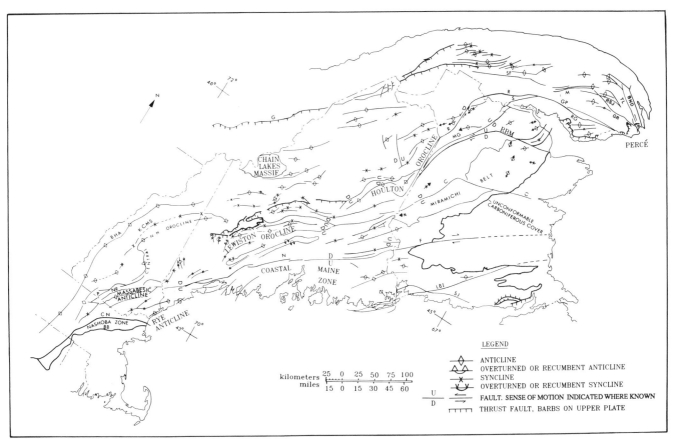

Figure 5. Structural Map of the northern Appalachians, excluding Newfoundland. Sources: Zen (1983b), Osberg and others (1985), Osberg (1978), Bothner and others (1984), Lyons and others (1982), Fyffe (1982), Hall and Robinson (1982), Ludman (1978), Eusden (1988), Malo and Beland (1989), Keppie (1982). Identified faults and structures: BB—Bloody Bluff; BHA—Bronson Hill Anti-clinorium; BNO—Bras Nord-Ouest; C—Catamaran; CN—Clinton-Newbury; F—Fredericton; G—Guadeloupe; GP—Grand Pabos; GR—Grande River; KCMS—Kearsage Central Maine Synclinorium; LBI—Lubec-Belle Isle; M—Marcil; MG—McKenzie Gulch; N—Norumbega; NR—Nonesuch River; P—Pinnacle; R—Restigouche; RMB—Rocky Brook-Mill Stream; RG—Riviere Garin; RSJ—Riviere St. Jean; S—Sellarsville; SF—Sainte-Florence; SJ—St. John's; TL—Troisieme Lac.

also consider the Isleboro–Rockport belt (Figs. 4 and 6) as a faulted part of the basement of the same unit.

The Benner Hill belt (Figs. 4 and 6) lies along the southwest margin of the St. Croix–Ellsworth belt of our definition and is separated from that belt and from the Isleboro–Rockport exposures (Fig. 6) by a series of thrusts. The Benner Hill sequence (Hussey, 1985) consists of middle grade metamorphosed fossiliferous Ordovician rusty and dark schists, laminated quartzites, and metasandstones. The precise correlation of the lithologies within the belt is at present unknown; in view of the fact that parts of the St. Croix–Ellsworth belt lie to the north and east of the Benner Hill sequence, even if it is a suspect terrane, the terrane must have been proximate to the St. Croix lithotectonic unit.

The Casco Bay belt (CB) is separated from the St. Croix belt by the St. George fault (Figs 4 and 6, St.GF.). The Casco Bay belt is included by Berry and Osberg (1989) with the South-Central

Maine belt. The Casco Bay belt rocks occur both in the Casco Bay area and in the Richmond belt. The sequence of rocks forming the belt, although yielding a general succession, is of unknown age (Hussey, 1988). It consists of essentially three units: the Cushing Formation; the Cape Elizabeth Formation, forming the lowest part of the Casco Bay Group; and a third unit, the upper part of the Casco Bay Group. This unit is located in an isolated syncline (Pankiwskyj, 1978) and is not recognized elsewhere in the belt. The Cushing Formation contains extensive volcanic and volcanogenic rocks, some of which are welded tuffs and can possibly be correlated with the late Proterozoic felsic volcanics of the Avalon superterrane. The Cape Elizabeth Formation may be unconformable on the Cushing (Hussey, 1986, 1988, p. 24), as the uppermost part of the former includes thick lenses of calc-silicate rocks as well as amphibolites and some quartzite and marble. All these rocks are variably deformed and metamor-

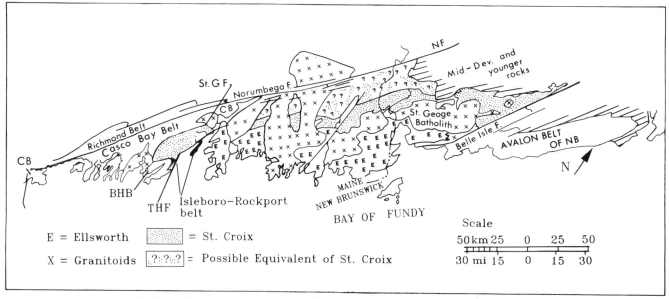

Figure 6. Coastal Maine belt and Avalon of New Brunswick, showing major subdivisions. Modified from Berry and Osberg (1989). Abbreviations: BHB—Benner Hill belt; CB—rocks of Casco Bay belt; NF—Norumbega fault; St.GF—St. George fault; THF—Turtle Head Fault.

phosed; the deformation is in places polyphase. Hussey (1988) suggested that this deformation in all its phases may correlate with that recognized in the Kearsarge–Central Maine synclinorium or the Fredericton–Central Maine belt and is therefore a reflection of the Acadian orogeny. On the other hand, if this deformation is earlier, it may be of late Proterozoic to early Ordovician age. The axial traces of both F_1 (recumbent) and F_2 (upright) folds trend northeast-southwest in the north-central part of the area and deflect to north-south direction in the south (Fig. 5). Farther to the southwest the belt is cut off by a thrust, the Nonesuch River fault (Fig. 4, NRF), that Hussey and Newberg (1978) considered as premetamorphic. This thrust, if it exists, separates the Casco Bay belt from the Merrimack trough sequence.

The Merrimack trough sequence (Fig. 4) has been recognized by Lyons and others (1982) as an entirely separate belt. In southeastern Maine and southeastern New Hampshire, its succession has been more or less established by Hussey (1962, 1985), Hussey and others (1986), and Swanson and Carrigan (1984). The rocks of the Merrimack trough are variably deformed metasedimentary formations with strong mylonitization of two periods and have been folded three times. The earliest folds are recumbent, and the latest are upright with variable axial plane cleavage. According to Gaudette and others (1984), the later 473±37-Ma Exeter pluton is postdeformational as well as postmetamorphic. This dating is consistent with middle Ordovician ages of the Newburyport Quartz-diorite and the Appledore Diorite dated by Zartman and Naylor (1984), and such data indicate that the deformation of the Merrimack Group rocks is pre-475 Ma. This

is so because there is little evidence for Taconian metamorphism and deformation in this area; thus both the rocks and these effects are in all probability Proterozoic. The oldest rocks are probably those of migmatitic complexes such as the Massabesic Gneiss, Rye Formation, and so on that form the cores of large anticlines. According to Hussey (1990, personal communication) there is a possibility that parts of the Merrimack and the Casco Bay groups are equivalent.

The Merrimack Group forms a belt that includes rocks that have been indicated as Paxton in Massachusetts (Zen, 1983b) and Hebron in Connecticut, where it is juxtaposed with the Bronson Hill belt (Fig. 4) across the Bigelow Brook fault complex. Stratigraphic and structural relations in this region are at present uncertain, and correlations in Massachusetts have to be at least partially revised, as indicated by Tucker and others (1988). A significantly new contribution that they make, and one whose implications are at present unclear, is that the core rocks of the Bronson Hill belt in Massachusetts show a Proterozoic single zircon crystallization age (606 +13/–9 Ma) and Alleghanian metamorphic effects (ca. 293 Ma). In the same part of the Bronson Hill belt, there are effects of the "Acadian" metamorphism and magmatism, yielding 367 to 350 Ma ages.

The rocks of the Merrimack trough as well as the adjacent Kearsarge–Central Maine synclinorium are characterized in their southwestern extremity by the so-called Hebron Gneiss, which is a mylonitized, partly melanged and recrystallized gneiss intruded in places by pegmatites of Permian age. In Massachusetts the Paxton Formation is seen as the continuation of the Hebron and also has been correlated by Barosh and Pease (1981) with the

Berwick Formation of the Merrimack Group. Thus in this chapter we extend the Merrimack Trough southward beyond its conventional limits into southern Connecticut (Fig. 4).

The structure of both central Maine belts (Fig. 4) as well as the coastal Maine belts shows remarkable arcuate deflection of structural trends, especially from southern Maine near the Lewiston orocline (Figs. 5) to eastern Maine near the Houlton orocline. We suggest that these oroclinal flexures may represent north- to northwest-directed thrusts that pass laterally into northerly trending strike-slip fault boundaries in the underlying basement blocks. The presence of exposed Proterozoic blocks in the Casco Bay and the St. Croix–Ellsworth belts of the Coastal Maine zone (Fig. 6) lends plausibility to this suggestion. At present most data suggest that rocks of the Avalon belt underlie most of the St. Croix–Ellsworth belt as far as the Turtle Head fault (Figs. 4 and 6, THF). If the Merrimack Trough sequence is Proterozoic with perhaps some cover of Silurian-Devonian formations (cf. Rast and Skehan, 1990), it can be equated in time with the Avalon St. Croix-Ellsworth belts. We suggest that the latter belt formed colliding blocks responsible for the local manifestations of the Acadian orogeny. The Merrimack trough in Massachusetts and Connecticut is separated from the Boston Avalon (area 6, Fig. 2) by the Putnam-Nashoba belt (Fig. 4, NB) that stretches from eastern Massachusetts almost to Long Island Sound in Connecticut. This belt does not show clear evidence for Acadian deformation, which also appears to be absent from the Boston Avalon block. The Putnam-Nashoba belt is bounded on the east and west sides by the Lake Char and Clinton-Newbury faults respectively that can be traced southwestward almost to Long Island Sound (Wintsch and Aleinikoff, 1987). Moreover, the belt is characterized by apparently Lower Paleozoic (Taconian?) metamorphism. The rocks of this belt have, at present, a problematic stratigraphy that cannot be easily fitted with other units of the northern Appalachian orogenic belt.

Connections with Newfoundland

In order to establish correlations with Newfoundland (area 2, Fig. 2), we must consider in the Gulf of St. Lawrence (Figs. 4 and 7) structural inflections first suggested by Marillier and others (1989). The complexity of deformation is here enhanced by the existence of a Carboniferous Magdalen basin shown in Figures 4 and 7. To the east of Magdalen Island the depocenter of this basin is over 12 km deep. The basin does not completely disappear before reaching Anticosti Island, and Carboniferous strata have been detected in the estuary of the St. Lawrence River (Sanford and Grant, 1990). Deeper seismic, gravity, and aeromagnetic profiles indicate that the basin is underplated by the mantle (Marillier and Verhoef, 1989), thus implying that the basin is an artifact of post-Acadian thinning of the crust. On the basis of deep seismic reflections, a Grenville basement wedge can be traced southeast under most of the Magdalen basin. Marillier and Verhoef (1989) argued convincingly that the Gulf of St. Lawrence

promontory is caused by the thinning of the lower crust during Carboniferous time rather than being the result of earlier Appalachian-related tectonic events. Marillier and Verhoef (1989) provided a seismic profile showing their interpretation of the extent of Grenville crust under the Gulf of St. Lawrence. To the west, south, and east transitional lower crust of unknown composition is recognized under central Newfoundland, parts of Cape Breton Island, and central New Brunswick and presumably continues into central Maine. On the basis of geological and geophysical data, Zen (1983a) has located the Taconian margin of the North American craton in Maine and New Brunswick, approximately along the southeastern margin of the Grenville wedge of our Figure 7. In Figure 7, section A-A′ is our schematic interpretation of the situation described by Ludman and others (1990) and based on analogues to the east noted above. According to Ludman and others, the Avalon crust in central Maine forms a wedge (Fig. 7, profile A–A′) that stretches from the southeast, tapers northwestward beyond the Norumbega fault zone, and underlies the Fredericton belt and the Miramichi block, closely approaching the Grenville crust. However, the shallower crustal units, including the Acadian belt described in this chapter, may be allochthonous on the deep lower crust (Marillier and Durling, 1990), although on the basis of geochemical evidence this interpretation is controversial (Whalen and Hegner, 1990).

The Avalon cover of the Avalon crust comes to the surface in southern New Brunswick, Cape Breton Island of Nova Scotia, and the Avalon Peninsula of Newfoundland. This major block (Fig. 4), which itself can be divided into smaller blocks collectively constituting the Avalon superterrane, is usually devoid of Silurian-Devonian strata except at the margins (Mascarene Group in New Brunswick, Arisaig in Nova Scotia, and perhaps parts of the Gander Group and Burgeo in Newfoundland). In Newfoundland, however, it is separated from the Avalon superterrane by a large fault that, despite a projection suggested by Barr and Raeside (1986), cannot clearly be identified on the 8b-5A seismic line of Marillier and others (1989).

Central Newfoundland (Dunnage zone of Williams, 1978) (Fig. 4) presents a novel Acadian problem. In this region Silurian strata conformably overlie the Ordovician, and both were deformed by Devonian orogenic forces, conventionally referred to as the Acadian. This point was forcibly made by Karlstrom and others (1982) and reiterated by Van der Pluijm (1987), who stressed the idea that the formation of the Taconian volcanic arc was followed by the onlap of late Ordovician and Silurian seas. These sediments and volcanic rocks were deformed in late Silurian to early Devonian time in a collision episode and then were affected by Devonian or possibly Carboniferous strike-slip northeast-southwest–trending faults. These faults, according to Currie and Piasecki (1989), were accompanied by severe thrusting directed to the northwest. Van der Pluijm and Van Staal (1988) suggested that the Dunnage and Gander belts of Newfoundland (Fig. 4) should be combined into a single Central Mobile Belt (Fig. 4) in which there is a continuity of deformation

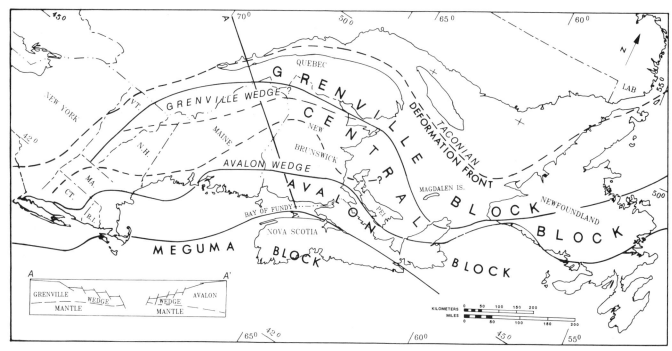

Figure 7. Lower crustal relationships in the northern Appalachians (major blocks). Sources: Marillier and others (1989), Stewart and others (1986), Ludman and others (1990), Tucker and others (1988).

throughout Upper Ordovician and Silurian time as thrusting and transcurrent faulting continued into later Silurian and Devonian time. They projected the Central Mobile Belt into central Maine. It seems to us that even if the details have to be worked out locally, such a model represents a viable extension and amplification of the one suggested by Rast and Stringer (1980, p. 241). Whalen (1989), moreover, suggested that because of the abundance of deformed Lower and Middle Silurian granites and volcanic rocks not only in south-central Newfoundland but also in the Canadian northern Appalachians, the Acadian orogeny was Late Silurian.

Connections with Nova Scotia

The main part of the northern Appalachians, stretching from Long Island Sound to Newfoundland, is separated from the so-called Meguma zone of southern Nova Scotia (area 4, Fig. 2; Fig. 4) by the Avalon superterrane (Area 3, Fig. 2). In Nova Scotia the deformed Cambrian-Ordovician sediments that are unconformably overlain by Silurian-Devonian, and then Carboniferous strata, are deformed by Acadian (Devonian) structures and intruded by Devonian to Carboniferous granites. The Silurian rocks are largely graptolitic slates succeeded by coarser clastic rocks, but the Devonian beds are paralic with fossiliferous layers.

Dallmeyer and Keppie (1987) approached the problem of the dating of the Acadian orogeny in the Meguma by detailed investigation of $^{40}Ar/^{39}Ar$ mineral ages. Keppie in 1982 proposed the main deformation to be early to mid-Devonian. The

main deformation produced upright folds and axial plane cleavage as well as associated low-grade metamorphism. The rocks were then overprinted by a regional low-grade metamorphism and late Devonian granites. Dallmeyer and Keppie (1987) enhanced this chronology by suggesting that the early regional metamorphism occurred about 400 to 410 Ma. This metamorphism was followed by a series of shallow granitic intrusions ranging in age from ca. 375 to 315 Ma and culminating in the intrusion of upper Devonian to Carboniferous (ca. 315 to 325 Ma) granites. Dallmeyer and Keppie (1987) also suggested sequences of deformation associated with these thermal events that straddle the Devonian-Carboniferous boundary and even may have continued into the Permian.

It therefore seems that although in the main part of the northern Appalachians there is some continuity from the Ordovician (Taconian) through to the Devonian (Acadian) tectonism, in the Meguma zone the deformation was more or less continuous from the Devonian (Acadian) to Permian (Alleghanian-Variscan) tectonism. This observation has led us (Rast and Skehan, 1989) to suggest that although the main orogenic belt is equivalent to the Caledonides of Britain, the Meguma belt is equivalent to the Acadian-Ligerian belt of western and central France (Fig. 1).

Lateral correlations with the British Caledonides

The idea that the Acadian orogen of the northern Appalachians prior to the Mesozoic opening of the Atlantic Ocean extended into Ireland and Britain (Fig. 1) was proposed by Bertrand in 1887 (p. 442), by Bailey (1928), and by Schuchert (1930). However, Schuchert suggested the possibility of the Acadian orogen of eastern North America being equivalent to the Armorican peninsula of France. Bertrand hesitated to invoke continental drift to explain the above relationships, but Bailey did so unequivocally.

The Acadian orogenic belt of the northern Appalachians has been interpreted as a product of collision of the North American continent with the so-called Avalonian superterrane (volcanic island arc). A collisional interpretation was first proposed by Bird and Dewey (1970) for the Taconian orogen, a concept that they applied to the Acadian in the same paper, implying that a suture existed between the Silurian-Devonian sedimentary sequence associated with the North American craton and that of what we now refer to as the Avalon superterrane. Although modifications have been introduced lately, the collisional hypothesis implies that an oceanic domain, called the Iapetus Ocean, existed between the continents and the superterrane.

In the aftermath of Bird and Dewey's (1970) interpretation of the Acadian orogenic belt, the correlation of the northern Appalachians with the British Caledonides became progressively more widely accepted. McKerrow (1988), for example, reinterpreted radiometric ages of the British Caledonides to correlate with the Acadian of North America. On this basis he suggested that the British-Canadian Acadian occurred in the middle of the Emsian (mid-Devonian). However, the geochronologic scale that he used places Emsian as an interval of 401 to 392 Ma, whereas Palmer (1983) placed it as 394 to 397 Ma, and the recent International Union of Geological Sciences (IUGS) scale has it as 390 to 395 Ma. Thus the correlation based on stratigraphic formations with specific isotopic age boundaries at present is not entirely valid.

Furthermore, McKerrow considered the whole of the British Isles, although there are some indications of cross-trend episodicity within this region. The deformation of the Southern Uplands of Scotland is Silurian (Barnes and others, 1989), whereas in northern England and Wales it appears to be Lower Devonian. Soper and Kneller (1990) suggested that the changeover takes place abruptly across the Iapetus suture of the British Isles. The Paleozoic strata southeast of this suture in eastern Ireland also yielded Devonian ages for late orogenic to postorogenic plutons (Murphy, 1987). Of course, these models depend on the presupposition that the Iapetus suture represents a compressional interface. It can also be interpreted as a transcurrent fault, and several models take into account this assumption. The Caledonian orogen of the British Isles is then interpreted (Hutton, 1987) as a series of elongated northeast-southwest–trending terranes separated by left-lateral transcurrent faults.

The duration of mid-Paleozoic movements (Caledonian-Acadian) in the British Isles is prolonged, and the evidence for these is variable. The earliest events are inferred from sedimentary evidence. In most parts of Britain there is a Late Silurian abrupt change of facies into red rocks, parts of which are undoubtedly subaerial as far west as Dingle, Ireland (Holland, 1987). Tradi-

tionally this change has been associated with the generation of the Early Devonian Old Red Sandstone (ORS) continent. Of the many sedimentological studies on the Lower ORS, that by Haughton and Farrow (1989) is particularly useful. In it they indicated that clasts in the Old Red Sandstone of the Midland Valley terrane suggest derivation from the Grampian terrane, which cannot have been juxtaposed against the Midlands Valley Terrane in end-Ordovician times by sinistral transpression or transtension. In the South, however, the Avalonian Midland terrane docked against Laurasia in mid-Devonian Acadian time. The repetitive continuity of these movements thus spanned the time between the Late Ordovician and mid-Devonian, causing rejuvenation of major faults. Barnes and others (1989) explained the apparent diachronism by southward-migrating deformation. In the absence of concrete evidence at present we find this difficult to accept.

The early manifestations of Caledonian orogeny in Britain possibly correspond to the Salinic disturbance reported from the northern Appalachians (Boucot, 1962; Roy, 1980). In the latter region the disturbance, although originally interpreted as a result of uplift and erosion, may be more significant. Roy (1980) has speculated on the possibility of the early recumbent folds in Maine being Salinic. Similar indications of recumbency of Silurian strata in west-central New Brunswick (Rast and others, 1980) may be attributed to the Salinic deformation. It is also quite possible that the very early deformation described by Moench (1970) from the Rangeley area is a manifestation of Salinic disturbance. Thus we suggest that the closing of the Iapetus Ocean began to take place in Llandovery time as a Salinic event.

One of the comparisons between the northern Appalachians and the Caledonides is based on the emplacement of syntectonic to posttectonic plutons. In the northern Appalachians of either the United States (Sinha, 1988) or Canada (Fyffe and others, 1981a), Acadian calc-alkaline plutonic rocks are widely distributed and range from Silurian to Devonian in age. This is exactly the situation in the British Isles.

Lateral correlations with the south-central Appalachians

The central and southern Appalachians constitute a very large region, which we divide into areas 7, 8, and 9 (Fig. 2). The northern Appalachian Acadian orogenic belt is difficult to trace through into the central and southern Appalachians. An east-west line of separation that can be drawn south of Long Island marks a substantial right lateral offset of the Appalachian orogen (Fig. 8). This line has been assumed to be a fault (Rast and Skehan, 1984; Rast, 1988; Getty and Gromet, 1988) or possibly an east-northeast–trending anticlinorium whose northern limb is exposed along the Connecticut coast (Skehan and Rast, 1990). On the mainland the fault cannot be easily traced south and southwest of New York, but it may be represented by the Martic line of the central Appalachians and in part covered by later Triassic deposits. To the south of the Martic line at present only indirect indica-

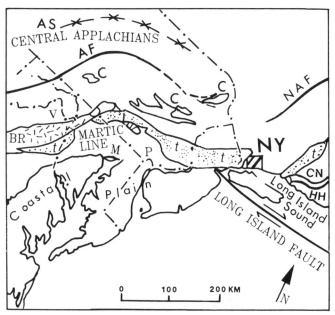

Figure 8. Position of branches of the Martic Line and its interpreted continuation as the Long Island fault. Modified after Rast (1984), Getty and Gromet (1988), Hatcher and others (1989). Features identified by abbreviation: AF—Appalachian front; AS—Appalachian syncline; BR—Blue Ridge (hachured pattern); C—Carboniferous; CN—Clinton-Newbury fault; HH—Honey Hill (correlated with Bloody Bluff), fault; M—Maryland; NAF—Northern Appalachian front; NY—New York; P—Pennsylvania; t—Triassic basins (dotted pattern); V—Virginia.

tions of the Acadian orogen have been recorded. Tectonics of this highly populated region are badly known, and at present the evidence for a fault cannot be pursued any further.

In the southern Appalachians the principal lithotectonic and geochronometric evidence for Acadian orogeny exists in Alabama (Tull and Telle, 1989) and also in the Piedmont near the Brevard fault zone (Dallmeyer, 1988). Ferrill and Thomas (1988) have suggested that in the Appalachians the Acadian tectonic regime was transpressional rather than purely compressional and that in the southern Appalachians movement was concentrated along large thrusts and transcurrent faults rather than developing overall ductile strain. Similarly, Sinha and others (1988) suggested that there is no geochemical evidence for Devonian metamorphism in mylonites from the Brevard zone, and therefore Acadian movements were absorbed in large faults and were not accompanied by regional metamorphism. On the basis of $^{40}Ar/^{39}Ar$ analyses of rocks in the Inner Piedmont and Alto allochthon, however, Dallmeyer (1988) pointed out that the possible Acadian metamorphism occurred prior to the formation of mylonites. Glover (1989) pointed out that stratigraphic ages of the clastic wedge in the Valley and Ridge of Virginia extended from the Middle Devonian (385 Ma) to early Mississippian (360 Ma), indicating the effects of the Acadian orogeny. The clastic wedge, in his interpretation, originated from the uplift and deformation in the Virginia Piedmont.

In the southern Appalachians there are three possible reasons for the difficulties in detecting and correlating Acadian orogenic events. The first is that Silurian and Devonian fossils are extremely rare in the southern Appalachians. The recent discovery of Silurian fossils in the foothills of the Appalachians by Unrug and Unrug (1990), on line with the Alabama outcrops, suggests the possibility of the existence of a former Acadian successor basin at the western edge of the Blue Ridge (Tull and Groszos, 1990). If this is so, then much of the western Blue Ridge and the Inner Piedmont must have been affected by Acadian deformational events that led to the supply of the sediments to these basins. In support of this notion Osberg and others (1989) advanced evidence for the so-called Concord plutonic suite with some thirty plutons about 400 Ma in age in the Charlotte belt as well as the large 400- to 415-Ma calc-alkaline granites in South Carolina. All these granitoids and associated mafic intrusions are situated within the equivalents of the Avalon superterrane (Carolina terrane) and have been thought to indicate that this terrane docked to North America in Acadian times. The Inner Piedmont belt has also gneissic granitoids varying in age between 460 and 370 Ma. These granitoids range from syntectonic to posttectonic. Posttectonic (ca. 390 Ma) pegmatites occur in the Spruce Pine area of the Blue Ridge, where they intrude high-grade metamorphic rocks and have no thermal aureoles, thus indicating broad contemporaneity with the Acadian metamorphism.

A second explanation of the difficulties of recognizing Acadian orogenic effects in the southern and central Appalachians is the pervasive Alleghanian thrusting. Thus the Acadian orogeny in the South also may have been compressional but is largely overprinted and confused by the effects of Alleghanian stacking.

In New England and Canada at present, Alleghanian thrusting is regarded as restricted to the southeast part of the region. However, in th past few years indications of deformations (Brookins, 1967; Naylor, 1969; Zartman and others, 1970; Zartman, 1988; Skehan and Murray, 1980; Skehan, 1983; Lyons and others, 1982; Geiser and Engelder, 1983; Wintsch and Aleinikoff, 1987; Tucker and others, 1988; Guidotti, 1989b; Guidotti and Cheney, 1989, Gromet, 1989, 1991) that we interpret as probably Alleghanian have been noted in areas farther to the northwest of those where Alleghanian thrusting has been demonstrated. Thus such thrusting may be more widespread in the northern Appalachians than is generally considered to be the case.

A third difficulty arises from the perception that in the south-central Appalachians there may have been a complete continuity between Taconian and Acadian movements and metamorphism, rendering all correlations more difficult than in the northern Appalachians. This suggestion is based on the interpretation of paleomagnetic studies. It has been found by Vick and others (1987) that the Carolina terrane docked to the developing American craton in Ordovician time. It seems possible that this event dates the initiation of an episode of docking that was completed in Silurian-Devonian time, while throughout the interval from mid-Ordovician to Lower Devonian the orogen was continuously metamorphosed and intruded by plutons. No doubt this lengthy orogenic development could be separated into shorter periods of specific thermotectonic activity, but these must be considered as stages of a general process. A similar interpretation has been advanced by Keppie and Dallmeyer (1987) for the gradation of Devonian-Carboniferous ages of crystalline metamorphic mica in Nova Scotia.

Recent evaluations of the geology of the Carolinas suggest that there is still doubt as to the significance of the Acadian orogeny in the southern Appalachians (Horton and Zullo, 1991).

GRANITES AND METAMORPHISM

Devonian metamorphism and intrusive granites are usually attributed to the Acadian orogeny. Until fairly recently most granites in the northern Appalachians were considered as either posttectonic or, more rarely, syntectonic intrusions. Unfortunately, in the existing literature granites and regional metamorphism tend to be treated separately, although recently this tendency has been changing (Guidotti, 1989a, 1989b). The recognition of regionally developed, diverse migmatitic rocks has been an important contribution to the interpretation of metamorphism (Robinson and others, 1989).

Relation of "Acadian" granites to metamorhism

Figure 9 shows the distribution of "Acadian" granites divided into Silurian, Devonian, and Carboniferous-Permian categories. These plutons are plotted on the map that shows the distribution of "Acadian" metamorphic zones in New England and Canada, the distribution of postmetamorphic cover rocks, and the northernmost limits of Alleghanian metamorphic overprint (Zartman, 1988). It is significant that granitoid intrusions falling in the time interval of 415 to 365 Ma, which we may call the Acadian *sensu stricto,* themselves can be divided into synmetamorphic and postmetamorphic. The former include the Bethlehem and Kinsman plutons, which on Rb/Sr dates by Lyons and Livingston (1977) are 405±78 Ma and 411±19 Ma respectively. Postmetamorphic plutons, such as the Mt. Katahdin and Lucerne intrusives, have Rb/Sr ages of 388±5 Ma and 371±21 Ma respectively (Loiselle and others, 1983a; Loiselle and others, 1983b).

In some cases a prolonged intrusive history has been recorded, as for example the multiple St. George batholith of New Brunswick, which according to Fyffe and others (1981b) can be divided into five intrusive episodes. These range from 406±7 Ma to 337±15 Ma; the youngest can be considered to be satellitic stocks or offshoots of the main granite. Whalen (1989) suggested that associations between intrusive bodies and extrusive volcanic sequences, which in Canada are predominantly Silurian, occur principally between the Humber and Avalon zones (the Central Mobile Belt, Fig. 4). These have been described by Williams and others (1989b). Yet in the same area and partly overlapping both the Humber and Avalon zones there are several completely posttectonic granites such as the Ackley (ca. 354 Ma) (Fig. 9). Similar posttectonic granitoids occur all the way to Connecticut.

Detailed work in the southern and central Appalachians on specifically Devonian granitoids is rare, but Kish (1990) has recently referred to some fifteen intrusions, which he classified into synkinematic to postkinematic tectonic settings. Their ages range between 435 and 354 Ma. Despite such a wide range, Kish found the data supportive of an age of 399±17 Ma for regional metamorphism in the Talladega belt of Alabama and Georgia. Kish pointed out that in general the ages accord well with the Devonian granites in New England. However, the mild metamorphism of the country rock is in many places Alleghanian.

In the northern Appalachians of New Brunswick and much of northern and eastern Maine (Fig. 9), granitoids intrude deformed Silurian-Devonian strata of moderate to low metamorphic grade. In southern Newfoundland, however, the grade of the country rocks is in places high, and it is here that the metamorphism is Silurian. Farther south, in the Appalachians of New Hampshire, Vermont, and central Massachusetts, high-grade metamorphism is Devonian (Chamberlain and Robinson, 1989; Schumacher and others, 1989). Farther south in Connecticut and parts of Rhode Island metamorphism is variable and is principally Alleghanian. These three metamorphic culminations appear to be independent of each other, although in Maine there are two successive regional metamorphic events attributed to the Acadian (Guidotti, 1989b).

Guidotti and Cheney (1989) pointed out that in western Maine both Devonian metamorphic isograds are interrupted by a post-Devonian regional metamorphism, which according to Aleinikoff and others (1985) may be of the same age as the Sebago pluton (Fig. 9). Nevertheless Guidotti and Cheney considered this metamorphism as late Acadian. However, since the pluton and associated metamorphic aureole lie within the boundary of Zartman's Permian disturbed belt where granitoids of Carboniferous age are common, we consider the Sebago and its aureole to have been produced during early Alleghanian stages.

An area of Late Paleozoic (Permian-Carboniferous) mild metamorphism also occurs in southern New Brunswick, where it has been described by Rast and Grant (1973) as associated with overthrusts at the edge of the Avalon belt and is independent of mild Acadian metamorphism that affects the Mascarene belt (Fig. 4).

In addition, mid-Paleozoic metamorphism affected the Meguma, but as already pointed out, the deformation and metamorphism range from Devonian to Carboniferous, and there is no evidence for Silurian metamorphism.

The irregular distribution of metamorphism and particularly its temperature highs can be traced throughout the northern Appalachians. Although in places this is structurally conditioned (Chamberlain and Karabinos, 1987), elsewhere there is no obvious relationship between macrostructures and regional metamorphism (Chamberlain and Rumble, 1988). Furthermore, within the larger metamorphic hot spots detected from mineral isograds, there are much smaller localized zones of particularly high temperature and the occurrence of kinzigitic gneisses and graphitic veins of granulite facies for which a temperature of

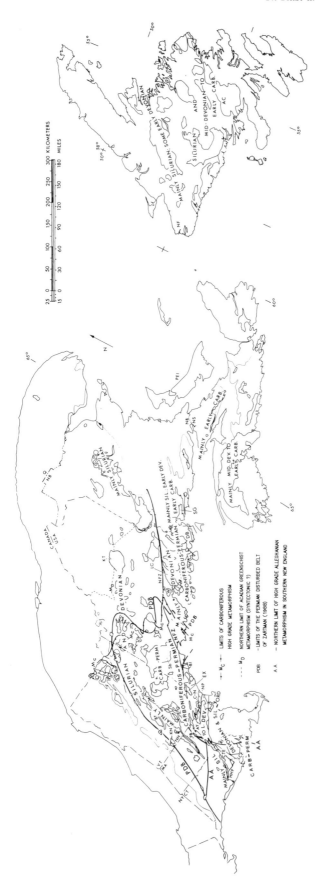

about 700°C has been estimated. Chamberlain and Rumble (1988) suggested that this occurred because of focusing of heat by hydrothermal fluid advection through fractures and channels. It is possible that the more extensive regional metamorphism was also focused but probably through upward rise of granitic and other magmas. This may be activated either by a compressive mechanism that is associated with syntectonic intrusions and migmatization or by posttectonic relaxation of pressure, permitting gravity-induced rise of later granites.

Relationships of plutonism to large structure

In the United States most magmatic events associated with the areas of metamorphism are commonly considered as Acadian *sensu stricto*. In Canada, Whalen (1989) pointed out the widespread Silurian granitic magmatism, although this is also known from parts of the United States (Ayuso and others, 1988; Sinha, 1988). The Devonian granites in the northern Appalachians of the United States can be divided into two groups: the early Devonian of about 400-Ma age (cf. Hubacher and Lux, 1987) and the mid- to late-Devonian ca. 380 to 350 Ma (Table 2). In northern Maine the former granites are taken to represent the main phase (mid-Devonian) of Acadian orogeny, although they crosscut the folds and associated cleavage that in the Canadian Appalachains are inferred to be of Silurian age (Dunning and others, 1988; Dunning and others, 1990). Moreover, although Hubacher and Lux (1987) suggested that the upright folding and cleavage in northeast Maine is Devonian, there is evidence for an unconformity separating Silurian and Devonian rocks near Presque Isle, Maine. This unconformity has been taken by Roy (1980) to be the Salinic disturbance, which is also equivalent to the Ludlovian change of facies in the British Isles and possibly the Late Silurian orogeny of Newfoundland (O'Brien and others, 1991) the Caledonian main phase of Greenland (Henriksen, 1985), and the Scandian of Scandinavia. Thus, it appears that the Salinic event is probably one of the most widespread in the North Atlantic. Whether the movements that have caused it were tectonic or eustatic at present is impossible to demonstrate with complete certainty. However, tectonic movements are very likely, since the Silurian development of red facies led to the formation of the Old Red Sandstone continent (Goldring and Langen-

Figure 9. Distribution of Acadian plutons and regional high-grade metamorphism in the northern Appalachians. Modified after Sinha (1988), Guidotti (1989a, b), Williams (1978), Robinson and others (1988), and Keppie (1982). Plutons, including those mentioned in text: AC—Ackley; AY—Ayer; BI—Biddeford; BT—Bethlehem; CTE—Canterbury; DB—DeBlois; EX—Exeter; FB—Fitchburg; KN—Kinsman; KT—Katahdin; LC—Lucerne; NP—Newburyport; SB—Sebago; SG—St. George; W—Waldo. Other features identified by abbreviation: C-N—Clinton-Newbury fault zone; HHF—Honey Hill fault; LCF—Lake Char fault; NFZ—Norumbega fault zone. American states and Canadian provinces as in Figure 2.

strassen, 1979). The effects of Devonian movements produced a widespread elevation of the continent (Neuman and others, 1989), to be followed by the spread of a Mississippian (Hercynian) ocean in the British Isles, in Maritime Canada, and possibly to the east of the southern Appalachians. This ocean was separated from an epicontinental sea by the Appalachian chain. The sea to the west of the Appalachian chain contains a distinctly endemic fauna (Sando and others, 1975).

The analysis of structural, metamorphic, and plutonic patterns in the Newfoundland Appalachians (Williams and others, 1989a, b) yields evidence for Late Silurian (Salinic) deformation and granites. These were succeeded by transcurrent movements and late granitoids attributed to the Acadian (Devonian) deformation, although as pointed out by Whalen (1989), this is not a unique interpretation. Strong (1980) has indicated the association of Appalachian granites and transpressive (strike-slip) regimes. On a similar basis Coyle and Strong (1987) proposed that granites of the Canadian and Maine Appalachians as well as those of the British Isles originate along the trace of the Iapetus suture—separating continental and oceanic crusts—where it has been affected by deep transcurrent faults that tapped the various levels of the lower crust. The trend-parallel faults, according to Currie and Piasecki (1989), were sinistral, but as is pointed out by Van der Pluijm and van Staal (1988), most of the faults identifiable on the ground are dextral, as they are in the Gaspé peninsula. We suggest that the faults are indeed dextral and originated by the distortion of the block of the Central Mobile belt (Fig. 4), including both the Gander and Dunnage zones of Williams (1978), between the Laurentian (Humber) and Avalon continents. This suggestion allows a synthesis of plate tectonic movements from Silurian to Devonian times.

PLATE TECTONIC INTERPRETATION

The Lower to Middle Paleozoic evolution of the Iapetus Ocean has frequently been conceived as a two-sided system (Bird and Dewey, 1970; Bradley, 1983), with either the Laurentian and Baltic continents facing each other or, as is thought more recently, the Laurentian and Gondwanan continents being opposites. Because it is difficult with a two-sided system to place two opposing continents precisely, there has been much recent disagreement as to the relative positions of Laurentia and Gondwana (Mason, 1988; Murphy and Nance, 1989). A more cogent system, proposed by Soper and Hutton (1984) and further expanded by Hutton (1989), is one that envisions a three-continent—Laurentia, Baltica, and Gondwana or the associated Armorica interaction—as is clear in Figure 1 and is diagrammed in Figure 10a. The interaction takes place across a triple junction.

In this scheme, the Iapetus Ocean has two branches—one separating Greenland and the Scandinavian part of Baltica and the other separating the main Laurentian plate from Gondwana. The former we refer to as the northeastern Iapetus and the latter as the southwestern Iapetus, although it is likely that most of the system lay in the southern hemisphere (Van der Voo, 1988).

The sequence of events subsequent to the Taconian orogeny follows.

1. Post-Taconian movements of the continental plates across the triple junction (Fig. 10a) led to the approach of Greenland and Baltica. The southwestern Iapetus ocean (SIO) at this stage consisted of several small oceans and intervening islands as volcanic and continental blocks. Some of these blocks were part of the Proterozoic Avalon island arc, whereas others were new volcanic islands (Neuman and others, 1988) and continental fragments (Chain Lakes massif).

Some of the island blocks in the northern Appalachians adhered to, or approached, the Laurentian craton. These indentor belts have distinct lithostratigraphies and vary in extent. The largest are the Merrimack trough and the Avalon, and the smallest is the Casco Bay (Fig. 4). Belts such as the Benner Hill and the Mascarene represent sediments caught and deformed between the indentor belts. In Britain, sinistral strike-slip deformation of the former Iapetus terranes occurred. It should be noted that the individual belts of the Acadian orogen can be considered in a tectonic framework as terranes or blocks in the Iapetus Ocean.

2. The approach of Baltica and Greenland resulted in a mid- to late-Silurian collision that produced divergent overthrusting in the East Greenland and Scandinavian Caledonides (Fig. 10b). We interpret this collision as generating the Salinic disturbance in the northern Appalachians, parts of Britain, and Armorica. The dominant fault motions in the SIO were dextral. However, in Britain, situated at the triple junction (Fig. 1, TJ; Fig. 10), sinistral deformation of Iapetan sediments occurred (Hutton, 1989). At this stage, Avalonian blocks could have been divided into two major fragments—the eastern and western Avalon (Neugebauer, 1989).

In Scandinavia the late Silurian Scandian orogeny (Table 1) (Dallmeyer, 1990) may represent the most complete manifestation of structural upheaval associated with the Baltica-Greenland collision. The southernmost manifestation of the collisional effects is recognized in Newfoundland (Dunning and others, 1990) and in New Brunswick, with high-grade metamorphic rocks surrounding granitoids of late Silurian age. It is significant, and Dunning and others (1990) also thought this, that the orogeny was Salinic. They proposed the restriction of the term *Acadian* to New England. However, we suggest the retention of the term *Caledonian* in Europe.

3. The next complex episode was Early to Middle Devonian (410 to 380 Ma) collision and final closure of any basins of deposition (Fig. 10c). This episode in the northern Appalachians is traditionally known as the Acadian orogeny. Former terranes from the Iapetus Ocean at this stage became compressional indentors. One indentor at this time was the Merrimack trough terrane (Fig. 4), which was probably a part of western Avalonia, in an area where the Iapetan crust was still oceanic. This collision involved high thermotectonic energy, resulting in the high-grade metamorphic spots of New Hampshire (Chamberlain and Rumble, 1988), Massachusetts (Schumacher and others, 1989), and northern Connecticut (Robinson and others, 1989).

We interpret the rather abrupt margin of high-grade metamorphism across Maine, which has been interpreted by Guidotti (1989a, b) as a position where flat-lying bodies of granites are replaced by more diapiric plutons, as a margin of the small ocean that occupied the site of New Hampshire, central Massachusetts, an Connecticut. Here the earliest recumbent folds and thrusts form a divergent system and are affected by later episodes of upright folding and doming (Schumacher and others, 1989; Eusden, 1988). This is the area where the main phase of the Acadian orogeny is conventionally given an age of 380 Ma, although the highest-grade metamorphism is earlier and has been attributed to the penetration of metamorphic fluids (Chamberlain

and Rumble, 1988). It is suggested that it is this Devonian collision that had as its indentor the Merrimack trough terrane as well as other blocks that had already adhered to it. The essentially compressive to transpressive collision at first produced dominantly sinistral faults, but these were succeeded by dextral faulting, which emplaced the Putnam-Nashoba zone (Fig. 4) of area 6 (Fig. 2) from an unknown location.

In eastern Avalonia, and particularly the Armorican Peninsula, a 380-Ma collision resulted in the Ligerian orogeny (Audren and Triboulet, 1989) where anatectic granites followed the emplacement of nappes. As in Nova Scotia, these events are episodically succeeded by Carboniferous phases of deformation and later

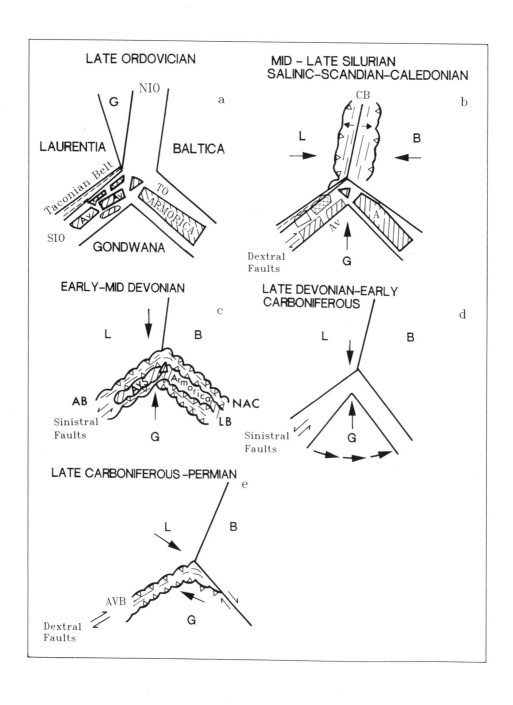

granite emplacement. Thus we consider the Armorican–Nova Scotian Acadian-Ligerian orogenic belt (Fig. 1) as yet another manifestation of mid-Paleozoic tectonics. The lateral equivalent of this belt has to be searched for in the southern Appalachians, where the Carolina block collided with, and was probably obducted onto, the thin wedge of the margin of Laurentia.

4. In the southern Appalachians it is thought (Hatcher, 1989) that despite the possibly earlier Ordovician collision, there was also a later Devonian–Early Mississippian collisional orogeny (Fig. 10d). The Avalon (Carolina) block (Fig. 1) acted as an indentor, possibly in a transpressive regime with a dextral strike-slip component and in places succeeded by oblique sinistral movements. The dextral displacement is reflected in subparallel shear zones, some being recorded from southwest Georgia as synchronous with the peak of metamorphism (Hooper and Hatcher, 1990), whereas others are later, and all are crosscut by sinistral faults. The peak of metamorphism in the area is conventionally taken as 350 Ma, later than the northwestern transport in the area, dated 360 Ma (Glover, 1989). Glover suggested that these Acadian movements involved the subduction of oceanic crust and melange below the Carolina block. Glover (1989) and Ferrill and Thomas (1988) suggested that oblique collision was the cause of the Acadian orogeny. Moreover, they proposed a

single, irregular Avalonian indentor, which produced successive orogenic highs by collision with the Laurentian craton.

We propose collision of separate blocks, each causing a thickening of the crust, high metamorphism, granite emplacement, and the generation of separate clastic wedges. The transpressive events in this system, we suggest, were later than the main collision, which had produced what Secor and others (1986) have named the Delmar deformation, resulting in cover rocks verging from northwest to southeast. The precise age of deformation is uncertain, although at least in part it appears to be pre-415 Ma. Between the areas considered by Hooper and Hatcher (1990) and Secor and others (1986) lies a large territory unmapped in detail. The Ocmulgee fault that separates the Avalon terrane from the Inner Piedmont of Georgia is considered a suture (Hooper and Hatcher, 1990) along which the accretion of lower-grade Avalon terrane to higher-grade Laurentia took place. The fault is a steep dextral shear zone, and the rocks of the adjacent Piedmont contain kyanite and are, therefore, middle- to high-pressure gneisses. The thermal peak of metamorphism in the Piedmont terrane is ca. 365 Ma (Odom and others, 1982), presumably Acadian, and may be equivalent to the post-Delmar metamorphism of South Carolina. In fact, Secor and others (1986) indicated that this belt of metamorphism extends from Virginia to Georgia, parallel to and of the same trend and length as the Carolina terrane. We suggest that this terrane was an indentor that caused a considerable thickening of the Appalachian crust, causing high metamorphic temperature and pressure.

Thus Acadian metamorphic highs date collisions of large fragments (blocks) of the Avalon superterrane with Laurentia, and transpressive events (first dextral and then sinistral) are essentially successors to these collisions and are caused by other plate tectonic events, including rotation. The cause of the Acadian movements *sensu stricto* can be explained by changes in the motion of the Gondwana-Armorica plates that also cause oblique closure of the Tornquist ocean (sea) (Fig. 10a; cf. Ziegler, 1982, p. 22). Similarly, the metamorphic high of Nova Scotia was also collisional, but the indentor must have been to the southeast (in present-day coordinates) of the Meguma and possibly was part of the Iberian microplate. The latter also was responsible for the collision with the Armorican plate to generate the Ligerian orogeny of the Armorican massif (cf. Lefort, 1989, p. 117). However, it can also be maintained that the indentor against the Meguma was Gondwana (cf. Lefort, 1989, p. 124).

5. The last Paleozoic tectonic episode in the North Atlantic area was the Alleghanian (Fig. 10e). This episode has left a pronounced overprint in southeastern New England (Gromet, 1989) and resulted in the collision of the composite Boston Avalon, Hope Valley alaskitic block that wedged under the Merrimack and Nashoba terranes (Fig. 4) and the Acadian orogenic belt of southern New England (New Hampshire to Connecticut, Fig. 4). Some of the effects of this episode have been described recently by Zartman (1988). From our point of view the Boston Avalon terrane (Rast and Skehan, 1990; Hepburn and others, 1987) had not suffered any Acadian (mid-Devonian) orogenic

---

Figure 10. Schematic plate tectonic development of mid- to late-Paleozoic orogenic belts in the Atlantic region. The figure illustrates the interaction of the large Paleozoic plates and smaller blocks. Thick arrows show direction of plate motion or rotation. Fault motion arrows indicate relative stresses within or between plates. Thrusts are shown with teeth on upper plate. a, Late Ordovician scheme. The Taconian orogenic belt is already adherent to Laurentia. Triple junction is stationary, post–Middle Ordovician arc collision. Av—Avalon blocks; G—Greenland; NIO—Northeastern Iapetus Ocean; SIO—Southwestern Iapetus Ocean; TO—Tornquist Ocean (sea?). Cratons—unornamented; Avalon terranes—hachured; non-Avalon terranes—cross-hatched; Taconian belt—dashed. b, Mid- to late-Silurian. Northeast Iapetus is closed, and Scandinavia and Greenland, as well as in part British Caledonides, form as mountain chains. Salinic movements in the northern Appalachians. Small oceanic domains persist in the southwestern Iapetus, where they separate Avalonian and other terranes. A—Armorica; Av—Avalonian blocks, hachured; B—Baltica; CB—Caledonian orogenic belt (thin line ornament); G—Gondwana; L—Laurentia. c, Early to Middle Devonian, formation of Acadian-Ligerian collisional belts. Note that the Acadian is between the Avalon superterrane (AvS) and Laurentia, and the Ligerian is between the Avalon superterrane and Gondwana. In North America both have been referred to as Acadian. We suggest that the equivalent of the Ligerian in North America be referred to as the Scotian. AB—Acadian belt; LB—Ligerian belt; NAC—North European Caledonides. Collision of Baltica and Laurentia distorts the Iapetan lithosphere and generates on the Laurentian side a major fragmented block with accompanying sinistral faults. Symbols as in d, Late Devonian–early Carboniferous. Drift of Gondwana toward Laurentia-Baltica and rotation accompanied by sinistral faults. The direction of relative rotation that resulted in production of sinistral faults is shown by arrows. Other symbols as before e, Late Carboniferous–Permian collision of Gondwana with Laurentia-Baltica. Alleghanian orogeny. AVB—Alleghanian-Variscan belt.

episode and, therefore, collided with Laurentia later than the Merrimack trough or the Carolina terranes. Thus in Silurian-Devonian-Carboniferous times what is now called the Avalonian composite terrane was represented by a series of blocks that, upon collision with Laurentia and its miogeocline, produced orogenic episodes of somewhat different ages and at somewhat different times. Most of the blocks collided in Devonian times, producing the Acadian orogeny, whereas the Boston Avalon block arrived in Carboniferous–Early Permian time at approximately the same time as Gondwana and Eurasia accreted together.

The accretion of either smaller blocks or supercontinents may have been accompanied by rotations or oblique motions. Thus on a large scale three possible circumstances were likely: (1) head-on collision involving high compression across the suture, (2) oblique collision resulting in either sinistral or dextral transpression in the area of the suture, and (3) transcurrent motion resulting in transcurrent faults at the suture. The head-on collision of major plates or microplates with the major plates produced linear orogenic belts, with localized effects if the microplates were relatively small and a more extensively maintained trend of deformation if the microplates were long, as for example an island arc. Our modified model envisions the Scandinavian and Greenland Late Silurian orogens as having been produced by a continent-continent collision (Fig. 10b) in a system (Fig. 10a) that involved three major continental masses and also smaller, but quite considerable, microplates. The effect of colliding Laurentia and Baltica, producing Laurasia, has been to distort the Iapetan lithosphere and to generate on the Laurentian side a major fragmented block with accompanying sinistral faults (Fig. 10c). The fragments of Avalon terrane that existed at this time would have participated in collision followed by dissection by faulting (Acadian orogeny). There is some paleomagnetic evidence of Early Devonian movement (Van der Voo, 1988) in which Gondwana moved away from Armorica, but during the Ligerian episode Gondwana moved toward Armorica in such a way that some rotation took place, producing dominantly sinistral faults (Fig. 10d). Lastly (Fig. 10e), it was in Carboniferous times that Gondwana reapproached Laurasia, again leading to the Alleghanian-Variscan orogenic episodes and involving rotations as well as collisions (Wintsch and Lefort, 1984).

Thus the attempts at discussing sedimentation, deformation, metamorphism, and granite emplacement in terms of worldwide orogenic cycles from a plate tectonic point of view, are unsound. Even notions such as time-transgressive episodes can only apply to individual pairs of plates or microplates. Terms such as cycles should apply only to specific local episodes but must not be assumed as repetitions of alternating movements. Salinic, Caledonian, Scandian, and Acadian are all sound concepts locally, provided that we do not extend them beyond the spatial and temporal limits of their definitions. One can accurately correlate stratigraphies, but not orogenies.

ACKNOWLEDGMENTS

This research was supported in part by a grant from the Jesuit Community of Boston College and in part by the Hudnall Endowment of the University of Kentucky. We thank Frances Ahearn, Edward M. Myskowski, and Patricia C. Tassia for assistance in various stages of the preparation of drafts of the manuscript. We are grateful to Professors Richard Naylor and Rolfe Stanley, who provided very helpful reviews of the original manuscript and suggestions for its improvement. David C. Roy graciously provided a final review of the revised manuscript and made suggestions that we believe have materially abbreviated and focused the manuscript. Any errors, however, are the responsibility of the authors.

REFERENCES CITED

Aleinikoff, J. N., Moench, R. H., and Lyons, J. B., 1985, Carboniferous U-Pb age of the Sebago batholith, southwestern Maine: Geological Society of America Bulletin, v. 69, p. 990–996.

Ashwal, L. D., Leo, G. W., Robinson, P., Zartman, R. E., and Hall, D. J., 1979, The Belchertown quartz monzodiorite pluton, West-Central Massachusetts; a syntectonic Acadian intrusion: American Journal of Science, v. 299, p. 936–969.

Audren, C., and Triboulet, C., 1989, Pressure-temperature-time deformation paths in metamorphic rocks and tectonic processes, as exemplified by the Variscan orogeny in South Brittany, France, *in* Daly, J. S., Cliff, R. A., and Yardley, B.W.D., eds., Evolution of metamorphic belts: Geological Society of London Special Publication 43, p. 441–446.

Autran, A., and Cogné, J., 1980, La zone interne de l'orogene varisque dans l'ouest de la France et sa place dans le developpement de la chaine hercynienne: 26th International Geological Congress in Paris, Colloque C. 6, p. 90–111.

Autran, A., and Guillot, P. L., 1975, L'evolution orogenique et metamorphique du Limousin au Paléozoique (Massif Central francais): Paris, Académie des Sciences, Comptes Rendus Hebdomadaires des Séances, t. 280, D, p. 1649–1672.

Ayuso, R. A., Horan, M., and Criss, R. E., 1988, Pb and O isotopic geochemistry of granitic plutons from northern Maine, *in* Sinha, A. K., Hewitt, D. A., and Tracy, R. J., eds., Frontiers in petrology: A special volume of the American Journal of Science, v. 288-A, p. 421–460.

Bailey, E. B., 1928, The Paleozoic mountain systems of Europe and America: Report to the British Association for the Advancement of Science, p. 57–76.

Barnes, R. P., Lintern, B. C., and Stone, P., 1989, Timing and regional implications of deformation in the Southern Uplands of Scotland: Geological Society of London Journal, v. 146, p. 905–908.

Barosh, P. J., and Pease, M. H., 1981, Correlation of the Oakdale and Paxton Formations with their equivalents from eastern Connecticut to southern Maine: Geological Society of America Abstracts with Programs, v. 13, p. 122.

Barr, S. M., and Raeside, R. P., 1986, Precarboniferous tectonostratigraphic subdivisions of Cape Breton Island, Nova Scotia: Maritime Sediments and Atlantic Geology, v. 22, p. 252–263.

Bates, R. L., and Jackson, J. A., eds., 1987, Glossary of geology: Falls Church, Virginia, American Geological Institute, 788 p.

Beland, J., 1974, La tectonique des Appalaches du Quebec: Geoscience Canada, v. 1, p. 26–32.

Bertrand, M., 1887, La Chaine des Alpes, et la formation du continent européen:

Bulletin de la Société Géologique de France, Série 3, v. 15, p. 423–457.

Bevier, M. L., 1988, Timing of Paleozoic plutonism in the Miramichi terrane, New Brunswick: Geological Society of America Abstracts with Programs, v. 20, p. 7.

Bevier, M. L., and Whalen, J. B., 1988, Implications of Silurian magmatism for the tectonic evolution of the Canadian Appalachians: Geological Society of America Abstracts with Programs, v. 20, p. A216.

Bird, J. M., and Dewey, J. F., 1970, Lithosphere plate–continental margin tectonics and the evolution of the Appalachian orogen: Geological Society of America Bulletin, v. 81, p. 1031–1060.

Boone, G. M., and Boudette, E. L., 1989, Accretion of the Boundary Mountains Terrane within the northern Appalachian orthotectonic zone, *in* Horton, J. W., Jr., and Rast, N., eds., Melanges and olistostromes of the U.S. Appalachians: Geological Society of America Special Paper 228, p. 17–42.

Bothner, W. A., Boudette, E. L., Fagan, T. J., Gaudette, H. E., Laird, J., and Olszewski, W. J., 1984, Geologic framework of the Massabesic Anticlinorium and Merrimack Trough, southeastern New Hampshire, *in* Hanson, L. S., ed., Geology of the Coastal Lowlands, Boston, MA to Kennebunk, ME: New England Intercollegiate Geologic Conference, Salem State College, Salem, Massachusetts, p. 186–206.

Boucot, A. J., 1962, Appalachian Siluro-Devonian, *in* Coe, K., ed., Some aspects of the Variscan Fold Belt: 9th Inter-University Geological Congress: Manchester, England, Manchester University Press, p. 155–163.

—— , 1968, Silurian and Devonian of the northern Appalachians, *in* Zen, E-an, White, W. S., Hadley, J. B., and Thompson, J. B., Jr., eds., Studies of Appalchian geology: Northern and maritime: New York, Wiley Interscience, p. 83–94.

Bradley, D. C., 1983, Tectonics of the Acadian orogeny in New England and adjacent Canada: Journal of Geology, v. 91, p. 381–400.

Brookins, D. G., 1967, Rb-Sr age evidence for Permian metamorphism of the Monson Gneiss, west-central Massachusetts: Geochimica and Cosmochimica Acta, v. 31, p. 281–283.

Cady, W. M., 1960, Stratigraphic and geotectonic relationships in northern Vermont and southern Quebec: Geological Society of America Bulletin, v. 71, p. 531–576.

Cawood, P. A., and Williams, H., 1988, Acadian basement thrusting, crustal delamination, and structural styles in and around the Humber Arm allochthon, western Newfoundland: Geology, v. 16, p. 370–373.

Chamberlain, C. P., and Karabinos, P., 1987, The influence of deformation on the pressure-temperature paths of metamorphism: Geology, v. 15, p. 42–44.

Chamberlain, C. P., and Robinson, P., eds., 1989, Styles of Acadian metamorphism with depth in the Central Acadian High, New England (a field trip honoring J. B. Thompson, Jr.): University of Massachusetts Department of Geology and Geography Contribution No. 63, 140 p.

Chamberlain, C. P., and Rumble, D., 1988, Thermal anomalies in a regional metamorphic terrane; an isotopic study of the role of fluids: Journal of Petrology, v. 29, p. 1215–1232.

Choukroune, P., Roure, F., Pinet, B., and ECORS Pyrenees Team, 1990, Main results of the ECORS Pyrenees profile: Tectonophysics, v. 173, p. 411–423.

Coyle, M., and Strong, D. F., 1987, Geology of the Springdale Gorup: A newly recognized Silurian epicontinental-type caldera in Newfoundland: Canadian Journal of Earth Sciences, v. 24, p. 1135–1148.

Currie, K. L., and Piasecki, M.A.J., 1989, Kinematic model for southwestern Newfoundland based upon Silurian sinistral shearing: Geology, v. 17, p. 938–941.

Dallmeyer, R. D., 1988, Late Paleozoic tectonothermal evolution of the western Piedmont and eastern Blue Ridge, Georgia: Controls on the chronology of terrane accretion and transport in the southern Appalachian orogen: Geological Society of America Bulletin, v. 100, p. 702–713.

—— , 1990, ^{40}Ar/^{39}Ar mineral age record of a polyorogenic evolution within the Seve and Köli nappes, Trøndelag, Norway: Tectonophysics, v. 179, p. 199–226.

Dallmeyer, R. D., and Keppie, J. D., 1987, Polyphase late Paleozoic tectono-thermal evolution of the southwestern Meguma Terrane, Nova Scotia: Evidence from ^{40}Ar/^{39}Ar mineral ages: Canadian Journal of Earth Sciences, v. 24, p. 1242–1254.

Dennis, J. G., ed., 1967, International tectonic dictionary, English terminology: Tulsa, American Association of Petroleum Geologists, 196 p.

de Sitter, L. U., 1964, Structural geology: New York, McGraw-Hill Book Co., 551 p.

Donohoe, H. V., and Pajari, G. E., 1974, The age of the Acadian deformation in Maine–New Brunswick: Maritime Sediments, v. 9, p. 78–82.

Dostal, J., Wilson, R. A., and Keppie, J. D., 1989, Geochemistry of Siluro-Devonian Tobique volcanic belt in northern and central New Brunswick (Canada): Tectonic implications: Canadian Journal of Earth Sciences, v. 26, p. 1282–1296.

Dunning, G. R., and Cousineau, P. A., 1990, U/Pb ages of single zircons from Chain Lakes massif and a correlative unit in ophiolitic melange in Quebec: Geological Society of America Abstracts with Programs, v. 22, p. 13.

Dunning, G. R., Krogh, T. E., O'Brien, S. J., Colman-Sadd, S. P., and O'Neill, P., 1988, Geochronologic framework for the central mobile belt in southern Newfoundland and the importance of Silurian orogeny: Geological Association of Canada, Mineralogical Association of Canada, and Canadian Geophysical Union Joint Annual Meeting, Programs with Abstracts, v. 13, p. A34.

Dunning, G. R., Barr, S. M., Raeside, R. P., and Jamieson, R. A., 1990, U-Pb zircon, titanite, and monazite ages in the Bras d'Or and Aspy terranes of Cape Breton Island, Nova Scotia; Implications for igneous and metamorphic history: Geological Society of America Bulletin, v. 102, p. 322–330.

Eusden, J. D., Jr., 1988, Stratigraphy, structure, and metamorphism across the "dorsal zone," central New Hampshire, *in* Bothner, W. A., ed., Guidebook for field trips in southwestern New Hampshire, southeastern Vermont, and north-central Massachusetts: New England Intercollegiate Geological Conference, Keene State College, Keene, New Hampshire, p. 40–59.

Ferrill, B. A., and Thomas, W. A., 1988, Acadian dextral transpression and synorogenic sedimentary successions in the Appalachians: Geology, v. 16, p. 604–617.

Fyffe, L. R., 1982, Taconian and Acadian structural trends in central and northern New Brunswick, *in* St-Julien, P., and Beland, J., eds., Major structural zones and faults of the northern Appalachians: Geological Association of Canada Special Paper 24, p. 117–130.

Fyffe, L. R., and Fricker, A., 1987, Tectonostratigraphic terrane analysis of New Brunswick: Maritime Sediments and Atlantic Geology, v. 23, p. 113–122.

Fyffe, L. R., Pajari, G. E., Jr., and Cherry, M. E., 1981a, The Acadian plutonic rocks of New Brunswick: Maritime Sediments and Atlantic Geology, v. 17, p. 23–36.

Fyffe, L. R., Pajari, G. E., and Cormier, R. F., 1981b, Rb-Sr geochronology of New Brunswick: Geological Society of America Abstracts with Programs, v. 13, p. 133.

Fyffe, L. R., Barr, S. M., and Bevier, M. L., 1988, Origin and U-Pb geochronology of amphibolite facies metamorphic rocks, Miramichi Highlands, New Brunswick: Canadian Journal of Earth Sciences, v. 25, p. 1674–1686.

Gaudette, H. E., Bothner, W. A., Laird, J., and Olszewski, W. J., 1984, Late Precambrian/early Paleozoic deformation and metamorphism in southeastern New Hampshire—Confirmation of an exotic terrane: Geological Society of America Abstracts with Programs, v. 16, p. 516.

Geiser, P., and Engelder, T., 1983, The distribution of layer-parallel shortening fabrics in the Appalachian foreland of New York and Pennsylvania: Evidence for two non-coaxial phases of the Alleghanian orogeny, *in* Hatcher, R. D., Jr., Williams, H., and Zietz, I., Contributions to the tectonics and geophysics of mountain chains: Geological Society of America Memoir 158, p. 161–175.

Getty, S. R., and Gromet, L. P., 1988, The southern terminus of the Hope Valley shear zone and Alleghanian polyphase deformation of basement rocks in southeastern New England: Tectonics, v. 7, p. 1325–1338.

Glover, L., III, 1989, Tectonics of the Virginia Blue Ridge and Piedmont, in Glover, L. III, Evans, N. H., Patterson, J. G., and Brown, W. R., eds., Tectonics of the Virginia Blue Ridge and Piedmont—Culpeper to Richmond, Virginia (28th International Geological Congress, field trip guidebook T363): Washington, D.C., American Geophysical Union, p. 1–29.

Goldring, R., and Langenstrassen, F., 1979, Open shelf and near shore clastic facies in the Devonian, in House, M. R., ed., The Devonian system: Special Papers in Palaeontology 23 Palaeontological Association, p. 81–97.

Gromet, L. P., 1989, Avalonian terranes and late Paleozoic tectonism in southeastern New England; Constraints and problems, in Dallmeyer, R. D., ed., Terranes in the Circum-Atlantic Paleozoic orogens: Geologic Society of America Special Paper 230, p. 193–212.

—— , 1991, Late Paleozoic plutonism and mobilization (?) of Avalonian basement in the southern New England Appalachians: Geological Society of America Abstracts with Programs, v. 23, p. 176.

Guidotti, C. V., 1989a, Metamorphism in Maine: An overview, in Tucker, R. D., and Marvinney, R.G., eds., Studies in Maine geology: Vol. 3: Igneous and metamorphic geology: Augusta, Maine Geological Survey, p. 1–17.

—— , 1989b, Mineralogic and textural evidence for polymetamorphism along a traverse from Oquossoc to Phillips to Weld, Maine, in Berry, A. W., Jr., ed., Guidebook for field trips in southern and west-central Maine: New England Intercollegiate Geologic Conference, Farmington, Maine, p. 144–164.

Guidotti, C. V., and Cheney, J. T., 1989, Metamorphism in western Maine; an overview, in Chamberlain, C. P., and Robinson, P., eds., Styles of Acadian metamorphism with depth in the Central Acadian High, New England (a field trip honoring J. B. Thompson, Jr.): University of Massachusetts Department of Geology and Geography Contribution No. 63, p. 17–37.

Hall, L. M., and Robinson, P., 1982, Stratigraphic-tectonic subdivisions of southern New England, in St-Julien, P., and Beland, J., eds., Major structural zones and faults of the northern Appalachians: Geological Association of Canada Special Paper 24, p. 15–41.

Harland, W. B., 1985, Caledonide Svalbard, in Gee, D. G., and Sturt, B. A., eds., The Caledonide Orogen—Scandinavia and Related Areas: New York, John Wiley & Sons, p. 999–1016.

Hatch, N. L., Jr., 1982, The Taconian Line in western New England and its implications to Paleozoic tectonic history, in St-Julien, P., and Beland, J., eds., Major structural zones and faults of the northern Appalachians: Geological Association of Canada Special Paper 24, p. 67–85.

—— , 1987, Lithofacies, stratigraphy, and structure in the rocks of the Connecticut Valley trough, eastern Vermont, in Westerman, D. S., ed., Guidebook for field trips in Vermont, vol. 2: New England Intercollegiate Geologic Conference, Norwich University, Northfield, Vermont, p. 192–212.

—— , 1988, Some revisions to the stratigraphy and structure of the Connecticut Valley trough, eastern Vermont: American Journal of Science, v. 288, p. 1041–1059.

Hatch, N. L., Jr., Moench, R. H., and Lyons, J. B.,1983, Silurian–Lower Devonian stratigraphy of eastern and south-central New Hampshire: Extensions from western Maine: American Journal of Science, v. 283, p. 739–761.

Hatcher, R. D., Jr., 1989, Tectonic synthesis of the U.S. Appalachians, in Hatcher, R. D., Jr., Thomas, W. A., and Viele, G. W., eds., The Appalachian-Ouachita orogen in the United States (The Geology of North America, vol. F-2): Boulder, Colorado, Geological Society of America, p. 511–536.

Hatcher, R. D., Jr., Osberg, P. H., Drake, A. A., Jr., Robinson, P., and Thomas, W. A., 1989, Tectonic map of the U.S. Appalachians, in Hatcher, R. D., Jr., Thomas, W. A., and Viele, G. W., eds., The Appalachian-Ouachita orogen in the United States (The Geology of North America, vol. F-2): Boulder, Colorado, Geological Society of America, Plate 1.

Haughton, P.D.W., and Farrow, C. M., 1989, Compositional variations in Lower Old Red Sandstone detrital garnets from the Midland Valley of Scotland and the Anglo-Welsh Basin: Geological Magazine, v. 126, p. 373–396.

Henriksen, N., 1985, The Caledonides of central East Greenland 70-76N, in Gee, D. G., and Sturt, B. A., eds., The Caledonide orogen—Scandinavia and related areas: New York, John Wiley & Sons, p. 1095–1113.

Hepburn, J. C., Hill, M., and Hon, R., 1987, The Avalonian and Nashoba terranes, eastern Massachusetts; An overview: Maritime Sediments and Atlantic Geology, v. 23, p. 1–12.

Holland, C. H., 1987, Stratigraphical and structural relationships of the Dingle Group (Silurian), County Kerry, Ireland: Geological Magazine, v. 124, p. 33–42.

Hooper, R. J., and Hatcher, R. D., Jr., 1990, Ocmulgee fault: The Piedmont-Avalon terrane boundary in central Georgia: Geology, v. 18, p. 708–711.

Horton, J. W., Jr., and Zullo, V. A., 1991, The geology of the Carolinas; Carolina Geological Society 50th Anniversary Volume: Knoxville, University of Tennessee Press, 406 p.

Hsu, K. J., 1989, Time and place in Alpine orogenesis—The Fermor Lecture, in Coward, M. P., Dietrich, D., and Park, R. G., eds., Alpine tectonics: Geological Society of London Special Publication 45, p. 421–443.

Hubacher, P. A., and Lux, D. R., 1987, Timing of Acadian deformation in northeastern Maine: Geology, v. 15, p. 80–83.

Hueber, F. M., Bothner, W. A., Hatch, N. L., Jr., Finney, S. C., and Aleinikoff, J. N., 1990, Devonian plants from southern Quebec and northern New Hampshire and the age of the Connecticut Valley trough: American Journal of Science, v. 290, p. 360–395.

Hussey, A. M., II, 1962, The geology of southern York County, Maine: Maine Geological Survey Special Geologic Studies Series 4, 67 p.

—— , 1985, Bedrock geology of the Bath and Portland 2 degree map sheets, Maine: Maine Geological Survey Open File 85-87, 82 p., map scale 1:250,000.

—— , 1986, Stratigraphic and structural relationships between the Cushing, Cape Elizabeth, Bucksport, and Cross River Formations, Portland-Boothbay area, Maine, in Newberg, D. W., ed., Guidebook for field trips in southwestern Maine: New England Intercollegiate Geological Conference, Bates College, Lewiston, Maine, p. 164–183.

—— , 1988, Lithotectonic stratigraphy, deformation, plutonism, and metamorphism, greater Casco Bay region, southwestern Maine, in Tucker, R. D., and Marvinney, R. G., eds., Studies in Maine geology: Vol. 1: Structure and stratigraphy: Augusta, Maine Geological Survey, p. 17–34.

Hussey, A. M. II, and Newberg, D. W., 1978, Major faulting in the Merrimack synclinorium between Hollis, New Hampshire, and Biddeford, Maine: Geological Society of America Abstracts with Programs, v. 10, p. 48.

Hussey, A. M. II, Bothner, W. A., and Thompson, J. A., 1986, Geological comparisons across the Norumbega fault zone, southwestern Maine, in Newberg, D. W., ed., Guidebook for field trips in southwestern Maine: New England Intercollegiate Geological Conference, Bates College, Lewiston, Maine, p. 53–79.

Hutton, D.H.W., 1987, Strike-slip terranes and a model for the evolution of the British and Irish Caledonides: Geological Magazine, v. 124, p. 405–425.

—— , 1989, Pre-Alleghanian terrane tectonics in the British and Irish Caledonides, in Dallmeyer, R. D., ed., Terranes in the Circum-Atlantic Paleozoic orogens: Geological Society of America Special Paper 230, p. 47–57.

Karlstrom, K. E., Van der Pluijm, B. A., and Williams, P. F., 1982, Structural interpretation of the eastern Notre Dame Bay area, Newfoundland—Regional post–Middle Silurian thrusting and asymmetrical folding: Canadian Journal of Earth Sciences, v. 19, p. 2325–2341.

Keppie, J. D., 1982, The Minas geofracture, in St-Julien, P., and Beland, J., eds., Major structural zones and faults of the northern Appalachians: Geological Association of Canada Special Paper 24, p. 263–280.

Keppie, J. D., and Dallmeyer, R. D., 1987, Dating transcurrent terrane accretion: An example from the Meguma and Avalon composite terranes in the northern Appalachians: Tectonics, v. 6, p. 831–847.

Keppie, J. D., and Dallmeyer, R. D., comps., 1989, Tectonic map of pre-Mesozoic terranes in circum-Atlantic Phanerozoic orogens: Sponsored by International Geological Correlation Programme, Project #233, available from Department of Mines and Energy, P.O. Box 1087, Halifax, Nova Scotia, Canada B3J 2X1.

Kish, S. A., 1990, Timing of middle Paleozoic (Acadian) metamorphism in the southern Appalachians: K-Ar studies in the Talladega belt, Alabama: Geology, v. 18, p. 650–653.

Lefort, J.-P., 1989, Basement correlation across the North Atlantic: Berlin, Springer-Verlag, 148 p.

Loiselle, M. C., Eriksson, S., Wones, D. R., and Sinha, A. K., 1983a, Timing and emplacement of post-Acadian plutons in central and eastern Maine: Geological Society of America Abstracts with Programs, v. 15, p. 187.

Loiselle, M. C., Hon, R., and Naylor, R. S., 1983b, Age of the Katahdin batholith, Maine: Geological Society of America Abstracts with Programs, v. 15, p. 146.

Ludman, A., 1978, Stratigraphy and structure of Silurian and pre-Silurian rocks in the Brookton-Princeton area, eastern Maine, *in* Ludman, A., ed., Guidebook for field trips in southeastern Maine and southwestern New Brunswick: New England Intercollegiate Geologic Conference, Queens College, Flushing New York, p. 145–161.

—— , 1986, Timing of terrane accretion in eastern and east-central Maine: Geology, v. 14, p. 411–414.

Ludman, A., Hopeck, J., Costain, J. K., Domoracki, W. J., Coruh, C., and Doll, W. E., 1990, Seismic reflection evidence for the NW limit of Avalon in east-central Maine: Geological Society of America Abstracts with Programs, v. 22, p. 32.

Lyons, J. B., and Faul, H., 1968, Isotope geochronology of the northern Appalachians, *in* Zen, E-an, White, W. S., Hadley, J. B., and Thompson, J. B., Jr., eds., Studies of Appalachian geology: Northern and maritime: New York, Wiley Interscience, p. 305–318.

Lyons, J. B., and Livingston, D. E., 1977, Rb-Sr age of the New Hampshire Plutonic Series: Geological Society of America Bulletin, v. 88, p. 1808–1812.

Lyons, J. B., Boudette, E. L., and Aleinikoff, J. N., 1982, The Avalonian and Gander Zones in Central Eastern New England, *in* St-Julien, P., and Beland, J., eds., Major structural zones and faults of the northern Appalachians: Geological Association of Canada Special Paper 24, p. 43–66.

Malo, M., and Beland, J., 1989, Acadian strike-slip tectonics in the Gaspé region, Quebec Appalachians: Canadian Journal of Earth Sciences, v. 26, p. 1764–1777.

Marillier, F., and Durling, P., 1990, Structural trends and terrane boundaries in the southwestern Gulf of Maine from seismic data: Geological Society of America Abstracts with Programs, v. 22, p. 53.

Marillier, F., and Verhoef, J., 1989, Crustal thickness under the Gulf of St. Lawrence, northern Appalachians, from gravity and deep seismic data: Canadian Journal of Earth Sciences, v. 26, p. 1517–1532.

Marillier, F., and six others, 1989, Crustal structure and surface zonation of the Canadian Appalachians: Implications of deep seismic reflection data: Canadian Journal of Earth Sciences, v. 26, p. 305–321.

Mason, R., 1988, Did the Iapetus Ocean really exist? Geology, v. 16, p. 823–826.

McKerrow, W. S., 1988, The development of the Iapetus Ocean from the Arenig to the Wenlock, *in* Harris, A. L., and Fettes, D. J., eds., The Caledonide-Appalachian orogen: Geological Society of London Special Publication 38, p. 405–412.

Moench, R. H., 1970, Premetamorphic down-to-basin faulting, folding, and tectonic dewatering, Rangeley area, western Maine: Geological Society of America Bulletin, v. 81, p. 1463–1496.

—— , 1989, Metamorphic stratigraphy of the northwestern part of the Kearsarge–Central Maine synclinorium, western Maine, *in* Moench, R. H., and St.-Julien, P., 1989, Northern Appalachian transect: Southeastern Quebec, Canada through Western Maine, U.S.A. (28th International Geological Congress, field trip guidebook T358): Washington, D.C., American Geophysical Union, 52 p.

Moench, R. H., and Pankiwskyj, K. A., 1988a, Geologic map of western interior Maine, with contributions by Boone, G. M., Boudette, E. L., Ludman, A., Newell, W. R., and Vehrs, T. I.: U.S. Geological Survey Miscellaneous Investigations Map I-1692, scale 1:250,000.

—— , 1988b, Definition, problems, and reinterpretation of early premetamorphic faults in western Maine and northeasten New Hampshire, *in* Tucker, R. D., and Marvinney, R. G., eds., Studies in Maine geology: Vol. 1: Structure and stratigraphy: Augusta, Maine Geological Survey, p. 35–50.

Murphy, J. B., 1987, Petrology of Upper Ordovician–Lower Silurian rocks of the Antigonish Highlands, Nova Scotia: Canadian Journal of Earth Sciences, v. 24, p. 752–759.

Murphy, J. B., and Nance, D. R., 1989, Model for the evolution of the Avalonian-Cacomian belt: Geology, v. 17, p. 735–738.

Naylor, R. S., 1969, Age and origin of the Oliverian domes, central-western New Hampshire: Geological Society of America Bulletin, v. 80, p. 405–428.

—— , 1971, Acadian orogeny: An abrupt and brief event: Science, v. 172, p. 558–560.

Neugebauer, J., 1989, The Iapetus model: A plate tectonic concept for the Variscan belt of Europe: Tectonophysics, v. 169, p. 229–256.

Neuman, R. B., 1967, Bedrock geology of the Shin Pond and Stacyville Quadrangles, Penobscot County, Maine: U.S. Geological Survey Professional Paper 524-I, 37 p.

Neuman, R. B., Van der Voo, R., and Van der Pluijm, B. A., 1988, The Ordovician Iapetus Ocean; Latitudinal stability for the bordering cratons and mobility for the terranes in the Northern Appalachian Central Mobile Belt: Geological Association of Canada, Mineralogical Association of Canada, and Canadian Geophysical Union Joint Annual Meeting, Program with Abstracts, v. 13, p. A90.

Neuman, R. B., Palmer, A. R., and Dutro, J. T., Jr., 1989, Paleontological contributions to Paleozoic paleogeographic reconstructions of the Appalachians, *in* Hatcher, R. D., Jr., Thomas, W. A., and Viele, G. W., ed., The Appalachian-Ouachita orogen in the United States (The Geology of North America, vol. F-2): Boulder, Colorado, Geological Society of America, p. 375–384.

O'Brien, S. J., O'Brien, B. H., O'Driscoll, C. F., Dunning, G. R., Holdsworth, R. E., and Tucker, R., 1991, Silurian orogenesis and the NW limit of Avalonian rocks in the Hermitage flexure, Newfoundland Appalachians: Geological Society of America Abstracts with Programs, v. 23, p. 109.

Odom, A. L., Hatcher, R. D., Jr., and Hooper, R. J., 1982, A premetamorphic tectonic boundary between contrasting Appalachian basements, southern Georgia Piedmont: Geological Society of America Abstracts with Programs, v. 14, p. 579.

Osberg, P. H., 1978, Synthesis of the geology of the northeastern Appalachians, U.S.A., *in* IGCP Project 27, Caledonian-Appalachian orogen of the North Atlantic Region: Geologic Survey of Canada Paper 78-13, p. 137–147.

—— , 1988, Geologic relations within the shale-wacke sequence in south-central Maine, *in* Tucker, R. D., and Marvinney, R. G., eds., Studies in Maine geology: Vol. 1: Structure and stratigraphy: Augusta, Maine Geological Survey, p. 51–73.

Osberg, P. H., Hussey, A. M. II, and Boone, G. M., eds., 1985, Bedrock geologic map of Maine: Augusta, Maine Geological Survey, scale 1:500,000.

Osberg, P. H., Tull, J. F., Robinson, P., Hon, R., and Butler, J. R., 1989, The Acadian orogen, *in* Hatcher, R. D., Jr., Thomas, W. A., and Viele, G. W., eds., The Appalachian-Ouachita orogen in the United States (The Geology of North America, vol. F-2): Boulder, Colorado, Geological Society of America, p. 179–232.

Palmer, A. R., 1983, The Decade of North American Geology 1983 time scale: Geology, v. 11, p. 503–504.

Pankiwskyj, K. A., 1978, Bedrock geology in the Coopers Mills–Liberty area, Maine, *in* Guidebook for field trips 3 and 4: Brunswick, The Geological Society of Maine, p. 5–8.

Pickering, K. T., Bassett, M. G., and Siveter, D. J., 1988, Late Ordovician–early Silurian destruction of the Iapetus Ocean; Newfoundland, British Isles, and Scandinavia—a discussion: Transactions of the Royal Society of Edinburgh: Earth Sciences, v. 79, p. 361–382.

Pin, C., and Peucat, J. J., 1986, Ages des épisodes de metamorphisme paléozoique dans le Massif central et le Massif armorican: Société géologique de France Bulletin, ser. 8, v. 2, p. 461–469.

Pique, A., and Michard, A., 1989, Moroccan Hercynides: A synopsis. The Pa-

leozoic sedimentary and tectonic evolution at the northern margin of West Africa: American Journal of Science, v. 289, p. 286–330.

Potter, R. R., Jackson, E. V., and Davies, J. L., 1968, Geological map of New Brunswick: New Brunswick Department of Natural Resources, scale 1:50,000.

Rankin, D. W., 1968, Volcanism related to tectonism in the Piscataquis volcanic belt, an island arc of Early Devonian age in north-central Maine, *in* Zen, E-an, White, W. S., Hadley, J. B., and Thompson, J. B., Jr., eds., Studies of Appalachian geology: Northern and maritime: New York, Wiley Interscience, p. 355–369.

Rast, N., 1984, The Alleghenian orogeny in eastern North America, *in* Hutton, D.H.W., and Sanderson, D. J., eds., Variscan tectonics of the North Atlantic region: Geological Society of London Special Publication 14, p. 197–217.

——, 1988, The evolution of the Appalachian chain, *in* Bally, A. W., and Palmer, A. R., eds., The Geology of North America: An overview (The Geology of North America, vol. A): Boulder, Colorado, Geological Society of America, p. 323–348.

Rast, N., and Grant, R. H., 1973, The Variscan front in southern New Brunswick, *in* Rast, N., ed., Geology of New Brunswick: New England Intercollegiate Geological Conference, 65th Annual Meeting, Field Guide to Excursions, p. 4–11.

Rast, N., and Skehan, S. J., J. W., 1984, Avalon zone in relation to Alleghanian deformation: Geological Society of America Abstracts with Programs, v. 16, p. 56.

——, 1989, The splitting of the mid-Paleozoic orogenic belt in the northern Appalachians and western Europe—An example of continental accretion: Geological Society of America Abstracts with Programs, v. 21, p. 227.

——, 1990, The Late Proterozoic geologic setting of the Boston Basin, *in* Socci, A. D., Skehan, J. W., and Smith, G. W., eds., Geology of the Composite Avalon Terrane of southern New England: Geological Society of America Special Paper 245, p. 235–247.

Rast, N., and Stringer, P., 1974, Recent advances and the interpretation of geological structure of New Brunswick: Geoscience Canada, no. 2, part 4, p. 15–25.

——, 1980, A geotraverse across a deformed Ordovician ophiolite and its Silurian cover, northern New Brunswick, Canada: Tectonophysics, v. 69, p. 221–245.

Rast, N., Lutes, G. G., and St. Peter, C., 1980, The geology and deformation history of the southern part of the Matapedia Zone and its relationship to the Miramichi Zone and Canterbury Basin, *in* Roy, D. C., and Naylor, R. S., eds., A guidebook to the geology of northeastern Maine and neighboring New Brunswick: New England Intercollegiate Geologic Conference, Presque Isle, Maine, p. 191–201.

Riva, J., and Malo, M., 1988, Age and correlation of the Honorat Group, southern Gaspé Peninsula: Canadian Journal of Earth Sciences, v. 25, p. 1618–1628.

Roberts, D., 1988, Timing of Silurian to middle Devonian deformation in the Caledonides of Scandinavia, Svalbard, and E. Greenland, *in* Harris, A. L., and Fettes, D. J., eds., The Caledonian-Appalachian orogen: Geological Society of London Special Publication 38, p. 429–435.

Robinson, P., Tracy, R. J., Santallier, D. S., Andreasson, P.-G., and Gil-Ibarguchi, J., 1988, Scandian-Acadian-Caledonian sensu strictu metamorphism in the age range 430–360 Ma, *in* Harris, A. L., and Fettes, D. J., eds., The Caledonian-Appalachian orogen: Geological Society of London Special Publication 38, p. 453–467.

Robinson, P., Tracy, R. J., Hollocher, K., Berry, H. N. IV, and Thomson, J. A., 1989, Basement and cover in the Acadian metamorphic high of central Massachusetts, *in* Chamberlain, P. C., and Robinson, P., eds., 1989, Styles of Acadian metamorphism with depth in the Central Acadian High, New England (a field trip honoring J. B. Thompson, Jr.): University of Massachusetts Department of Geology and Geography Contribution 63, p. 69–140.

Rodgers, J., 1970, The tectonics of the Appalachians: New York, Wiley Interscience, 271 p.

Roy, D. C., 1980, Tectonics and sedimentation in northeastern Maine and adjacent New Brunswick, *in* Roy, D. C., and Naylor, R. S., eds., A guidebook to

the geology of northeastern Maine and neighboring New Brunswick: New England Intercollegiate Geologic Conference, Presque Isle, Maine, p. 1–21.

Sando, W. J., Bamber, E. W., and Armstrong, A. K., 1975, Endemism and similarity indices: Clues to the zoogeography of North American Mississippian corals: Geology, v. 3, p. 661–664.

Sanford, B. V., and Grant, A. C., 1990, Bedrock geological mapping and basin studies in the Gulf of St. Lawrence: Geological Survey of Canada Paper 90-1B, p. 33–42.

Schuchert, C., 1930, Orogenic times of the northern Appalachians: Geological Society of America Bulletin, v. 41, p. 701–724.

Schumacher, J. C., Schumacher, R., and Robinson, P., 1989, Acadian metamorphism in central Massachusetts and south-western New Hampshire: Evidence for contrasting P-T trajectories, *in* Daly, J. S., Cliff, R. A., and Yardley, B.W.D., eds., Evolution of metamorphic belts: Geological Society of London Special Publication 43, p. 453–460.

Secor, D. T., Jr., Snoke, A. W., Bramlett, K. W., Costello, O. P., and Kimbrell, O. P., 1986, Character of the Alleghanian orogeny in the southern Appalachians: Pt 1. Alleghanian deformation in the eastern Piedmont of South Carolina: Geological Society of America Bulletin, v. 97, p. 1319–1328.

Sinha, A. K., 1988, Plutonism in the U.S. Appalachians, *in* Sinha, A. K., Hewitt, D. A., and Tracy, R. J., eds., Frontiers in petrology: A special volume of the American Journal of Science, v. 288-A, p. ix–xii.

Sinha, A. K., Hewitt, D. A., and Rimstidt, J. D., 1988, Metamorphic petrology and strontium isotope geochemistry associated with the development of mylonites: An example from the Brevard fault zone, North Carolina, *in* Sinha, A. K., Hewitt, D. A., and Tracy, R. J., eds., Frontiers in petrology: A special volume of the American Journal of Science, v. 288-A, p. 115–147.

Skehan, S.J., J. W., 1983, Geological profiles through the Avalonian terrane of southeastern Massachusetts, Rhode Island, and eastern Connecticut, U.S.A., *in* Rast, N., and Delany, F., eds., Profiles of orogenic belts: American Geophysical Union Geodynamic Series, v. 10, p. 275–300.

——, 1988, Evolution of the Iapetus Ocean and its borders in pre-Arenig times: A synthesis, *in* Harris, A. L., and Fettes, D. J., eds., The Caledonian-Appalachian orogen: Geological Society of London Special Publication 38, p. 185–229.

Skehan, S.J., J. W., and Murray, D. P., 1980, A model for the evolution for the eastern margin (EM) of the northern Appalachians, *in* Wones, D. R., ed., Proceedings; The Caledonides in the U.S.A.: IGCP Project 27, Virginia Polytechnic Institute and State University Memoir 2, p. 67–72.

Skehan, S.J., J. W., and Piqué, A., 1989, Comparative evolution of Avalon Zone of New England and of Panafrican Zone of West Africa from Proterozoic through Late Paleozoic time: Washington, D.C., 28th International Geological Congress, Abstracts v. 3 of 3, p. 130.

Skehan, S.J., J. W., and Rast, N., 1990, Pre-Mesozoic evolution of Avalon terranes of southern New England, *in* Socci, A. D., Skehan, S.J., J. W., and Smith, G. W., eds., Geology of the Composite Avalon Terrane of southern New England: Geological Society of America Special Paper 245, p. 235–247.

Soper, N. J., and Hutton, D.H.W., 1984, Late Caledonian sinistral displacements in Britain: Implications for a three-plate collision model: Tectonics, v. 3, p. 781–794.

Soper, N. J., and Kneller, B. C., 1990, Cleaved microgranite dykes of the Shap swarm in the Silurian of N.W. England: Geological Journal, v. 25, p. 161–170.

Spear, F. S., and Harrison, T. M., 1990, Post-Acadian uplift history of central New England from detailed petrologic and ^{40}Ar/^{39}Ar studies: Geological Society of America Abstracts with Programs, v. 22, p. 71.

Stewart, D. B., and eight others, 1986, The Quebec–Western Maine seismic reflection profile: Setting and first year results, *in* Barazangi, M., and Brown, L., eds., Reflection seismology: The continental crust (Geodynamics Series, vol. 14): Washington, D.C., American Geophysical Union, p. 189–199.

Stringer, P., 1974, Acadian post folding cleavage in northern Appalachians. Two phases of Acadian deformation in northeast New Brunswick: Geological

Association of Canada–Mineral Association of Canada Joint Annual Meeting, abstract and 3rd circular, Program, p. 10, Author abstracts, p. 90.

——, 1975, Acadian slaty cleavage noncoplanar with fold axial surfaces in the northern Appalachians: Canadian Journal of Earth Sciences, v. 12, p. 949–961.

Strong, D. F., 1980, Granitoid rocks and associated mineral deposits of eastern Canada and western Europe, *in* Strangway, D. W., ed., The continental crust and its mineral deposits: Geological Association of Canada Special Paper 20, p. 741–769.

Suess, E., 1908, The face of the earth, vol. III, translated by W. J. Sollas: Oxford, Clarendon Press, 400 p.

Swanson, M. T., and Carrigan, J., 1984, Ductile and brittle structures within the Rye Formation of coastal Maine and New Hampshire, *in* Hanson, L. S., ed., Geology of the Coastal Lowlands, Boston, MA to Kennebunk, ME: New England Intercollegiate Geologic Conference, Salem State College, Salem, Massachusetts, p. 165–185.

Thomas, W. A., 1977, Evolution of Appalachian-Ouachita salients and recesses from reentrants and promontories in the continental margin: American Journal of Science, v. 277, p. 1233–1278.

Tucker, R. D., Robinson, P., and Hollocher, K., 1988, U-Pb zircon, titanite, and monazite dating in "basement" rocks of the Bronson Hill anticlinorium, central Massachusetts: Geological Society of America Abstracts with Programs, v. 20, p. 216.

Tull, J. F., and Groszos, M. S., 1990, Nested Paleozoic "successor" basins in the southern Appalachian Blue Ridge: Geology, v. 18, p. 1046–1049.

Tull, J. F., and Telle, W. R., 1989, Tectonic setting of olistostromal units and associated rocks in the Talladega slate belt, Alabama Appalachians, *in* Horton, J. W., Jr., and Rast, N., eds., Melanges and olistostromes of the U.S. Appalachians: Geological Society of America Special Paper 228, p. 247–270.

Unrug, R., and Unrug, S., 1990, Paleontological evidence of Paleozoic age for the Walden Creek Group, Ocoee Supergroup, Tennessee: Geology, v. 18, p. 1041–1045.

Van der Pluijm, B. A., 1987, Timing and spatial distribution of deformation in the Newfoundland Appalachians: A "multi-stage collision" history: Tectonophysics, v. 135, p. 15–24.

Van der Pluijm, B. A., and Van Staal, C. R., 1988, Characteristics and evolution of the central mobile belt, Canadian Appalachians: Journal of Geology, v. 96, p. 535–547.

Van der Voo, R., 1988, Paleozoic paleogeography of North America, Gondwana, and intervening displaced terranes: Comparisons of paleomagnetism with paleo-climatology and biogeographical patterns: Geological Society of America Bulletin, v. 100, p. 311–324.

Vick, H. K., Channell, J.E.T., and Opdyke, N. D., 1987, Ordovician docking of the Carolina slate belt; Paleomagnetic data: Tectonics, v. 6, p. 573–583.

Westerman, D. S., 1987, Structures in the Dog River fault zone between Northfield and Montpelier, Vermont, *in* Westerman, D. S., ed., Guidebook for field trips in Vermont, vol. 2: New England Intercollegiate Geologic Conference, Norwich University, Northfield, Vermont, p. 109–132.

Whalen, J. B., 1989, The Topsails igneous suite, western Newfoundland: An Early Silurian subduction-related magmatic suite? Canadian Journal of Earth Sciences, v. 26, p. 2421–2434.

Whalen, J. B., and Hegner, E., 1990, Geochemical and isotopic signatures of granites across the Canadian Appalachian orogen: Tectonic implications: Geological Society of America Abstracts with Programs, v. 22, p. 78.

Williams, H., 1978, Tectonic lithofacies map of the Appalachian orogen: Memorial University of Newfoundland Map no. 1, scale 1:1,000,000.

——, 1984, Miogeoclines and suspect terranes of the Caledonian-Appalachian orogen: Tectonic patterns of the North Atlantic region: Canadian Journal of Earth Sciences, v. 21, p. 887–901.

Williams, H., Colman-Sadd, S. P., and Swinden, H. S., 1988, Tectonic-stratigraphic subdivisions of central Newfoundland, *in* Current Research, part B: Geological Survey of Canada Paper 88-1B, p. 91–98.

Williams, H., Dickson, W. L., Currie, K. L., Hayes, J. P., and Tuach, J., 1989a, Preliminary report on a classification of Newfoundland granitic rocks and their relations to tectonostratigraphic zones and lower crustal blocks: Geological Survey of Canada Paper 89-1B, p. 47–53.

Williams, H., Piasecki, M.A.J., and Colman-Sadd, S. P., 1989b, Tectonic relationships along the proposed central Newfoundland Lithoprobe transect and regional correlations: Geological Survey of Canada Paper 89-1B, p. 55–66.

Williams, H. S., 1895, Geological biology: An introduction to the geological history of organisms: New York, Henry Holt & Co., 395 p.

Wintsch, R. P., and Aleinikoff, J. N., 1987, U-Pb isotopic and geologic evidence for Late Paleozoic anatexis, deformation, and accretion of the Late Proterozoic Avalon terrane, south central Connecticut: American Journal of Science, v. 287, p. 107–126.

Wintsch, R. P., and Lefort, J.-P., 1984, A clockwise rotation of Variscan strain orientation in SE New England and regional implications, *in* Hutton, D.H.W., and Sanderson, E., eds., Variscan tectonics of the North Atlantic region: Geological Society of London Special Publication 14, p. 245–252.

Zartman, R. E., 1988, Three decades of geochronologic studies in the New England Appalachians: Geological Society of America, v. 100, p. 1168–1180.

Zartman, R. E., and Naylor, R. S., 1984, Structural implications of some radiometric ages of igneous rocks in southeastern New England: Geological Society of America Bulletin, v. 95, p. 522–539.

Zartman, R. E., Hurley, P. M., Krueger, H. W., and Giletti, B. J., 1970, A Permian disturbance of K-Ar radiometric ages in New England; Its occurrence and cause: Geological Society of America Bulletin, v. 81, p. 3359–3374.

Zen, E-An, 1983a, Exotic terranes in the New England Appalachians—Limits, candidates, and ages: A speculative essay, *in* Hatcher, R. D., Jr., Williams, H., and Zietz, I, Contributions to the tectonics and geophysics of mountain chains: Geological Society of America Memoir 158, p. 55–81.

Zen, E-an, ed., 1983b, Bedrock geologic map of Massachusetts: U.S. Geological Survey, scale 1:250,000.

Ziegler, P. A., 1982, Geological atlas of western and central Europe: Amsterdam, Elsevier Scientific Publishing Co., 130 p.

MANUSCRIPT ACCEPTED BY THE SOCIETY JUNE 8, 1992

Geological Society of America
Special Paper 275
1993

Paleogeography, accretionary history, and tectonic scenario: A working hypothesis for the Ordovician and Silurian evolution of the northern Appalachians

Ben A. van der Pluijm, Rex J.E. Johnson, and Rob Van der Voo
Department of Geological Sciences, University of Michigan, 1006 C. C. Little Building, Ann Arbor, Michigan 48109-1063

ABSTRACT

Paleomagnetic and tectonostratigraphic data for the northern Appalachians record Silurian closure of a major ocean, the Iapetus Ocean, that was bordered by the Laurentian craton and the Avalonian microcontinent. In Ordovician times this ocean consisted of at least two basins (Iapetus I and II) and extended from a paleolatitude of 10 to 20°S (Laurentian margin) to ca. 50°S (Avalonian margin); Gondwana was located yet farther south. Paleomagnetic data from the Middle Ordovician Robert's Arm, Chanceport, and Summerford groups in north-central Newfoundland, which represent intraoceanic arcs and ocean islands, yield paleolatitudes of 30 to 33°S. In contrast, pillow lavas of the upper part of the Late Cambrian to Lower Ordovician Moreton's Harbour Group, which are currently juxtaposed to the Chanceport Group along the Lobster Cove–Chanceport Fault, acquired their remanence at 11°S in a marginal island arc setting. Subaerial deposits of the mid-Silurian Botwood Group that unconformably overlie marine sequences in northeastern Newfoundland yield a primary magnetization with a paleolatitude of 24°S, which is indistinguishable from the Early Silurian position of the southeast-facing Laurentian margin. Silurian closure of Iapetus is supported by the timing of thrusting and folding and by the age of angular unconformities in the Central Mobile Belt.

The combined paleomagnetic and tectonostratigraphic data present a working hypothesis for the geometry and tectonic evolution of the northern Appalachians. In Early Ordovician times, a volcanic arc and a back-arc basin (Iapetus I) were located near the Laurentian margin. Following Middle Ordovician obduction of ophiolites onto the Laurentian margin when Iapetus I closed, convergence of Avalon and Laurentia by northward subduction continued until closure of Iapetus II was complete by the Late Silurian.

In view of these data, the traditional subdivision of Ordovician "Taconic" and Devonian "Acadian" orogenies needs to be revised. Maintaining the terminology of orogenic phases, one either has to expand the time interval of the Acadian orogeny to include the Silurian or add an orogenic phase (Caledonian?) in between Taconic and Acadian. In either case, Early to Middle Paleozoic closure of Iapetus should be viewed in terms of a progressive deformation history with peak deformation pulses rather than temporally discrete orogenies.

van der Pluijm, B. A., Johnson, R.J.E., and Van der Voo, R., 1993, Paleogeography, accretionary history, and tectonic scenario: A working hypothesis for the Ordovician and Silurian evolution of the northern Appalachians, *in* Roy, D. C., and Skehan, J. W., eds., The Acadian Orogeny: Recent Studies in New England, Maritime Canada, and the Autochthonous Foreland: Boulder, Colorado, Geological Society of America Special Paper 275.

INTRODUCTION

The Appalachian-Caledonian orogen was one of the first pre-Mesozoic mountain belts to be interpreted in terms of continent-continent collision (e.g., Wilson, 1966; Dewey, 1969; Bird and Dewey, 1970; Church and Stevens, 1971; and many subsequent contributions) and more recently has been analyzed within the concept of displaced terranes (e.g., Williams and Hatcher, 1983; Keppie, 1985, 1989; Fyffe and Fricker, 1987; Barr and Raeside, 1989). Along the orogen, the northern Appalachian segment between Massachusetts and Newfoundland is one of the more interesting examples of collision and terrane accretion processes and consequently has received much attention (e.g., Rast and Skehan, 1983; Zen, 1983; Neuman, 1984; Dunning and Krogh, 1985; Jacobi and Wasowski, 1985; O'Hara and Gromet, 1985; Ludman, 1986; Ayuso, 1986; van der Pluijm, 1987; van Staal, 1987; Williams and Hiscott, 1987; Chandler and others, 1987; Pickering and others, 1988; van der Pluijm and van Staal, 1988; Williams and others, 1988; Zartman, 1988; Boone and Boudette, 1989; Marillier and others, 1989; Stockmal and others, 1990; van Staal and others, 1990; Ayuso and Bevier, 1991). The picture that emerges is one of continent-continent collision between the Laurentian craton and the Avalon terrane, with the latter forming the coastal strip that ranges from eastern Massachusetts (the Boston Basin) to the type area in eastern Newfoundland. In between Laurentia and Avalon, however, several displaced terranes are found in the Central Mobile Belt of the northern Appalachians.

Orogenic episodes of the northern Appalachians have been recognized since the 1930s (e.g., Schuchert, 1930), and in the application of modern plate tectonics to the Appalachians these orogenies have been equated with tectonic accretion. The most widely recognized orogenies of the northern Appalachians are the Taconic, Acadian, and Alleghanian events, which traditionally have been considered to be Ordovician, Devonian, and (Permo-) Carboniferous in age, respectively (e.g., Williams and Hatcher, 1983; Keppie, 1985, 1989; Rast and Skehan, this volume). Locally, evidence for an orogeny that may predate the Taconic, the Penobscot event, has been reported (Neuman, 1967; Boone and Boudette, 1989; Neuman and Max, 1989). Up to the early 1980s, the Taconic orogeny was considered the main orogenic event of the northern Appalachians (e.g., Williams, 1979; Williams and Hatcher, 1983). However, in the past several years the importance of Silurian orogenic activity has become increasingly recognized. In this chapter we will mainly concentrate on the post-Taconic history of the northern Appalachians, which is related to the Acadian orogeny.

The Acadian phase of the New England and Canadian Appalachians, which is characterized by pervasive folding of Lower to Middle Paleozoic units and is accompanied by widespread plutonism, is widely considered to be a Devonian event (e.g., Rodgers, 1968; Williams, 1979; Williams and Hatcher, 1983; Zen, 1983; Hatcher, 1988). In more recent years, however, the timing of this orogenic event has received renewed attention.

Silurian ages were obtained for regional deformation (e.g., Karlstrom and others, 1982; Colman-Sadd and Swinden, 1984; Chorlton and Dallmeyer, 1986; van der Pluijm, 1986; Currie and Piasecki, 1989; Keppie, 1989; Lafrance, 1989; Elliott and others, 1991; Waldron and Milne, 1991) as well as metamorphic/plutonic activity (Coyle and Strong, 1987; Whalen and others, 1987; Whalen, 1989; Bevier and Whalen, 1990; Dunning and others, 1990). Paleomagnetic and tectonostratigraphic results from north-central Newfoundland that are presented in this paper further demonstrate the significance of Silurian tectonic activity. They indicate that the Early to Middle Paleozoic orogenic history of the northern Appalachians is characterized by progressive deformation associated with oblique accretion of a variety of tectonic elements.

The Newfoundland Appalachians are subdivided into the miogeocline (Humber zone), the Central Mobile Belt, and the Avalon terrane (Fig. 1; Williams, 1979; Williams and Hatcher, 1983; van der Pluijm and van Staal, 1988; Williams and others, 1988). The Humber zone recorded the formation and subsequent destruction of the northern cratonic (Laurentian) margin of Iapetus. A lower Paleozoic shelf facies (Bradley, 1989) is structurally overlain by Lower Ordovician ophiolite slices, such as the Humber Arm and Hare Bay allochthons, that were emplaced in Middle Ordovician times (Stevens, 1970; Church and Stevens, 1971; Williams, 1975). Subsequent Middle Paleozoic deformation modified this sequence (Cawood and Williams, 1988). The Grenvillian basement that underlies the Humber zone outcrops in several places in western Newfoundland, and off-shore deep-seismic data indicate that the basement extends a considerable distance under the neighboring Central Mobile Belt (Keen and others, 1986; Marillier and others, 1989). The Central Mobile Belt (CMB) preserves the main remnants of the Iapetus Ocean, and based on structural as well as geophysical grounds the CMB appears to be largely allochthonous (Karlstrom and others, 1982; Karlstrom, 1983; Colman-Sadd and Swinden, 1984; Stockmal and others, 1987; van der Pluijm, 1987; Marillier and others, 1989). Characteristic tectonic elements in the northern CMB include Lower Paleozoic magmatic arcs, ocean islands, ophiolites, and deep-marine sedimentary sequences as well as early Middle Paleozoic subaerial deposits. The present distribution of these elements and their relationships favor elimination of the previously used Dunnage and Gander two-fold subdivision of the CMB (compare Williams, 1979, and Williams and Hatcher, 1983, with van der Pluijm, 1987, and Williams and others, 1988). The surface boundary between the CMB and the Humber zone is marked by ophiolite assemblages (the Baie Verte–Brompton Line; Williams and St-Julien, 1982). The Avalon terrane, which represents the southern bordering continental block of Iapetus, was a stable tectonic element after Proterozoic times (O'Brien and others, 1983; Rast and Skehan, 1983; Krogh and others, 1988). The boundary of the Avalon terrane with the CMB in northeastern Newfoundland is a major fault, the Dover Fault, which exhibits a complex movement history (e.g., Kennedy and others, 1982). Deep-seismic reflection data across this boundary

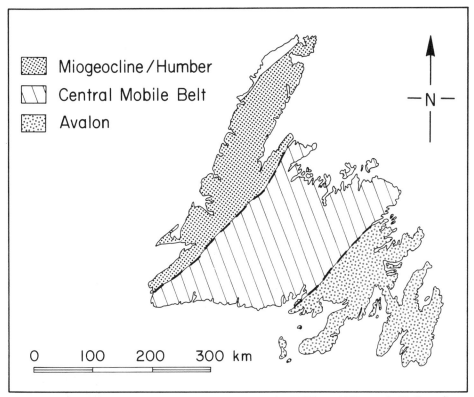

Figure 1. Subdivision of the Newfoundland Appalachians (after Williams, 1979; van der Pluijm and van Staal, 1988).

seem to indicate that the Dover fault penetrates the mantle (Keen and others, 1986), which probably reflects major strike-slip motion that was superimposed on the Silurian compressional history.

In many studies of the northern Appalachians, the Notre Dame Bay area of the north-central CMB in Newfoundland (Fig. 2) has received considerable attention, largely because of its excellent coastal exposure. A wealth of stratigraphic (e.g., Kay, 1967; Hibbard and Williams, 1979; McKerrow and Cocks, 1978, 1981; Dean, 1978; Kean and others, 1981; Arnott and others, 1985; van der Pluijm and others, 1987), sedimentologic (e.g., Helwig and Sarpi, 1969; Horne, 1970; Watson, 1981; Arnott, 1983; Pickering, 1987), geochemical (e.g., Strong, 1977; Strong and Dickson, 1978; Jacobi and Wasowski, 1985; Wasowski and Jacobi, 1985; Swinden and others, 1989; Fyffe and Swinden, 1991), and structural studies (e.g., Helwig, 1967, 1970; Horne, 1969; Kennedy, 1975; Nelson, 1981; Karlstrom and others, 1982; van der Pluijm, 1986; Kusky and others, 1987; Blewett and Pickering, 1988; Elliott and others, 1989; Lafrance, 1989; Williams and Piasecki, 1990) have been carried out in this region. The results presented in this chapter are largely derived from this part of the northern Appalachians.

PALEOMAGNETIC DATA: PALEOGEOGRAPHY

The existence of an ocean between Laurentia, comprising North America and Greenland plus northern Great Britain, and western Gondwana, consisting of South America and Africa, has

been well documented (e.g., Wilson, 1966; Dewey, 1969; Cocks and Fortey, 1982; Fortey and Cocks, 1988). The name Iapetus was proposed for this ocean by Harland and Gayer (1972). Early paleomagnetic studies were hampered by the occurrence of widespread remagnetizations (Van der Voo, 1988, 1989) and therefore contributed relatively limited data to Appalachian paleogeographic reconstructions compared to other Phanerozoic orogenic belts. Over the past few years, however, improved techniques and sampling methods have significantly expanded these data, in particular for the northern Appalachians.

Paleomagnetic results from the Avalon terrane indicate intermediate to high paleolatitudes similar to those of northern Gondwana (Johnson and Van der Voo, 1986; Van der Voo, 1988; Torsvik and Trench, 1991; Trench and Torsvik, 1991), implying that this continental block was located along the southern margin of Iapetus. During Ordovician time, the Appalachian margin of Laurentia was oriented approximately east-west and located at low paleolatitudes of 10 to 20°S (Briden and others, 1988; Van der Voo, 1988). The northern margin of Avalon, on the other hand, was at paleolatitudes of ca. 50°S. Biogeographic data for the Iapetus-bordering blocks (Cocks and Fortey, 1982; Neuman, 1984; Fortey and Cocks, 1988) are in good agreement with these paleomagnetic data (Van der Voo, 1988). Moreover, paleontologic evidence from several units in the CMB of the Appalachian orogen suggests the presence of seamounts in Iapetus, as represented by the Celtic fauna (Neuman, 1984, 1988);

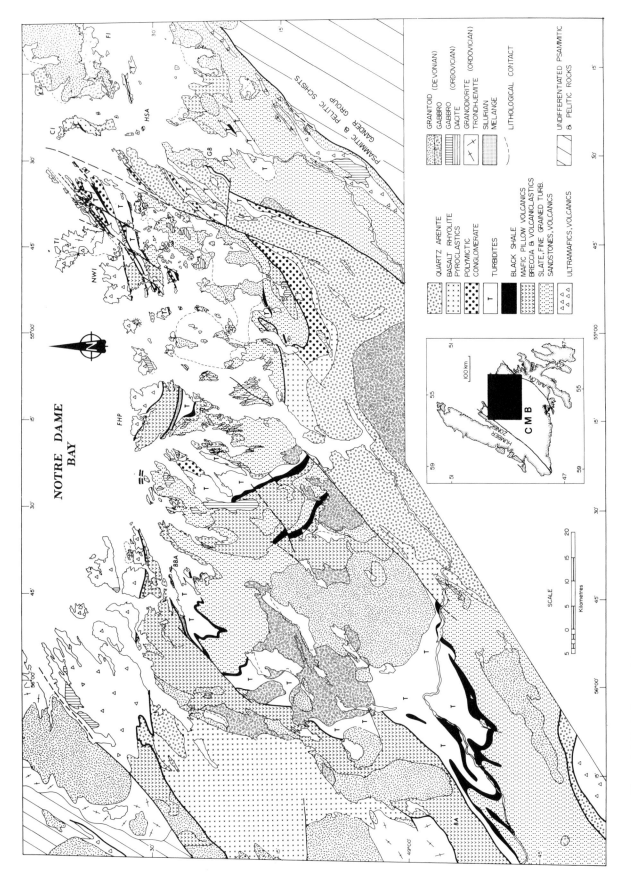

Figure 2. Lithologic map of the Notre Dame Bay area in northeastern Newfoundland. The map was compiled from Williams (1964), Helwig (1967), Nelson (1979), Currie and others (1980a), Kean and others (1981) and additional mapping by K. E. Karlstrom, B. A. van der Pluijm, and P. F. Williams. Abbreviations used on the map are: BA—Buchans area; BBA—Badger Bay area; CI—Change Islands; FHP—Fortune Harbour Peninsula; FI—Fogo Island; GB—Gander Bay; HSA—Hamilton Sound area; NWI—New World Island; TI—Twillingate Island.

however, the location of these islands as well as other Ordovician tectonic elements of Iapetus relative to the bordering continents was largely unknown.

We present paleomagnetic results from three key lithostratigrahic units of the CMB of northeastern Newfoundland. They are labeled in Figure 2 as (1) ultramafics, volcanics, (2) mafic pillow volcanics, breccia, and volcaniclastics, and (3) basalt, rhyolite, pyroclastics. These units, which comprise many local formation names, are grouped by us as (1) *island arc and back-arc basin,* (2) *intra-oceanic volcanics,* and (3) *subaerial cover,* respectively. Samples from a large number of sites were collected using a portable, gas-powered drill in the field or from oriented hand specimens that were drilled in the lab. Stratification planes were determined from interbedded sediments or flow tops. Thus, inclinations can be measured with respect to the paleohorizontal and because inclinations are not affected by block rotations (i.e., rotation without internal distortion), paleolatitudes can be determined. Detailed paleomagnetic data for these units, including thermal demagnetization and magnetic mineralogy, are presented elsewhere (Gales and others, 1989; Johnson and others, 1991; Van der Voo and others, 1991). In this chapter only the mean directions before and after tectonic correction are presented (Fig. 3). In Table 1 we summarize the Ordovician and Silurian paleopoles for these units, in addition to other paleomagnetic results used in this paper.

Island arc and back-arc basin

The Upper Cambrian to Lower Ordovician Moreton's Harbour Group of northeastern Newfoundland forms part of a volcanic arc that was built onto oceanic crust (Strong and Payne, 1973; Kean and Strong, 1975; Williams and Payne, 1975). Samples were collected from the upper units of the Moreton's Harbour Group, the Little Harbour and Western Head formations, that are assigned a Lower Ordovician age based on regional correlation (Dean, 1978). Stepwise thermal demagnetization of pillow basalts revealed a characteristic magnetization that is carried by magnetite (Johnson and others, 1991). The site means for 23 sites in field coordinates as well as after tectonic correction are shown in Figure 3a. A statistically significant, positive fold test indicates a primary age for the magnetization. This is further supported by a positive contact test and the occurrence of dual polarities. The mean magnetic direction of flows and intrusives combined is 171°/22° (Declination/inclination), which corresponds to a paleopole at 29°S, 315°E. Deutsch and Rao (1977) obtained a similar paleopole position (32°S, 310°E) for a limited suite from the Moreton's Harbour Group. From these data we calculate a paleolatitude of 11°S, which is indistinguishable from the location of the Appalachian margin of Laurentia in Early Ordovician times (Table 1). This indicates that a volcanic arc was located near the northern margin of Iapetus.

Intra-oceanic volcanics

The Chanceport Group of western New World Island is currently juxtaposed to the Moreton's Harbour Group along the

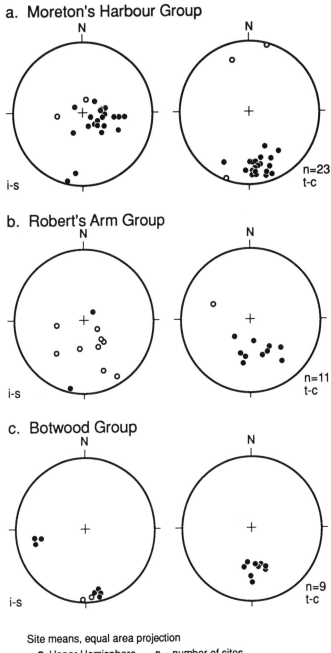

Figure 3. Paleomagnetic data for the Moreton's Harbour, the Robert's Arm, and the Botwood groups. Site means in field coordinates (in-situ) and after tectonic correction are given; n is number of site means. Detailed analysis of the data is given in Johnson and others (1991), Van der Voo and others (1991), and Gales and others (1989), respectively.

steeply dipping Lobster Cove–Chanceport Fault (Dean and Strong, 1977) that separates units with marked lithologic and geochemical contrasts (Williams, 1979; Bostock, 1988; Swinden and others, 1989). This sequence of mafic volcanic rocks forms part of a belt that includes the Cottrell's Cove Group on the

TABLE 1. ORDOVICIAN AND SILURIAN PALEOPOLES OF AVALON AND UNITS IN THE CENTRAL MOBILE BELT OF NEWFOUNDLAND, AND LAURENTIAN MEAN PALEOPOLES FOR EARLY AND MIDDLE PALEOZOIC TIMES

Rock Unit/ Time Interval (Ma)	Pole Position Lat., Long (°)	k, K	α_{95}, A_{95}
Laurentia			
Du, Dm/Du (366–378)	29S, 291E	24	9
Dm, Dl (379–397)	23S, 290E	40	8
Su/Dl (398–414)	03S, 277E	26	7
Su, Sm (415–429)	19S, 304E	52	6
Ou/Sl, Ou, Om (430–467)	16S, 324E	19	8
Ol/m, Ol (468–505)	15S, 344E	50	18
Cu, Cm, C (506–542)	09S, 338E	25	12
Cl (543–575)	05S, 350E	22	15
Central Mobile Belt			
Botwood, Sm/l	16S, 313E	54	7
Robert's Arm, Om/l	06S, 320E	14	10
Moreton's Harbour, Ol	29S, 315E	23	7
Avalon			
Dunn Point, Sl/Ou	02N, 316E	79	4
Dunn Point, Sl/Ou	02S, 310E	69	4
Stapeley (U.K.), Ou/m	27N, 036E	89	5

Note: D = Devonian; S = Silurian; O = Ordovician; C = Cambrian; u = upper; m = middle; l = lower; k, K and α_{95}, A_{95} are statistical parameters associated with the means.

Fortune Harbour Peninsula and the Robert's Arm Group in the Badger Bay area (Fig. 2). Geochemical evidence (Swinden, 1987; Bostock, 1988) for the Robert's Arm Group indicates an island arc origin for this Middle Ordovician (473 +2/–2 Ma, U/Pb; Dunning and others, 1987) unit. Paleomagnetic results from the Robert's Arm Group are shown in Figure 3b. The means from 11 sites cluster significantly better after tectonic correction, with a mean direction of 161°/52° (D/I). A primary age for this magnetization is indicated by the positive fold test (Fig. 3b) and a direction in one site that is antipodal to that in other sites. The intermediate inclination that we obtain implies that the Robert's Arm mafic volcanics were formed at a paleolatitude of 33°S. Results from the correlative Chanceport Group are essentially identical (paleolatitude of 30°S; Van der Voo and others, 1991), and samples from mafic volcanics of the Summerford Group of New World Island (Fig. 2) also yield a direction that is indistinguishable from that of the Robert's Arm Group. The Summerford Group, which represents a seamount (Jacobi and Wasowski, 1985) preserving fossils of the Celtic fauna (Neuman, 1984), was located at a paleolatitude of 31°S in Middle Ordovician times (Van der Voo and others, 1991). Collectively these three groups represent one or more volcanic terranes that were formed away from the margins of Iapetus.

Subaerial cover

Volcanic and sedimentary units of the mid-Silurian Botwood Group stratigraphically overlie marine sequences in eastern Notre Dame Bay; we will give detailed geologic information on the Botwood Group and the nature of its contact with underlying units in a later section of this chapter. Mafic volcanics of the Botwood Group were studied using paleomagnetic techniques by Lapointe (1979) and Gales and others (1989). In Figure 3c the mean directions for nine sites before and after tilt-correction are shown, and it was concluded by Gales and others (1989) that this direction represents magnetization at the time of extrusion. The mean direction after tilt correction is 175°/43°, which corresponds to a paleopole at 16°S, 313°E and a paleolatitude of 24°S. The similar paleolatitude of the Laurentian margin shows that by mid-Silurian times elements of Iapetus that are overlain by the Botwood Group had accreted to North America and that they had become exposed to surface weathering. Buchan and Hodych (1989) reported anomalous paleomagnetic results from Lower Silurian red beds and volcanics of the King George IV Lake area in southwestern Newfoundland, that may be explained by very large right-lateral strike-slip motion.

Bordering continental block and synthesis

In Figure 4 we have summarized the available paleomagnetic data for the Iapetus-bordering continental blocks, showing the paleolatitudes of Laurentia and as a function of time. The Laurentian paleolatitudes are based on a compilation of all the data for North America and Scotland (Van der Voo, 1990); the corresponding paleopoles for the various time intervals are listed in Table 1. The paleopoles for units of Avalon were previously summarized in Van der Voo (1988). Two new results are added:

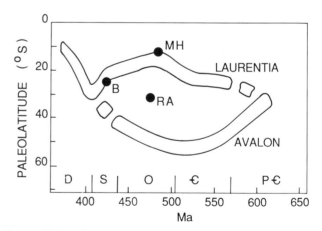

Figure 4. Paleolatitudinal variation of Laurentia, Avalon, and other geographical elements in Iapetus from late Precambrian to Devonian times. Based on data listed in Table 1 and Van der Voo (1988, 1989). Abbreviations: B—Botwood Group; MH—Moreton's Harbour Group; RA—Robert's Arm Group (modified from van der Pluijm and others, 1990).

a reexamination of the Upper Ordovician Dunn Point Formation in Nova Scotia (Johnson and Van der Voo, 1990) that confirms the previous conclusion of Van der Voo and Johnson (1985; cf. Seguin and others, 1987) and results from the Middle to Upper Ordovician Stapeley Formation in Great Britain (McCabe and Channell, 1990). The paleopoles for Laurentia and Avalon have been used to calculate the paleolatitudes of these blocks as they occur in Newfoundland. Figure 4 shows that the Avalonian and Laurentian continents have a maximum *latitudinal* width of ca. 35°. Given that the Laurentian margin was oriented approximately east-west in Ordovician times, this corresponds to a width of ca. 4,000 km for Iapetus.

The overlap of the paleolatitudes for the Moreton's Harbour Group and the Appalachian margin of Laurentia in Early Ordovician times dictates that an Ordovician volcanic arc was located near the northern margin of Iapetus (Fig. 4). The paleolatitudes for the early Middle Ordovician Robert's Arm, Chanceport, and Summerford groups, on the other hand, show that arcs and seamounts were located widely separate from both margins of Iapetus (Fig. 4). Finally, "Laurentian" paleolatitudes from the subaerial Botwood Group show that underlying Ordovician and Early Silurian tectonic elements of Iapetus had been accreted to Laurentia by the mid-Silurian.

TECTONOSTRATIGRAPHIC DATA: TIMING OF ACCRETION

The paleomagnetic results presented in this chapter enable us to reconstruct the location of various tectonic elements of Iapetus. The timing of accretion of these elements, however, is only indirectly constrained by these data. The ages and nature of regional deformation, overlap sequences, and unconformities, on the other hand, provide a direct means to determine the timing and significance of orogenic events. In the Notre Dame Bay area, widespread folding with an associated axial-plane cleavage affected all post-Middle Paleozoic (i.e., "post-Taconic") depositional units of the CMB (e.g., Karlstrom and others, 1982; van der Pluijm, 1986). Recent U/Pb geochronology of intrusives places a Late Silurian upper age limit on this folding (408 +2/−2 Ma; Elliott and others, 1991). Following a brief description of the pertinent geology of eastern New World Island, new tectonostratigraphic data from the Change Islands will be presented in this section that bear directly on the significance of the Acadian orogeny in the northern Appalachians.

On eastern New World Island, bedding-parallel faults that are responsible for the repetition of stratigraphy and the existence of discrete movement zones indicate the presence of thrusts (Fig. 5; van der Pluijm, 1986). Stacking of discrete stratigraphic packages occurred in an accretionary wedge setting and was probably accompanied by strike-slip motion. Melanges that are located at the base of these thrust sheets constrain the timing of emplacement. The Joey's Cove Melange (McKerrow and Cocks, 1987; Arnott, 1983; Reusch, 1983) at the base of sheet B (Fig. 5)

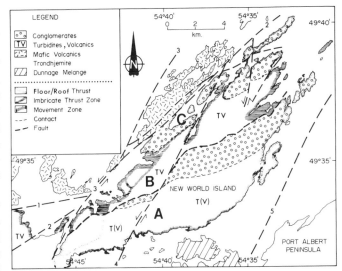

Figure 5. Simplified litho-structural map of eastern New World Island showing the presence of three thrust sheets, labeled A to C from bottom to top. Faults shown are: 1—Chanceport Fault; 2—Lukes Arm Fault; 3—Virgin-Village Fault; 4—Burnt Arm Fault; 5—Reach Fault (slightly modified from van der Pluijm, 1986).

contains clasts of all surrounding rock types (Ordovician volcanics, limestones, and shales) in a matrix of late Llandoverian sandstone (Fig. 6). A similar melange unit, the Byrne Cove Melange, is present at the base of sheet C on eastern New World Island (Fig. 5). These melanges are interpreted as synthrusting olistostromes and give an Early Silurian age for the part of the accretionary wedge exposed on New World Island (van der Pluijm, 1986).

On the Change Islands, which are located to the east of New

Figure 6. Field photograph of the Joey's Cove melange near Cobbs Arm, eastern New World Island. Blocks of predominantly Llandeilian limestone are contained in a late Llandoverian sandstone matrix.

World Island (Fig. 2), stratigraphic and structural relationships provide additional evidence for Silurian deformation. A simplified geologic map of Change Islands is shown in Figure 7, which illustrates that the area is dominated by deposits of the Botwood Group (Twenhofel and Shrock, 1937; Williams, 1964; Eastler, 1971). Hematite-rich volcanics and interbedded sandstones are overlain by micaceous quartz arenites. Columnar jointing and hematite weathering in the volcanics (Fig. 8a) and desiccation cracks in the sandstones (Fig. 8b) are present at some localities and indicate that subaerial or very shallow marine conditions prevailed during their deposition. Fossils from the Botwood Group give Llandoverian to Ludlovian ages (Berry and Boucot, 1970; Eastler, 1971), and a 422 +2/–2 Ma age (U/Pb; Elliott and others, 1991) from a bimodal dike suite that intrudes quartz

arenites on the Port Albert Peninsula (Fig. 2) places a mid-Silurian upper limit on the age of the Botwood Group. Its correlative in western Newfoundland, the Springdale Group (e.g., Kean and others, 1981), was dated as 429 +6/–5 Ma (U/Pb; Chandler and others, 1987). The generally shallowly dipping Botwood Group overlies a sequence of steeply dipping graywacke turbidites with some interbedded conglomerates. These marine deposits are correlatives of clastic deposits on eastern New World Island and areas to the west that yielded Ashgillian-Llandoverian ages (Dean, 1978; Kean and others, 1981; Arnott and others, 1985; van der Pluijm and others, 1987; Williams and O'Brien, 1991). Isolated occurrences of corals in graywackes on western Change Islands indicate a Llandoverian age for this unit (Eastler, 1971). The boundary between these graywackes/conglomerates and the

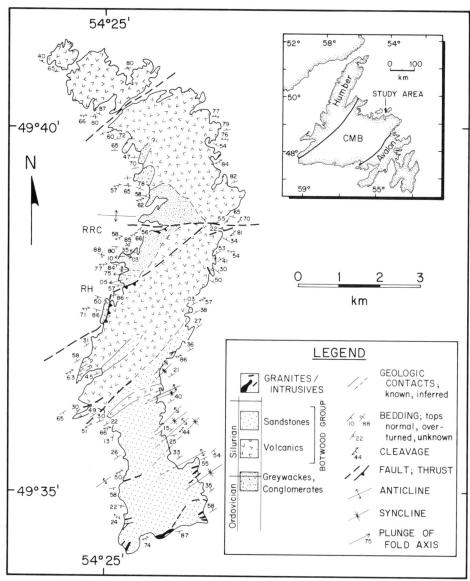

Figure 7. Geologic map of the Change Islands. Abbreviations are: RH—Randall's head; RRC—Red Rock Cove. Based on Eastler (1971) and mapping by J. E. Gales and B. A. van der Pluijm.

Figure 8. Columnar jointing in volcanics (a) and desiccation cracks in sandstone (b) of the Botwood Group. These features demonstrate the subaerial or very-shallow marine nature of this unit.

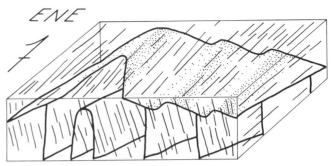

Figure 9. Schematic diagram of the structure of the Change Islands. The gently folded, shallowly dipping surface is an angular unconformity (locally faulted) that separates underlying graywackes from overlying deposits of the Botwood Group. The trace of the regional cleavage is marked on the faces of the block. Structures in the graywackes define a large isoclinal anticline (the Red Rock Cove anticline) that is crosscut by the regional cleavage. This regional cleavage, however, is axial planar to open folds in the Botwood Group.

Botwood Group is a shallowly dipping, northwest-directed thrust fault at its southernmost exposures (Fig. 7), similar to that on the Port Albert Peninsula to the southwest of the Change Islands (Karlstrom and others, 1982). At localities to the north, however, thrusting is absent.

Folds in graywackes at Randell's Head on the western side of Change Islands (RH, Fig. 7) are tight, upright structures that are crosscut by the regional cleavage (van der Pluijm, 1990). Superimposition of cleavage on these folds is demonstrated by the fold-cleavage relationships in Red Rock Cove (RRC, Fig. 7). The orientation of cleavage is essentially constant and cuts both limbs of this isoclinal anticline (Fig. 9). Folds in the Botwood Group, on the other hand, are relatively open, upright structures with the regional cleavage parallel to their axial planes (Fig. 9). These relationships show that parts of the deformation history of the graywackes must predate folding of the Botwood Group and regional cleavage formation. This early deformation is absent in the Botwood Group, and thus the boundary between Botwood units and underlying graywackes is an *angular* unconformity. This relationship is further supported by the abrupt change from a

deep marine depositional setting for the graywackes to a subaerial or very-shallow marine environment for the Botwood Group. The age of the angular unconformity is constrained by the age of the Botwood Group, i.e., mid-Silurian.

Tectonostratigraphic studies on New World Island and Change Islands demonstrate the importance of Early Silurian deformation. In these areas, thrusting and the formation of an accretionary wedge characterize the style of closure of Iapetus near the Laurentian margin. Superimposed regional folding and cleavage formation were completed by Late Silurian times.

DISCUSSION: TECTONIC SCENARIO

The combined paleomagnetic and tectonostratigraphic data enable us to propose a scenario for the Early to Middle Paleozoic tectonic evolution of the Newfoundland Appalachians that can be used as a working hypothesis for other areas in the Appalachian-Caledonian chain. In this discussion we distinguish five elements in north-central Newfoundland (Fig. 10): two continental blocks (Laurentia and Avalon), a volcanic arc with a marginal sea (a supra-subduction zone), the main ocean basin, and a subaerial cover sequence. The latter three elements are mainly found in the Central Mobile Belt, although segments of the marginal sea are currently located on the Laurentian margin (the allochthons). The three tectonostratigraphic elements of the CMB are represented in this chapter by the Moreton's Harbour, Robert's Arm–Chanceport–Summerford, and Botwood groups, respectively. Our subdivision contrasts with Williams (1979) but is not unlike that proposed more recently by Williams and others (1988), who subdivided the CMB into the Notre Dame subzone, the Exploits subzone, and the Gander zone. The boundary between the Notre Dame and Exploits subzones is marked by the Red Indian Line. Our results place the surface expression of this fundamental boundary to the north of the Robert's Arm–Chanceport belt, along the Lobster Cove–Chanceport Fault. We add that the CMB contains several other characteristic elements,

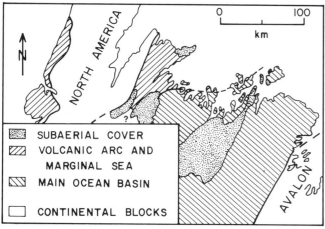

Figure 10. Paleogeographic elements of northern Newfoundland used in this chapter. The map is based on Williams and others (1988); note, however, that the boundary between the arc belt and the oceanic belt does not everywhere coincide with the location of the Red Indian Line of Williams and others (1988) (modified from van der Pluijm and others, 1990).

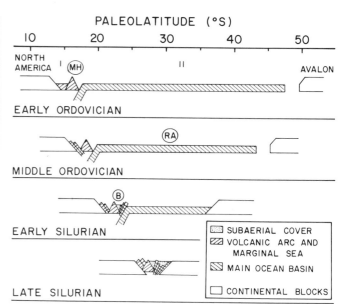

Figure 11. Schematic cross sections showing the paleolatitudinal progression of Iapetus-bordering continents and selected units of the CMB. Abbreviations used are: I—Iapetus I; II—Iapetus II; B—Botwood Group; MH—Moreton's Harbour Group; RA—Robert's Arm Group (modified from van der Pluijm and others, 1990).

such as the Annieopsquotch Complex (Dunning and Chorlton, 1985), the Gander River Ultramafic Belt (Blackwood, 1982), and the Pipestone Pond Complex (Colman-Sadd and Swinden, 1984), but in the absence of associated paleomagnetic data we will limit our discussion to the broad groupings shown in Figure 10. The nature of the boundary between the main ocean basin and the Avalon microcontinent in northeastern Newfoundland is obscured by the deep, steeply dipping Dover Fault (Keen and others, 1986). Therefore, we will not attempt to interpret its early history in our current scenario. However, from our work in neighboring areas it appears that in Early Paleozoic times at least part of the Avalonian margin is characterized by an arc/back-arc basin complex (Liss and others, 1991; see also van Staal and others, 1991). We have chosen to present our data in a sequence of highly schematic, north-south cross sections (Fig. 11). Such presentation of the data clearly ignores orogen-parallel plate motions (east-west displacements), but it is consistent with the limitation of paleomagnetic data that lack the necessary paleolongitudinal control to determine such likely displacements.

In Early Paleozoic times, the northern margin of Iapetus was located at low paleolatitudes. The northern Iapetus Ocean was characterized by a small ocean basin (marginal sea) with relatively young (Early Ordovician) oceanic crust, which separated the Laurentian margin from the volcanic arc. Van der Pluijm and van Staal (1988) introduced the terminology Iapetus I and II for the northerly and southerly basins (ancient coordinates), respectively. In our current use of these names we extend this terminology to reflect the temporal sequence of basin closure. Subsequent collision of the arc with North America, possibly as a result of plate boundary evolution (Dewey and Shackleton, 1984), resulted in closure of the marginal sea (Iapetus I) and the obduction of ophiolite. Portions of these ophiolites are now preserved in, for

example, the Humber Arm and Hare Bay allochthons, and their emplacement has traditionally characterized the Taconic orogeny (Williams, 1979). The interpretation of one (or more) small ocean basin(s) as the origin for these ophiolites is not new (e.g., Dewey and Bird, 1971; Church and Stevens, 1971; Kennedy, 1975; Williams and Payne, 1975; Kidd, 1977; Searle and Stevens, 1984; and more recently Jenner and others, 1991), but the view that these ophiolites represent the main Iapetus basin seems to have dominated interpretations since the landmark paper of Dewey (1969; e.g., Williams, 1979; Williams and Hatcher, 1983; Karson, 1984; Casey and others, 1985). In contrast, we contend that the short time interval that separates the formation of these Appalachian ophiolites and their obduction (10 to 20 m.y.; Dunning and Krogh, 1985; Spray, 1988; Jenner and others, 1991) is best explained by opening of a back-arc basin, which was rapidly followed by its closure. Paleomagnetic results from the Skinner Cove volcanics of the Humber Arm allochthon may lend further support to this interpretation. Beaubouef and others (1988) reported a low paleolatitude (ca. 12°S) for this unit, which indicates a near-Laurentian position. Northerly obduction of ophiolites on the Laurentian margin would suggest that minor southward underplating under the volcanic arc accompanied closure of the marginal sea. However, main subduction of the Iapetus Ocean occurred in a northerly direction, which is in contrast to most tectonic interpretations of the northern Appalachians that favor southerly subduction (e.g., Church and Stevens, 1971; Strong and others, 1974; Strong, 1977; Williams, 1979; Currie and others, 1980b; Jacobi, 1981; Colman-Sadd, 1982; Williams and Hatcher, 1983; Stockmal and others, 1987,

1990) over northerly subduction (e.g., Dewey, 1969; Bird and Dewey, 1970; Williams and Payne, 1975; this chapter).

Volcanic rocks of the Robert's Arm Group were formed in the central portion of Iapetus II and reached the margin of North America in Early Silurian times. This is recorded by the formation of a Lower Silurian accretionary wedge (van der Pluijm, 1986), Lower Silurian calc-alkaline volcanism (e.g., Whalen and others, 1987; Whalen, 1989) and a mid-Silurian angular unconformity (this chapter; see also Chandler and others, 1987; Bostock, 1988). By Late Silurian times, final closure of Iapetus was accomplished by the collision of Avalon with North America, which is recorded by widespread regional folding and cleavage formation and the deposition of the Botwood Group and correlatives.

CONCLUSION

The paleogeography and tectonostratigraphic history of parts of the northern Appalachians indicate that Iapetus closed by major Ordovician to Silurian subduction along its northern margin with Laurentia. Iapetus consisted of at least two basins with a combined latitudinal width of ca. 35°. Located in the north was a back-arc basin (Iapetus I) bordered by Laurentia and a volcanic arc. The main ocean basin (Iapetus II) was situated to the south of this arc and contained a variety of tectonic elements. Deformation in the Appalachians recorded the progressively outboard accretion of these geographically distinct elements (e.g., arcs and seamounts) to Laurentia.

The Silurian was a time of major deformation in the Central Mobile Belt of the Newfoundland Appalachians. It follows mid-Ordovician closure of Iapetus I and emplacement of allochthons on the Laurentian margin ("Taconic orogeny"). The Silurian history of the area described in this chapter and elsewhere represents the second Appalachian deformation stage, which has been called the "Acadian orogeny." However, in view of its traditionally Devonian age, one must either redefine the time interval of the Acadian or add a name (Caledonian?) in between Taconic and Acadian (see also Rast and Skehan, this volume). In either case, the early orogenic history of the northern Appalachians should be viewed in terms of progressive deformation, rather than as a series of temporally discrete orogenies. The timing and nature of these "orogenies" suggest that they represent times of peak deformation, or *pulses*, that occurred during continued convergence of Avalon and Laurentia and that they are not necessarily present everywhere along the orogen.

ACKNOWLEDGMENTS

Our Appalachian research is supported by the National Science Foundation, most recently under grants EAR 89-05811 and EAR 91-18021. Fieldwork on the Change Islands was carried out with Julie Gales; Laura Knutson collaborated on the paleomagnetic study of the Robert's Arm Group. Useful reviews were given by two anonymous referees; we thank David Roy and James Skehan for inviting us to participate in the symposium and this volume.

REFERENCES CITED

Arnott, R. J., 1983, Sedimentology of Upper Ordovician–Silurian sequences on New World Island, Newfoundland: Separate fault-controlled basins?: Canadian Journal of Earth Sciences, v. 20, p. 345–354.

Arnott, R. J., McKerrow, W. S., and Cocks, L.R.M., 1985, The tectonics and depositional history of the Ordovician and Silurian rocks of Notre Dame Bay, Newfoundland: Canadian Journal of Earth Sciences, v. 22, p. 607–618.

Ayuso, R. A., 1986, Lead-isotopic evidence for distinct sources of granite and for distinct basements in the northern Appalachians, Maine: Geology, v. 14, pp. 322–325.

Ayuso, R. A., and Bevier, M. L., 1991, Regional differences in Pb isotopic compositions of feldspars in plutonic rocks of the northern Appalachian mountains, U.S.A. and Canada: A geochemical method of terrane correlation: Tectonics, v. 10, pp. 191–212.

Barr, S. M., and Raeside, R. P., 1989, Tectono-stratigraphic terranes in Cape Breton Island, Nova Scotia: Implications for the configuration of the northern Appalachian orogen: Geology, v. 17, p. 822–825.

Beaubouef, R. T., Casey, J. F., and Hall, S. A., 1988, A paleomagnetic study of the Skinner Cove volcanic assemblage, western Newfoundland [abs.]: EOS Transactions of the American Geophysical Union, v. 69, p. 1158.

Berry, W.B.N., and Boucot, A. T., 1970, Correlation of the North American Silurian rocks: Geological Society of America Special Paper 102, p. 289.

Bevier, M. L., and Whalen, J. B., 1990, Tectonic significance of Silurian magmatism in the Canadian Appalachians: Geology, v. 18, p. 411–414.

Bird, J. M., and Dewey, J. F., 1970, Lithosphere plate–continental margin tectonics and the evolution of the Appalachian orogeny: Geological Society of America Bulletin, v. 81, p. 1031–1060.

Blackwood, R. F., 1982, Geology of Gander Lake (2D/15) and Gander River (2E/2) area: Newfoundland Department of Mines and Energy Mineral Development Division Report 82-4, p. 56.

Blewett, R. S., and Pickering, K. T., 1988, Sinistral shear during Acadian deformation in north-central Newfoundland, based on transecting cleavage: Journal of Structural Geology, v. 10, p. 125–127.

Boone, G. M., and Boudette, E. L., 1989, Accretion in the Boundary Mountains terrane within the northern Appalachian orthotectonic zone, *in* Horton, J. S., Jr., and Rast, N., eds., Melanges and olistostromes of the U.S. Appalachians: Geological Society of America Special Paper 228, p. 17–42.

Bostock, H. H., 1988, Geology and petrochemistry of the Roberts Arm Group, Notre Dame Bay, Newfoundland: Geological Survey of Canada Bulletin, v. 369, 84 p.

Bradley, D. C., 1989, Taconic plate kinematics as revealed by foredeep stratigraphy, Appalachian orogen: Tectonics, v. 8, p. 1037–1049.

Briden, J. C., and eight others, 1988, Palaeomagnetic constraint on the evolution of the Caledonian-Appalachian orogen, *in* Harris, A. L., and Fettes, D. J., eds., The Caledonian-Appalachian orogen: Geological Society Special Publication 38, p. 35–48.

Buchan, K. L., and Hodych, J. P., 1989, Early Silurian paleopole for redbeds and volcanics of the King George IV Lake area, Newfoundland: Canadian Journal of Earth Sciences, v. 26, p. 1904–1917.

Casey, J. F., Elthon, D. L., Siroky, F. X., Karson, J. A., and Sullivan, J., 1985, Geochemical and geologic evidence bearing on the origin of the Bay of Islands and coastal complex ophiolites of western Newfoundland: Tectonophysics, v. 116, p. 1–40.

Cawood, P. A., and Williams, H., 1988, Acadian basement thrusting, crustal delamination, and structural styles in and around the Humber Arm allochthon, western Newfoundland: Geology, v. 16, p. 370–373.

Chandler, F. W., Sullivan, R. W., and Currie, K. L., 1987, The age of the Springdale Group, western Newfoundland, and correlative rocks—Evidence for a Llandovery overlap assemblage in the Canadian Appalachians: Transactions of the Royal Society of Edinburgh, v. 78, p. 41–49.

Chorlton, L. B., and Dallmeyer, R. D., 1986, Geochronology of Early to Middle Paleozoic tectonic development in the southwest Newfoundland Gander zone: Journal of Geology, v. 94, p. 67–89.

Church, W. R., and Stevens, R. K., 1971, Early Paleozoic ophiolite complexes of the Newfoundland Appalachians as mantle-oceanic crust sequences: Journal of Geophysical Research, v. 76, p. 1460–1466.

Cocks, L.R.M., and Fortey, R. A., 1982, Faunal evidence for oceanic separations in the Palaeozoic of Britain: Journal of the Geological Society of London, v. 139, p. 465–478.

Colman-Sadd, S. P., 1982, Two-stage continental collision and plate driving forces: Tectonophysics, v. 90, p. 263–282.

Colman-Sadd, S. P., and Swinden, H. S., 1984, A tectonic window in central Newfoundland? Geological evidence that the Appalachian Dunnage Zone may be allochthonous: Canadian Journal of Earth Sciences, v. 21, p. 1349–1367.

Coyle, M., and Strong, D. F., 1987, Geology of the Springdale Group: A newly recognized Silurian epicontinental-type caldera in Newfoundland: Canadian Journal of Earth Sciences, v. 24, p. 1135–1148.

Currie, K. L., and Piasecki, M.A.J., 1989, Kinematic model for southwestern Newfoundland based upon Silurian sinistral shearing: Geology, v. 17, p. 938–941.

Currie, K. L., Pajari, G. E., Jr., and Pickerill, R. K., 1980a, Comments on the boundaries of the Davidsville Group, northeastern Newfoundland, in Current research: Geological Survey of Canada Paper 80-1A, p. 115–118.

Currie, K. L., Pickerill, R. K., and Pajari, G. E., Jr., 1980b, An early Paleozoic plate tectonic model of Newfoundland: Earth and Planetary Sciences Letters, v. 48, p. 8–14.

Dean, P. L., 1978, The volcanic stratigraphy and metallogeny of Notre Dame Bay, Newfoundland: Memorial University of Newfoundland Geological Report 7, 205 p.

Dean, P. L., and Strong, D. F., 1977, Folded thrust faults in Notre Dame Bay, central Newfoundland: American Journal of Science, v. 177, p. 97–108.

Deutsch, E. R., and Rao, K. V., 1977, New paleomagnetic evidence fails to support rotation of western Newfoundland: Nature, v. 266, p. 314–318.

Dewey, J. F., 1969, The evolution of the Caledonian/Appalachian orogen: Nature, v. 222, p. 124–129.

Dewey, J. F., and Bird, J. M., 1971, Origin and emplacement of the ophiolite suite: Appalachian ophiolites in Newfoundland: Journal of Geophysical Research, v. 76, p. 3179–3206.

Dewey, J. F., and Shackleton, R. M., 1984, A model for the evolution of the Grampian tract in the early Caledonides and Appalachians: Nature, v. 312, p. 115–121.

Dunning, G. R., and Chorlton, L. B., 1985, The Annieopsquotch ophiolite belt of southwest Newfoundland: Geology and tectonic significance: Geological Society of America Bulletin, v. 96, p. 1466–1476.

Dunning, G. R., and Krogh, T. E., 1985, Geochronology of ophiolites of the Newfoundland Appalachians: Canadian Journal of Earth Sciences, v. 22, p. 1659–1670.

Dunning, G. R., Kean, B. F., Thurlow, J. G., and Swinden, H. S., 1987, Geochronology of the Buchans, Roberts Arm, and Victoria Lake Groups and Masfield Cove Complex, Newfoundland: Canadian Journal of Earth Sciences, v. 24, p. 1175–1184.

Dunning, G. R., and six others, 1990, Silurian orogeny in the Newfoundland Appalachians: Journal of Geology, v. 98, p. 895–913.

Eastler, T. E., 1971, Geology of Silurian rocks, Change Islands and easternmost Notre Dame Bay, Newfoundland [Ph.D. thesis]: New York, Columbia University, 147 p.

Elliott, C. G., Barnes, C. R., and Williams, P. F., 1989, Southwest New World Island stratigraphy: New fossil data, new implications for the history of the Central Mobile Belt, Newfoundland: Canadian Journal of Earth Sciences, v. 26, p. 2062–2074.

Elliott, C. G., Dunning, G. R., and Williams, P. F., 1991, New U/Pb zircon age constraints on the timing and deformation in north-central Newfoundland and implications for early Paleozoic Appalachian orogenesis: Geological Society of America Bulletin, v. 103, p. 125–135.

Fortey, R. A., and Cocks, L.R.M., 1988, Arenig to Llandovery faunal distributions in the Caledonides, in Harris, A. L., and Fettes, D. J., eds., The Caledonian-Appalachian orogen: Geological Society of London Special Publication 38, p. 233–246.

Fyffe, L. R., and Fricker, A., 1987, Tectonostratigraphic terrane analysis of New Brunswick: Maritime Sediments and Atlantic Geology, v. 23, p. 113–122.

Fyffe, L. R., and Swinden, H. S., 1991, Paleotectonic setting of Cambro-Ordovician volcanic rocks in the Canadian Appalachians: Geoscience Canada, v. 18, p. 145–157.

Gales, J. E., van der Pluijm, B. A., and Van der Voo, R., 1989, Paleomagnetism of the Lawrenceton Formation volcanic rocks, Silurian Botwood Group, Change Islands, Newfoundland: Canadian Journal of Earth Sciences, v. 26, p. 296–304.

Harland, W. V., and Gayer, R. A., 1972, The Arctic Caledonides and earlier oceans: Geological Magazine, v. 109, p. 289–314.

Hatcher, R. D., Jr., 1988, The third synthesis: Wenlock to mid-Devonian (end of Acadian orogeny), in Harris, A. L., and Fettes, D. J., eds., The Caledonian-Appalachian orogen: Geological Society of London Special Publication 38, p. 499–504.

Helwig, J., 1967, Stratigraphy and structural history of the New Bay area, north-central Newfoundland [Ph.D. thesis]: New York, Columbia University, 211 p.

—— , 1970, Slump folds and early structures, northeastern Newfoundland Appalachians: Journal of Geology, v. 78, p. 172–187.

Helwig, J., and Sarpi, E., 1969, Plutonic-pebble conglomerates, New World Island, Newfoundland, and history of eugeosynclines, in Kay, M., ed., North Atlantic geology and continental drift: American Association of Petroleum Geologists Memoir 12, p. 443–466.

Hibbard, J., and Williams, H., 1979, Regional setting of the Dunnage Melange in the Newfoundland Appalachians: American Journal of Science, v. 279, p. 993–1021.

Horne, G. S., 1969, Early Ordovician chaotic deposits in the central volcanic belt of northeastern Newfoundland: Geological Society of America Bulletin, v. 80, p. 2451–2464.

—— , 1970, Complex volcanic-sedimentary patterns in the Magog Belt of northeastern Newfoundland: Geological Society of America Bulletin, v. 81, p. 1767–1788.

Jacobi, R. D., 1981, Peripheral bulge—A causal mechanism for the lower/middle Ordovician unconformity along the western margin of the northern Appalachians: Earth and Planetary Sciences Letters, v. 56, p. 245–251.

Jacobi, R. D., and Wasowski, J. J., 1985, Geochemistry and plate-tectonic significance of the volcanic rocks of the Summerford Group, north-central Newfoundland: Geology, v. 13, p. 126–130.

Jenner, G. A., Dunning, G. R., Malpas, J., Brown, M., and Brace, T., 1991, Bay of Islands and Little Port complexes, revisited: Age, geochemical and isotopic evidence confirm suprasubduction-zone origin: Canadian Journal of Earth Sciences, v. 28, p. 1635–1652.

Johnson, R.J.E., and Van der Voo, R., 1986, Paleomagnetism of the Late Precambrian Fourchu Group, Cape Breton Island, Nova Scotia: Canadian Journal of Earth Sciences, v. 23, p. 1673–1685.

—— , 1990, Pre-folding magnetization reconfirmed for the Late Ordovician/Early Silurian Dunn Point volcanics, Nova Scotia: Tectonophysics, v. 178, p. 193–205.

Johnson, R.J.E., van der Pluijm, B. A., and Van der Voo, R., 1991, Paleomagnetism of the Moreton's Harbour Group, Northeastern Newfoundland Appalachians: Evidence for an early Ordovician island arc near the Laurentian Margin of Iapetus: Journal of Geophysical Research, v. 96, p. 11689–11701.

Karlstrom, K. E., 1983, Reinterpretation of Newfoundland gravity data and arguments for an allochthonous Dunnage Zone: Geology, v. 11, p. 263–266.

Karlstrom, K. E., van der Pluijm, B. A., and Williams, P. F., 1982, Structural interpretation of the eastern Notre Dame Bay area, Newfoundland: Regional post–Middle Silurian thrusting and asymmetrical folding: Canadian Journal of Earth Sciences, v. 19, p. 2325–2341.

Karson, J. A., 1984, Variations in structure and petrology in the Coastal Complex, Newfoundland: Anatomy of an oceanic fracture zone, in Gass, I. G., Lippard, S. J., and Shelton, A. W., eds., Ophiolites and oceanic lithosphere:

Geological Society of London Special Publication 13, p. 131–144.

Kay, M., 1967, Stratigraphy and structure of northeastern Newfoundland and bearing on drift in the North Atlantic: American Association of Petroleum Geologists Bulletin, v. 51, p. 579–600.

Kean, B. F., and Strong, D. F., 1975, Geochemical evolution of an Ordovician island arc of the Central Newfoundland Appalachians: American Journal of Sciences, v. 275, p. 97–118.

Kean, B. F., Dean, P. L., and Strong, D. F., 1981, Regional geology of the Central Volcanic Belt of Newfoundland, *in* Swanson, E. A., Strong, D. F., and Thurlow, J. G., eds., The Buchans orebodies: Fifty years of geology and mining: Geological Association of Canada Special Paper 22, p. 65–78.

Keen, C. E., and seven others, 1986, Deep-seismic reflection profile across the northern Appalachians: Geology, v. 14, p. 141–145.

Kennedy, M. J., 1975, Repetitive orogeny in the northeastern Appalachians— New plate models based upon Newfoundland examples: Tectonophysics, v. 28, p. 39–87.

Kennedy, M. J., Blackwood, R. F., Colman-Sadd, S. P., O'Driscoll, C. F., and Dickson, W. L., 1982, The Dover–Hermitage Bay fault: Boundary between the Gander and Avalon zones, eastern Newfoundland, *in* St. Julien, P., and Beland, J., eds., Major structural zones and faults of the northern Appalachians: Geological Association of Canada Special Paper 24, p. 231–248.

Keppie, J. D., 1985, The Appalachian collage, *in* Gee, D. G., and Sturt, B. A., eds., The Caledonide Orogen: Part 2, Scandinavia and related areas: Chichester, United Kingdom, John Wiley & Sons, p. 1217–1226.

——, 1989, Northern Appalachian terranes and their accretionary history, *in* Dallmeyer, R. D., ed., Terranes in the circum-Atlantic Paleozoic orogens: Geological Society of America Special Paper 20, p. 159–192.

Kidd, W.S.F., 1977, The Baie Verte lineament, Newfoundland: Ophiolite complex floor and mafic volcanic fill of a small Ordovician marginal basin, *in* Talwani, M., and Pitman, W. C., eds., Island arcs, deep sea trenches and back-arc basins: American Geophysical Union, Maurice Ewing Series 1, p. 407–418.

Krogh, T. E., Strong, D. F., O'Brien, S. J., and Papezik, V. S., 1988, Precise U/Pb zircon dates from the Avalon terrane in Newfoundland: Canadian Journal of Earth Sciences, v. 25, p. 442–453.

Kusky, T. M., Kidd, W.S.F., and Bradley, D. C., 1987, Displacement history of the Northern Arion Fault and its bearing on the post-Taconic evolution of North-central Newfoundland: Journal of Geodynamics, v. 7, p. 105–133.

Lafrance, B., 1989, Structural evolution of a transpression zone in north central Newfoundland: Journal of Structural Geology, v. 11, p. 705–716.

Lapointe, P. L., 1979, Paleomagnetism and orogenic history of the Botwood Group and Mount Peyton batholith, Central Mobile Belt, Newfoundland: Canadian Journal of Earth Sciences, v. 16, p. 866–876.

Liss, M. J., van der Pluijm, B. A., and Van der Voo, R., 1991, Paleogeography of the Middle Ordovician Tetagouche Group, Northern New Brunswick: Paleomagnetic evidence [abs.]: EOS Transactions of the American Geophysical Union, v. 72, p. 106.

Ludman, A., 1986, Timing of terrane accretion in eastern and east-central Maine: Geology, v. 14, pp. 411–414.

Marillier, F., and six others, 1989, Crustal structure and surface zonation of the Canadian Appalachians: Implications of deep seismic reflection data: Canadian Journal of Earth Sciences, v. 26, p. 305–321.

McCabe, C., and Channell, J.E.T., 1990, Paleomagnetic results from volcanic rocks of the Shelve inlier, Wales: Evidence for a wide Late Ordovician Iapetus Ocean in Britain: Earth and Planetary Sciences Letters, v. 96, p. 458–468.

McKerrow, W. S., and Cocks, L.R.M., 1978, A Lower Paleozoic trench-fill sequence, New World Island, Newfoundland: Geological Society of America Bulletin, v. 89, p. 1121–1132.

——, 1981, Stratigraphy of eastern Bay of Exploits, Newfoundland: Canadian Journal of Earth Sciences, v. 18, p. 751–765.

Nelson, K. D., 1979, Geology of the Badger Bay–Seal Bay area, north-central Newfoundland [Ph.D. thesis]: Albany, State University of New York, 184 p.

——, 1981, Melange development in the Boones Point Complex, north-central

Newfoundland: Canadian Journal of Earth Sciences, v. 18, p. 433–442.

Neuman, R. B., 1967, Bedrock geology of the Shin Pond and Stacyville quadrangles, Penobscot County, Maine: U.S. Geological Survey Professional Paper 524-I, p. 11–137.

——, 1984, Geology and paleobiology of islands in the Ordovician Iapetus Ocean: Review and implications: Geological Society of America Bulletin, v. 95, p. 1188–1201.

——, 1988, Paleontological evidence bearing on the Arenig-Caradoc development of the Iapetus Ocean basin, *in* Harris, A. L., and Fettes, D. J., eds., The Caledonian-Appalachian orogen: Geological Society Special Publication 38, p. 269–274.

Neuman, R. B., and Max, M. D., 1989, Penobscottian-Grampian-Finnmarkian orogenies as indicators of terrane linkages, *in* Dallmeyer, R. D., ed., Terranes in the Circum-Atlantic Paleozoic Orogens: Geological Society of America Special Paper 230, p. 31–45.

O'Brien, S. J., Wardle, R. J., and King, A. F., 1983, The Avalon Zone: A Pan-African terrane in the Appalachian orogen of Canada: Geological Journal, v. 18, p. 195–222.

O'Hara, K. D., and Gromet, L. P., 1985, Two distinct late Precambrian (Avalonian) terranes in southeastern New England and their late Paleozoic juxtaposition: American Journal of Science, v. 285, pp. 673–709.

Pickering, K. T., 1987, Wet-sediment deformation in the Upper Ordovician Point Leamington Formation: An active thrust-imbricate system during sedimentation, Notre Dame Bay, north-central Newfoundland, *in* Jones, M. E., and Preston, R.M.F., eds., Deformation of sediments and sedimentary rocks: Geological Society of London Special Publication, p. 213–239.

Pickering, K. T., Bassett, M. G., and Siveter, D. J., 1988, Late Ordovician–early Silurian destruction of the Iapetus Ocean: Newfoundland, British Isles and Scandinavia—A discussion: Transactions of the Royal Society of Edinburgh, v. 79, p. 361–382.

Rast, N., and Skehan, J. W., 1983, The evolution of the Avalonian plate: Tectonophysics, v. 100, p. 257–286.

Reusch, D. N., 1983, The New World Island Complex and its relationship to nearby formations, north-central Newfoundland [M.Sc. thesis]: St. John's, Memorial University of Newfoundland.

Rodgers, J., 1968, The eastern edge of the North American continent during the Cambrian and Early Ordovician, *in* Zen, E-an, and others, eds., Studies of Appalachian geology: Northern and maritime: New York, John Wiley and Sons, p. 141–150.

Schuchert, C., 1930, Orogenic times of the northern Appalachians: Geological Society of America Bulletin, v. 41, p. 701–724.

Searle, M. P., and Stevens, R. K., 1984, Obduction in ancient, modern and future ophiolites, *in* Gass, I. G., Lippard, S. J., and Shelton, A. W., eds., Ophiolites and oceanic lithosphere: Geological Society of London Special Publication 13, p. 303–319.

Seguin, M. K., Rao, K. V., and Deutsch, R. R., 1987, Paleomagnetism and rock magnetism of Early Silurian Dunn Point volcanics, Avalon Zone, Nova Scotia: Physics of the Earth and Planetary Interiors, v. 46, p. 369–380.

Spray, J. G., 1988, Thrust-related metamorphism beneath the Shetland Islands oceanic fragment, northeast Scotland: Canadian Journal of Earth Sciences, v. 25, p. 1760–1776.

Stevens, R. K., 1970, Flysch sedimentation and Ordovician tectonics in west Newfoundland, *in* Lajoie, J., ed., Flysch sedimentology in North America: Geological Association of Canada Special Paper 7, p. 165–177.

Stockmal, G. S., Colman-Sadd, S. P., Keen, C. E., O'Brien, S. J., and Quinlan, G., 1987, Collision along an irregular margin: A regional plate tectonic interpretation of the Canadian Appalachians: Canadian Journal of Earth Sciences, v. 24, p. 1098–1107.

Stockmal, G. S., Colman-Sadd, S. P., Keen, C. E., Marillier, F., O'Brien, S. J., and Quinlan, G. M., 1990, Deep seismic structure and plate tectonic evolution of the Canadian Appalachians: Tectonics, v. 9, p. 45–62.

Strong, D. F., 1977, Volcanic regimes of the Newfoundland Appalachians, *in* Barager, W.R.A., Coleman, L. C., and Hall, J. M., eds., Volcanic regimes in Canada: Geological Association of Canada Special Paper 16, p. 61–90.

Strong, D. F., and Dickson, W. L., 1978, Geochemistry of Paleozoic granitoid plutons from contrasting tectonic zones of northeast Newfoundland: Canadian Journal of Earth Sciences, v. 15, p. 145–156.

Strong, D. F., and Payne, J. G., 1973, Early Paleozoic volcanism and metamorphism of the Moreton's Harbour–Twillingate area, Newfoundland: Canadian Journal of Earth Sciences, v. 10, p. 1363–1379.

Strong, D. F., Dickson, W. L., O'Driscoll, C. F., Kean, B. F., and Stevens, R. K., 1974, Geochemical evidence for an east-dipping Appalachian subduction zone in Newfoundland: Nature, v. 248, p. 37–39.

Swinden, H. S., 1987, Geology and mineral occurrences in the central and northern parts of the Robert's Arm Group, central Newfoundland, *in* Current research: Geological Survey of Canada Paper 87-1A, p. 381–390.

Swinden, H. S., Jenner, G. A., Kean, B. F., and Evans, D.T.W., 1989, Volcanic rock geochemistry as a guide for massive sulphide exploration in central Newfoundland, *in* Current Research: Geological Survey of Newfoundland Department of Mines Report 89-1, p. 201–219.

Torsvik, T. H., and Trench, A., 1991, The Ordovician history of the Iapetus ocean in Britain: New paleomagnetic constraints: Journal of the Geological Society, v. 148, p. 423–425.

Trench, A., and Torsvik, T. H., 1991, A revised Paleozoic apparent polar wander path for southern Britain (eastern Avalonia): Geophysical Journal International, v. 104, p. 227–233.

Twenhofel, W. H., and Shrock, R. R., 1937, Silurian strata of Notre Dame Bay and Exploits Valley, Newfoundland: Geological Society of America Bulletin, v. 48, p. 1743–1772.

van der Pluijm, B. A., 1986, Geology of eastern New World Island, Newfoundland: An accretionary terrane in the northeastern Appalachians: Geological Society of America Bulletin, v. 97, p. 932–945.

—— , 1987, Timing and spatial distribution of deformation in the Newfoundland Appalachians: A "multistage collision" history: Tectonophysics, v. 135, p. 15–24.

—— , 1990, Synchroneity of folding and crosscutting cleavage in the Newfoundland Appalachians?: Journal of Structural Geology, v. 12, p. 1073–1076.

van der Pluijm, B. A., and van Staal, C. R., 1988, Characteristics and evolution of the Central Mobile Belt, Canadian Appalachians: Journal of Geology, v. 96, p. 535–547.

van der Pluijm, B. A., Karlstrom, K. E., and Williams, P. F., 1987, Fossil evidence for fault-derived stratigraphic repetition in the northeastern Newfoundland Appalachians: Canadian Journal of Earth Sciences, v. 24, p. 2337–2350.

van der Pluijm, B. A., Johnson, R.J.E., and Van der Voo, R., 1990, Early Paleozoic paleogeography and accretionary history of the Newfoundland Appalachians: Geology, v. 18, p. 898–901.

Van der Voo, R., 1988, Paleozoic paleogeography of North America, Gondwana, and intervening displaced terranes: Comparisons of paleomagnetism with paleoclimatology and biogeographical patterns: Geological Society of America Bulletin, v. 100, p. 311–324.

—— , 1989, Paleomagnetism of North America: The craton, its margins and the Appalachians, *in* Pakiser, L. C., and Mooney, W. D., eds., Geophysical framework of the continental United States: Geological Society of America Memoir 172, p. 447–470.

—— , 1990, Phanerozoic paleomagnetic poles from Europe and North America and comparisons with continental reconstructions: Reviews of Geophysics, v. 28, p. 167–206.

Van der Voo, R., and Johnson, R.J.E., 1985, Paleomagnetism of the Dunn Point Formation (Nova Scotia): High paleolatitudes for the Avalon Terrane in the late Ordovician: Geophysical Research Letters, v. 12, p. 337–340.

Van der Voo, R., Johnson, R.J.E., van der Pluijm, B. A., and Knutson, L. C., 1991, Paleogeography of some vestiges of Iapetus: Paleomagnetism of the Ordovician Robert's Arm, Summerford and Chanceport Groups, Central Newfoundland: Geological Society of America Bulletin, v. 103, p. 1564–1575.

van Staal, C. R., 1987, Tectonic setting of the Tetagouche Group in northern New Brunswick: Implications for plate tectonic models of the northern Appalachians: Canadian Journal of Earth Sciences, v. 24, p. 1329–1351.

van Staal, C. R., Ravenhurst, C. E., Winchester, J. A., Roddick, J. C., and Langton, J. P., 1990, Post-Taconic blueschist suture in the northern Appalachians of northern New Brunswick, Canada: Geology, v. 18, p. 1073–1077.

van Staal, C. R., Winchester, J. A., and Bedard, J. H., 1991, Geochemical variations in Middle Ordovician volcanic rocks of the Northern Miramichi Highlands and their tectonic significance: Canadian Journal of Earth Science, v. 28, p. 1031–1049.

Waldron, J.W.F., and Milne, J. V., 1991, Tectonic history of the central Humber Zone, western Newfoundland Appalachians: Post-Taconian deformation in the Old Man's Pond area: Canadian Journal of Earth Sciences, v. 28, p. 398–410.

Wasowski, J. J., and Jacobi, R. D., 1985, Geochemistry and tectonic significance of the mafic volcanic blocks in the Dunnage Melange, north central Newfoundland: Canadian Journal of Earth Sciences, v. 22, p. 1248–1256.

Watson, M. P., 1981, Submarine fan deposits of the Upper Ordovician–Lower Silurian Milleners Arm Formation, New World Island, Newfoundland [Ph.D. thesis]: Oxford, Oxford University.

Whalen, J. B., 1989, The Topsails igneous suite, western Newfoundland: An Early Silurian subduction-related magmatic suite? Canadian Journal of Earth Sciences, v. 26, p. 2421–2434.

Whalen, J. B., Currie, K. L., and van Breemen, O., 1987, Episodic Ordovician-Silurian plutonism in the Topsails igneous terrane, western Newfoundland: Transactions of the Royal Society of Edinburgh, v. 78, p. 17–28.

Williams, H., 1964, Botwood map-area, Newfoundland: Geological Survey of Canada Preliminary Series Map 60-1963, scale 1:250,000.

—— , 1975, Structural succession, nomenclature, and interpretation of transported rocks in western Newfoundland: Canadian Journal of Earth Sciences, v. 12, p. 1874–1894.

—— , 1979, Appalachian orogen in Canada: Canadian Journal of Earth Sciences, v. 16, p. 792–807.

Williams, H., and Hatcher, R. D., Jr., 1983, Appalachian suspect terranes, *in* Hatcher, R. D., Jr., Williams, H., and Zietz, I., eds., Contributions to the tectonics and geophysics of mountain chains: Geological Society of America Memoir 158, p. 33–53.

Williams, H., and Hiscott, R. N., 1987, Definition of the Iapetus rift-drift transition in western Newfoundland: Geology, v. 15, p. 1044–1047.

Williams, H., and Payne, J. G., 1975, The Twillingate granite and nearby volcanic groups: An island arc complex in northeast Newfoundland: Canadian Journal of Earth Sciences, v. 12, p. 982–995.

Williams, H., and Piasecki, M.A.J., 1990, The Cold Spring melange and a possible model for the Dunnage-Gander zone interaction in central Newfoundland: Canadian Journal of Earth Sciences, v. 27, p. 1126–1134.

Williams, H., and St-Julien, P., 1982, The Baie Verte–Brompton line: Early Paleozoic continent ocean interface in the Canadian Appalachians, *in* St-Julien, P., and Beland, J., eds., Major structural zones and faults of the northern Appalachians: Geological Association of Canada Special Paper 24, p. 177–208.

Williams, H., Colman-Sadd, S. P., and Swinden, H. S., 1988, Tectonic-stratigraphic subdivisions of central Newfoundland, *in* Current Research: Geological Survey of Canada Paper 88-1B, p. 91–98.

Williams, S. H., and O'Brien, B. H., 1991, Silurian (Llandovery) graptolites from the Bay of Exploits, north-central Newfoundland, and their geological significance: Canadian Journal of Earth Sciences, v. 28, p. 1534–1540.

Wilson, J. T., 1966, Did the Atlantic close and then re-open?: Nature, v. 211, p. 676–681.

Zartman, R. E., 1988, Three decades of geochronologic studies in the New England Appalachians: Geological Society of America Bulletin, v. 100, p. 1168–1180.

Zen, E-an, 1983, Exotic terranes in the New England Appalachians—Limits, candidates, and ages: A speculative essay, *in* Hatcher, R. D., Jr., Williams, H., and Zietz, I., eds., Contributions to the tectonics and geophysics of mountain chains: Geological Society of America Memoir 158, p. 55–81.

MANUSCRIPT ACCEPTED BY THE SOCIETY JUNE 8, 1992

Printed in U.S.A.

Geological Society of America
Special Paper 275
1993

Comments on Cambrian-to-Carboniferous biogeography and its implications for the Acadian orogeny

A. J. Boucot
Department of Zoology, Oregon State University, Corvallis, Oregon 97331-2914

ABSTRACT

In considering the tectonic evolution of the Acadian orogen in New England and Maritime Canada, account should be taken of the fact that from the Early Cambrian through the Mississippian the Coastal Acadia marine faunas of eastern North America are very distinctive biogeographically from those occurring to the west and northwest within the Northern Appalachians and adjacent parts of the continent. The limited amount of boundary mixing can be largely ascribed to dispersible larval stages of a few taxa. This biogeographic situation suggests that the surface-current circulation pattern that maintained this type of reproductive isolation and biogeographic integrity from the Early Cambrian through the Mississippian implies a certain level of geographic remoteness as well, although specific distances cannot be derived from such data. Paleogeographies and plate tectonic concepts need to be consistent with the available biogeographic information.

Early-through-Late Cambrian biogeographic units have been recognized, since early in this century, in the Acadian orogen and adjacent regions. Baltic Realm (= Atlantic Realm, = Acado-Baltic Realm) faunas are restricted to the coastal regions of New England (Boston region), the Maritimes (St. John, New Brunswick), Nova Scotia, and eastern Newfoundland. On the other hand, Laurentian Realm (= Pacific Realm) faunas are restricted to a belt that extends from western Newfoundland down the valley of the St. Lawrence River and through Vermont and eastern New York.

Ordovician biogeography is less well documented, with earliest Ordovician (Tremadocian) Baltic Realm–type faunas also occurring in eastern Newfoundland, Nova Scotia, and Coastal Acadia, whereas on the North American Platform to the west, Laurentian Realm–type faunas occur. Baltic Realm faunas of the later Early Ordovician (Arenigian) are known in eastern Newfoundland. In central Newfoundland, the few shelly faunas are largely of the Laurentian Realm. There is no useful Middle to Upper Ordovician (upper Arenigian-Ashgillian) fauna from Coastal Acadia, whereas fauna of that age-span on the northwestern edge of the Northern Appalachians is strictly Laurentian Realm in character. During this time interval in central Newfoundland, central and northern New Brunswick, and northern Maine, the few shelly faunas presently available are largely of the Laurentian Realm, with a Baltic admixture in some cases, although more study is required for a definitive statement on this matter. In the latest Ashgillian (Hirnantian) there is at least one good occurrence of the globally very extensive cold water, Gondwana (=Malvinokaffric) Realm *Hirnantia* fauna in eastern Gaspé.

Silurian European Province faunas occur in Coastal Acadia, including parts of mainland Nova Scotia, southern New Brunswick, coastal Maine, and the Boston area. North American Province Silurian occurs farther to the west and northwest, including

Boucot, A. J., 1993, Comments on Cambrian-to-Carboniferous biogeography and its implications for the Acadian orogeny, *in* Roy, D. C., and Skehan, J. W., eds., The Acadian Orogeny: Recent Studies in New England, Maritime Canada, and the Autochthonous Foreland: Boulder, Colorado, Geological Society of America Special Paper 275.

northern New Brunswick, northern Maine, and the Connecticut River Valley region, as well as in areas farther to the west.

Marine Devonian units of Old World Realm type occur in Nova Scotia, coastal Maine, and southern New Brunswick. Eastern Americas Realm faunas, on the other hand, are present to the west and northwest in central and northern New Brunswick, northern Maine, eastern Quebec, and New Hampshire.

Mississippian marine European Province faunas are present in Newfoundland, Nova Scotia, and coastal New Brunswick. By contrast, both North American and Southeastern Province faunas are known well to the west of Greater Acadia in the Southern Appalachians and the Mid-Continent region. Later Paleozoic European and North American nonmarine biota are all of Euramerian Province type; that is, there is no evidence for provincialism within the Northern Appalachians.

These biogeographic data are of concern in tectonic analysis, particularly of the Acadian orogen, because belts yielding biogeographically similar organisms are unlikely to have been geographically remote from each other, and vice versa. Examples of boundary biogeographic mixing also indicate greater proximity—such examples are known in a few areas within the Northern Appalachians, particularly in eastern Quebec for the Early and Middle Devonian and in Nova Scotia for the Early Devonian. Early Devonian mixing is also present in northern Maine and adjacent New Brunswick. Boundary biogeographic mixing is also known in central Newfoundland and a few locales to the southwest in northern Maine and New Brunswick for the Middle and later Ordovician.

The timing of the Acadian orogeny deduced from datable fossils preserved above and below the post-Acadian unconformity is Middle Devonian, with a Givetian date for the maximum being most likely. This statement applies to all parts of the Northern Appalachians except for Newfoundland, where datable beds overlying Acadian-deformed strata are scarce.

INTRODUCTION

Geologists commonly think of paleontologic data as useful only in the dating and correlation of strata. Some geologists also routinely avail themselves of the useful information about depositional environments provided by certain fossil groups for some time intervals; that is, information establishing deposition in nearshore versus offshore, deep water versus shallow water, rough water versus quiet water, hypersaline versus hyposaline or normal salinity, and so forth. An additional category of geologically useful data is provided by the information of historical biogeography. What is historical biogeography, and how can it be geologically useful? Can historical biogeography provide evidence useful in better understanding the timing of the Acadian Orogeny?

Causes of biogeographical distinctiveness

Organisms, living and fossil, are not distributed globally in a completely random manner. The first-order regularity in their distribution is latitudinal. Higher latitudes today, and in the past, are characterized by a biota, both marine and nonmarine, consisting of fewer kinds of organisms than is the case at lower latitudes. By fewer kinds I refer to fewer genera and species and also fewer higher categories up to the class level. This statement merely reflects situations in which tropical-subtropical forests, seashores, and other environments yield far more diverse kinds of organisms than is the case at high latitudes. It should be borne in mind, however, that the latitudinal diversity gradient is in large part a function of the global climatic gradient. The climatic gradient is very high today, with glacial ice at both poles, but it was much lower during most of the geologic past. Superimposed on this primary, latitudinal gradient in organic diversity are longitudinal boundaries. The lower-latitude longitudinal barriers are such features as the Isthmus of Panama that has separated Atlantic and eastern Pacific tropical marine organisms from each other since the beginning of the Pliocene. At the same time such barriers provide a land bridge for some of the terrestrial biotas that permits them to migrate north and south. Another similar feature is the Middle East land barrier that has similarly separated Indo-Pacific from Atlantic Ocean marine environments since some time in the latest Oligocene–earlier Miocene. The longitudinal barriers come and go through geologic time.

Both the latitudinal and longitudinal barriers hinder reproductive communication between similar environments. For example, the abundant polar maritime birds, the alcids, have many habits similar to those of the penguins in terms of habitat and food supply, although, unlike penguins, the alcids can fly. Neither group can invade the territory of the other because of the reproductive barrier posed by the broad tropical-subtropical belt (Galapagos penguins flourish only because of the unusual, equatorial, cold-water perturbation provided by the Humboldt Current). Some barriers to reproductive communication may be very broad, such as the one cited, whereas others may involve relatively narrow transitions from a warm-water mass to a cold-water mass, such as the Gulf Stream and Labrador Current boundary off eastern North America or the warm-water Agulhas and cold-water Benguela currents off the east and west coasts, respectively, of southern Africa. These biogeographic facts of life do not, unfortunately, provide us with any measurable distances. Again, it should be remembered that the modern biogeographic distinctiveness of the New World Coral Reefs vis-à-vis the Old World Reefs does not provide us with any tools for measuring absolute distances in the past, any more than that provided by the Quaternary distinctiveness of the tropical North Atlantic marine benthos from that of the North Pacific across the Isthmus of Panama.

In view of examples such as those above, it should be clear that biogeographic data do not provide a tool for measuring the absolute width of the Iapetus Sea or Ocean (see Pickering and others, 1988, for a consideration of this problem). One simply cannot determine whether the Iapetus waterway was an ancient Mediterranean-size sea or whether it was an Atlantic-size or Pacific-size ocean. All that these data provide are a factual basis for stating that the two regions involved were far enough apart to have prevented a high level of reproductive communication. We can provide many hypothetical water masses with various shapes, sizes, and current circulation patterns that would be capable of explaining the biogeographic distinctions we observe.

However, for anyone concerned about the paleogeography of past tectonic plates and allied questions, biogeography cannot be ignored. When rocks presently adjacent to each other, as are those of Coastal Acadia and the remainder of the Northern Appalachians (Boucot, 1968, Figs. 6-2 and 6-3), yield fossils representing organisms that in life clearly existed in strongly contrasting biogeographical regions, a serious geological problem is posed. For example, it is geologically risky to permit a biogeographically cool- or cold-region fauna, such as that of the Cambrian Baltic Realm, to exist immediately adjacent to a warm-region fauna, such as that of the Laurentian Realm, unless there is an intermediate, faunally gradational region representing a transition from cold to warm environments. Such a transitional region, in this particular case, is unknown, although fossil-barren regions could, of course, contain transitional fauna. Using available data, one must therefore at least consider that present-day Coastal Acadia and the remainder of the fossiliferous Northern Appalachians were not nearly as close to each other at the time their sediments were laid down as they are today.

Early use of biogeographic analysis in the Acadian orogen

It has been recognized since the beginning of the twentieth century that the Cambrian, Early Ordovician, and Silurian rocks and fossils of Coastal Acadia (eastern Newfoundland, Nova Scotia, southeastern New Brunswick, coastal Maine, eastern Massachusetts, and Rhode Island) are very different from those of the rest of Greater Acadia present to the west. These differences were first noted by Walcott (1889), who was able to distinguish the Coastal Acadian Cambrian trilobites from those occurring to the west and northwest in eastern New York, Vermont, the St. Lawrence Valley region, and points west. This distinction ultimately became formalized in biogeographic terms as the "Atlantic Province" and "Pacific Province" respectively. These biogeographic terms are essentially equal to the "Baltic" and "Laurentian" realms of contemporary usage. Walcott used the terms *Atlantic Province* and *Pacific Province* in his 1909 paper (p. 201) and clearly had a biogeographic rather than a biostratigraphic lithofacies concept in mind. One can observe his thoughts developing on this problem of Cambrian provincialism by reading his publications on Cambrian problems from 1889 to after the turn of the century. Schuchert (1915) and others then postulated a hypothetical barrier landmass, Appalachia (which included a western peninsula called the New Brunswick Geanticline), between the provinces. The barrier landmass was located in a region that had not yielded any Cambrian fossils and thus might have formed a physical barrier to reproductive communication. Grabau (1921, p. 229–231), for example, used *Appalachia* for the same concept. Today it is recognized that there is little evidence favoring a major Cambrian local source area within the Northern Appalachians, whereas there is good evidence, beginning in the later Ordovician, for such a feature in the Central and Southern Appalachians. Too little was known at that time about Ordovician fossils in Coastal Acadia for much to be said about Ordovician biogeographic problems; in general, not much was said about Ordovician biogeography globally until after World War II.

Later than Walcott's efforts, Williams (1912)—working on the Coastal Acadian Silurian and earliest Devonian marine fossils—recognized a certain European affinity in them as contrasted with those to the west and northwest in Gaspé, northern New Brunswick and adjacent Maine as well as in New York and points south and west. Williams was possibly as much impressed with lithofacies differences as with the faunal differences, however, since he compared his Coastal Acadian materials with similar siliciclastic rocks from Wales and the Welsh Borderland rather than with carbonate facies Silurian present to the east in varied parts of Europe. As described below, all of the Coastal Acadian Silurian fauna are now assigned to the European Province of the North Atlantic Region of the North Silurian Realm (see Boucot, 1990, Fig. 1).

The small amount of marine Mississippian present in Coastal Acadia has been recognized for some time (Bell, 1929) to have European affinities. Sando (1992) can also be consulted for a recent treatment of the global marine Mississippian biogeography. For the Pennsylvanian it was early recognized that the coal swamp floras of eastern North America were very similar to those of Europe, and the Euramerian Province concept for a Pennsylvanian flora has been widely used for almost a century. The amount of Permian data is, even today, too limited to be of much biogeographic consequence.

CURRENT UNDERSTANDING OF CAMBRIAN-TO-CARBONIFEROUS BIOGEOGRAPHY

What is known today concerning the paleobiogeography of the Northern Appalachians that has some bearing on the Acadian orogeny? The first-order biogeographic fact is that Coastal Acadia, including eastern Newfoundland, Nova Scotia, southern New Brunswick (the Long Reach region and points to the southeast), coastal Maine, eastern Massachusetts, and Rhode Island, is very distinct in terms of its marine fossils from the regions to the west and northwest from the earliest Cambrian through the Early Devonian as detailed below. This distinction is not a merely local, Northern Appalachian feature. Coastal Acadia differs from the region to the west and northwest in a truly global manner, that is, this boundary is a globally significant feature rather than one of chiefly local concern. Marine data for the Mississippian are also consistent with this view. However, beginning with the Pennsylvanian one finds only nonmarine biotas within Greater Acadia, and these are all of similar type biogeographically. Greater Acadia, for biogeographic purposes, is defined here as the Northern Appalachians east of the Hudson River, southeast of the St. Lawrence River; it includes all of Maritime Canada.

I referred earlier to the fact that the Cambrian and Early Ordovician biogeographic units differ "in a truly global manner." This means that the fauna used in the biogeographical assessment are distinctive at the "realm level," the highest biogeographic rank. It is now customary to use a hierarchy of decreasing levels of biogeographic distinctiveness consisting of "realm," "region," "province," and "subprovince." It is now recognized that the younger Ordovician biogeographic distinctions are probably also at the realm level, that the Silurian units are at the province level, and that the Early and Middle Devonian units are again at the realm level followed by Mississippian distinctions at the province level. Following the Mississippian we have only Pennsylvanian data from the terrestrial environment that shows no evidence of biogeographic provincialism; the limited Permian data are permissively consistent with the same conclusion.

Cambrian

Cambrian biogeography is based chiefly on trilobites, because they are better known, are more abundant as individual specimens, and include more genera by far than all the other Cambrian fossils combined. Species are little used for the purposes of Paleozoic biogeography. Globally the marine Lower Cambrian is divided into two units, one characterized by olenellid trilobites and the other by redlichids; these two units correspond to realms. The Middle and Upper Cambrian may similarly be divided into four biogeographic units: Baltic ("European"), Siberian, Laurentian ("North American"), and Chinese. These four units are also considered to be realms (Boucot and Gray, 1987, Figs. 1, 2; brief discussion). The Baltic Realm represents cool to cold, although nonglacial, conditions, whereas the Laurentian and Chinese probably represent warm temperate conditions that

typically produced abundant limestones. These are followed by a Siberian Realm occurring to the north of a tropical-subtropical region indicated by the presence of bauxites in Soviet Central Asia. Palmer (1973, 1974) gave the best available summaries of Cambrian trilobite biogeography, and Dean (1976, 1985) provided additional Greater Acadia data.

Baltic Realm Cambrian uniqueness extends from Narragansett Bay (Skehan and others, 1978) north through the Boston area (skipping coastal Maine, which has no recognized Cambrian fossils) to the classic Saint John, New Brunswick region, then crosses over into Nova Scotia (Antigonish County; Landing and others, 1980) and includes the whole of Cape Breton Island. The Baltic Realm is also well represented on the Avalon Peninsula of eastern Newfoundland. This belt corresponds generally to what is now commonly referred to as the Avalon Composite Terrane. The distinctiveness of the Baltic Realm is also seen far to the southwest in the Carolina Slate Belt of South Carolina (Samson and others, 1990). The Cambrian faunas of this lengthy belt have been recognized for their high-level provincialism since Walcott's day. This belt stands in contrast to the Laurentian Realm (Dean, 1985) occurring in western Newfoundland, Quebec, Vermont, New York, and points both west and southwest in the Appalachians and on the North American Platform. The Cambrian trilobite biogeographic pattern persists, zone-by-zone, from the beginning of the period until its end. Thus, overall the Cambrian is an interval of high provincialism.

Dean, as indicated above, prefers the term *Laurentian* for the Cambrian fauna of the North American Platform, including the New York to Alabama Appalachians plus Greenland, Spitzbergen, northwestern Norway and northwestern Scotland areas. In central Newfoundland a single Middle Cambrian occurrence (Kay and Eldredge, 1968) is reinterpreted by Dean (1985, p. 27) as possibly being located "near the Gondwanaland site of Iapetus." This interpretation is mostly based, however, on a single genus (*Bailiella*), with the second genus (*Kootenia*) present in the collection being equivocal.

Ordovician

Ordovician marine biogeography as a whole has been little investigated when compared with that of the Cambrian, Silurian, and Devonian. The best available global summary is Jaanusson's (1979); Dean (1976, 1985) provided useful data from Greater Acadia, and Neuman (1984) has given a similarly useful summary largely devoted to the brachiopod data of the North Atlantic region. Ordovician biogeography has been based primarily on trilobites, followed by brachiopods, conodonts, and graptolites.

Greater Acadia biogeography of the Ordovician is complicated by the fact that fossils of this period are restricted chiefly to the western and northwestern areas in eastern New York, Vermont, the Saint Lawrence River Valley, and western Newfoundland. Ordovician fossils are relatively few and localities very scattered elsewhere. In addition, inland Ordovician fossil localities to the west and northwest of Coastal Acadia are not only few in number but have been discovered chiefly since World War II and still need additional study. The post–middle Arenig (Whiterock and younger) fossils are commonly biogeographically very complex. All too often they apparently represent a boundary-region of mixed genera with Baltic and North American affinities and even a few with Mediterranean affinities when allied British units are considered. The so-called Celtic biogeographic unit (Williams, 1973, Fig. 9) is actually a boundary mixture of Mediterranean and Baltic forms with a few relatively cosmopolitan and North American forms. In these, the proportions of the realm-level units vary from locality to locality and horizon to horizon. Unfortunately, a detailed biogeographic analysis of these mixed forms has not yet been carried out. There are very few Ordovician fossils within the Northern Appalachians of pre–Late Arenig (=pre-Whiterock) age, so that biogeographic information for the Lower Ordovician is inadequate. Neuman and Max (1989) summarized the evidence favoring an orogenic event, the Penobscotian Disturbance, that may be partly responsible for the relative rarity of these older fossils. Keeping all of this in mind, it is clear that Ordovician biogeography of the Greater Acadian region will certainly undergo substantial modifications in coming years as new collections are made, in contrast with the better-understood picture for the Cambrian, Silurian, Devonian, and Carboniferous.

In Coastal Acadia, earlier Ordovician (Tremadocian) Baltic Realm graptolite faunas, the dendroid faunas, have been known for a long time from the Saint John region of New Brunswick, Nova Scotia (including those from the upper part of the Meguma Series), and eastern Newfoundland. Laurentian Realm graptolites prevail outside Coastal Acadia to the west. Dean (1985) pointed out that the limited Coastal Acadian shelly Tremadocian fauna is of the "Gondwanaland Realm" type *sensu lato*. Coastal Acadia has not yet yielded biogeographically useful younger graptolitic or shelly Ordovician faunas except for an Arenigian occurrence in eastern Newfoundland (Dean, 1985). There the Gondwana Realm shallow-water trilobite, *Neseuretus,* occurs, representing widespread cool or cold water of higher latitudes. Additional collecting in the Cape George area of Antigonish County, Nova Scotia (following Neuman, 1968, p. 51; Boucot and others, 1974), might provide helpful data on Coastal Acadia provincialism. Further details on the Early Ordovician biogeography is very important, since younger Ordovician fossils are very uncommon in Coastal Acadia from the Avalon Peninsula of eastern Newfoundland to eastern Massachusetts.

The post–Lower Ordovician (post–middle Arenigian) shelly and graptolitic faunas occurring to the west and northwest of Coastal Acadia, in central Newfoundland, central and northern New Brunswick, and northern Maine, are both chiefly typical of a North American shelly type (Dean, 1976, 1985; Dean *in* Skinner, 1974, p. 43; Neuman, 1984) or of a Pacific Realm graptolitic type. There are, however, varying admixtures of Baltic Realm genera in time and space northwest of Coastal Acadia. It should be pointed out here that with the pelagic graptoloid graptolites of the Ordovician, the terms *Atlantic* and *Pacific* Realms, as contrasted with Laurentian and Baltic for the shelly faunas, may be more appropriate because of the far more widespread nature and far larger and fewer biogeographic units needed to describe Ordovician graptolite biogeography.

Silurian

The Coastal Acadian marine Silurian faunas have been termed "European" from at least the beginning of this century (Boucot, 1990). Modern investigations support this conclusion, although the level of distinctiveness is biogeographically much lower than that present in the Cambrian and Ordovician. The Cambrian biogeographic units in Greater Acadia, as discussed above, are distinctive at the realm level, the highest-level biogeographic units, whereas the Silurian units are assigned to the European and American Provinces of the "North Atlantic Region" (itself a subdivision of the North Silurian Realm). This lower-level distinctiveness does not necessarily translate into shorter distances of geographic separation within the Acadian orogen during the Silurian, however. The greater level of similarity of the Silurian fauna across Greater Acadia correlates with increased levels of reproductive communication, as judged by the far higher percentage of shared taxa. These shared taxa, in turn, may be ascribed to minor changes in geography capable of redirecting surface-current circulation patterns. Also recall that Cambrian biogeography is based on trilobites whose comparative dispersibility as compared with the brachiopods and ostracodes used in the Silurian is little known. Trying to figure out the dispersal capabilities of extinct groups of organisms is a very relative business, particularly since all of the Cambrian trilobites were well out of the way before any of the Silurian groups appeared on the scene. The phrase "well out of the way" refers to taxa at the superfamilial and ordinal levels, not just genera, species, or even families. We are really ignorant of the necessary paleobiological facts here. Inland from the European Province of Coastal Acadia, shelly Silurian faunas belong to the North American Province that extends from central and northern New Brunswick, Quebec, and Maine westward to almost the western margins of Silurian North America (Boucot, 1990, Fig. 1). The Coastal Acadian European Province shelly faunas were last seen just northwest of the Coastal Volcanic Belt (Boucot, 1968, Figs. 6-2, 6-3), and their provincial distinction has not yet been made in the older Silurian of Newfoundland west of the Avalon Peninsula.

It should be emphasized here, while commenting that Silurian graptolites are not too useful for biogeographic purposes within the Appalachian region, that each group of organisms, in any one major time interval, has its own characteristic level or provincialism. In the Silurian, for example, most graptolites are relatively cosmopolitan, most brachiopods and rugose corals moderately provincial, and most ostracodes and stalked echinoderms highly provincial. In this connection, it should be noted that the most abundant fossil groups used for biogeographic purposes in Coastal Acadia have been brachiopods and ostracodes; the highly provincial stalked echinoderms are essentially absent throughout Greater Acadia, possibly owing to the relative rarity of their favored carbonate environments.

Devonian

The lower-level biogeographic distinctiveness of the Silurian in Greater Acadia is rapidly replaced by a far higher Early Devonian realm-level provincialism that is comparable to that of the Cambrian and Ordovician. If one were to equate this increase in biogeographic provincialism with an increase in geographic separation, a very stately minuet carrying continental blocks back and forth more than once during the earlier Paleozoic would have to be inferred. There is no geologic evidence strongly favoring the closing of a broad seaway separating the New World from the Old World toward the end of the Silurian followed by a rapid opening during the Early Devonian. Changing levels of provincialism do not necessarily translate directly into distance, since many additional factors may be involved in any specific case.

Marine Middle Devonian is lacking in Coastal Acadia, but the earlier Middle Devonian is represented in Quebec (Lake Temiscouta and St. George areas; Boucot, 1968, 1975, 1988, Figs. 1, 2) and displays a mixture of North American (Eastern Americas Realm) and European (Rhenish-Bohemian Region of the Old World Realm) types. This Middle Devonian mixture is preceded by pure Eastern Americas Realm Lower Devonian, except for a few European taxa in the later Early Devonian of eastern Quebec. This Early Devonian mixing in eastern Quebec does not provide us with any direct measures of distance because we know nothing about the larval transport capabilities of the taxa involved. The later Middle Devonian of the Montreal region (St. Helen's Island; Boucot and others, 1986) is mixed like that of New York and the central Appalachians farther south. During later Middle Devonian time this mixing entails inclusion of taxa of both the Eastern Americas and Old World Realms, as well displayed in the Hamilton fauna of eastern North America. These mixed Middle Devonian fauna merely indicate that a previous barrier to reproductive communication responsible for maintaining the integrity of the Eastern Americas and Old World Realms in the Early and early Middle Devonian had been breached so as to permit thorough mixing. It is possible that geographic changes involving the Acadian orogeny were at least partly responsible for making the mixing possible. The Upper Devonian, as stated earlier, is everywhere very cosmopolitan.

Carboniferous through Permian

Mississippian marine biogeography overall has not received the attention it merits. Sando's (1992) treatment, however, is an exception. It has been recognized for some time (Bell, 1929) that the marine megafossils of Coastal Acadia have European affinities. Sando placed the corals in his European Province, whereas those in the nearest North American areas to the west (Southern Appalachians and Midcontinent area, which compose Sando's Southeastern Province) are said to be distinctive at the provincial

level. The terrestrial floras of the time are globally fairly cosmopolitan (Chaloner and Meyen, 1973). There are no known marine Pennsylvanian faunas involved here, and the floras all belong to a single phytogeographic unit, the so-called Euramerian Province (Chaloner and Meyen, 1973). Except for a limited part of Prince Edward Island, where terrestrial beds with a limited fossil content are known, Permian biogeographic evidence is lacking in Greater Acadia.

CHANGING GLOBAL CLIMATIC GRADIENTS AND PROVINCIALISM

At all times since the beginning of the Cambrian Period there is good evidence for first-order, latitudinal provincialism, just as there is today. However, the global climatic gradient goes up and down through time, with highest gradients being present at times when continental glaciation reaches sea level, as at present. The lowest gradients exist when equable conditions are present and warm temperate organisms reach high latitude. Such conditions prevailed in the Lower Eocene, for example, when crocodiles were present in northern Ellesmere Island, palms grew in west-central Greenland, nummulitic limestones were deposited on the Hatton-Rockall Plateau, and mangrove swamps flourished in the Paris Basin eastward from London. Any consideration of the geologic implications of biogeography in Greater Acadia must take these variations in climatic gradients into consideration.

The Lower Cambrian was a time of globally low climatic gradients, as indicated by the absence of cool-climate sediments and faunas, which helps explain the widespread presence of marine siliciclastics associated with marine redbeds in the Cambrian. The red pigment of such marine redbeds is presumably derived from a nonmarine regional source subjected to weathering in warm climates. As used here, the term *redbeds* does not necessarily include marine, commonly oolitic, iron ores that formed within the marine environment under conditions not necessarily related to warm climates (Boucot and Gray, 1986). Thus the famous Lower Ordovician Wabana oolitic iron ores of eastern Newfoundland and similar iron ores in the Mediterranean Ordovician of Europe are considered to be cool-water deposits despite the presences of some types of redbeds, that is, the oolitic iron ores.

The Coastal Acadian Early Cambrian, with its siliciclastics locally containing marine redbeds and lacking abundant, unweathered, detrital mica flakes, contrasts with the younger Cambrian siliciclastics without redbeds and containing abundant unweathered mica. In the Middle and Upper Cambrian there was a pronounced increase in the global climatic gradient, correlating with the absence of marine siliciclastics interbedded with such warm climate indicators as redbeds from the high latitude younger Cambrian of Coastal Acadia. For both the younger Cambrian and Ordovician there are, of course, lower latitude, warmer region strata present in North America to the west and northwest, well away from Coastal Acadia.

The same situation continues into the Ordovician. The very

high climatic gradient of the terminal Ordovician (later Ashgillian) is not actually represented by tillites in Coastal Acadia, but some conglomeratic rocks of the White Rock Formation in southwestern Nova Scotia have been interpreted to be of glacial-marine origin (Schenk, 1972). Undoubted Ashgillian tillites are widespread in North Africa (Beuf and others, 1971). The climatic maximum of the later Ashgillian is represented in eastern Gaspé by the occurrence of the cool water *Hirnantia* fauna of the Ashgillian stratigraphically between warmer-water sequences in both the older Ordovician and younger Early Silurian (Sheehan and Lesperance, 1981). Wright (1968, 1985) pointed out similar occurrences in Ireland and the Lake District of northern England, where the latest Ordovician *Hirnantia* fauna also lie above warmer-water Ordovician faunas. The common presence of a disconformity on the North American Platform as well as at many off-platform localities during the latest Ordovician may be largely due to Late Ordovician glacial drawdown. It is also possible that the disconformities may be related to Taconian orogenesis in the Appalachian system.

For the Silurian, both the European and North American Provinces present respectively in Coastal Acadia and to the west and northwest represent relatively warm water environments as contrasted with the Middle Cambrian through the Ordovician. In the Early Devonian, the situation shifts to one in which the Coastal Acadia Old World Realm faunas represent a warmer biota, whereas coeval faunas of the Eastern American Realm to the west and northwest appear to represent warm temperate conditions (Boucot, 1988). In the Mississippian, the widespread evaporites and normal marine limestones of the Windsor Group and coeval units in Coastal Acadia signify an arid, warm climate that was probably situated at lower, southern latitudes, during a time interval of relatively low, but not lowest global climatic gradient. The Pennsylvanian, on the other hand, was locally a time of relatively humid conditions, as suggested by the widespread coal swamps that may reflect tropical-subtropical latitudes to the north of a Southern Hemisphere arid belt.

BIOGEOGRAPHIC CONCLUSIONS

The most important conclusion concerning the biogeographic evolution of the Northern Appalachians is that Coastal Acadia—encompassing the Narragansett Bay and Boston regions, coastal Maine and southern New Brunswick, all of Nova Scotia, and the Avalon Platform of eastern Newfoundland—formed a Cambrian–through–Early Devonian biogeographic unit of Old World affinity, whereas parts of the orogen to the west and northwest of Coastal Acadia contain faunas of North American affinities. Before the Acadian orogeny (pre-Givetian), changing levels of provincialism overall in Greater Acadia are as follows: (1) Cambrian biogeographic differences are at the Realm level (highest rank), (2) Ordovician differences are not as clear as one might wish but are probably also at the Realm level, (3) Silurian differences are pretty well established at the Province level (relatively low), and (4) Early Devonian through early Middle Devo-

nian (Eifelian) differences are again at the realm level. Post-Acadian levels of provincialism in Greater Acadia are as follows: (1) Middle and Late Devonian biogeographic data are scanty, but they are consistent with a fairly cosmopolitan distribution pattern of little value in regional biogeographic discrimination; (2) Mississippian biogeographic differences are at the Province level; and (3) Pennsylvanian data show no differences, and Permian data are too limited for any definitive statements to be made. For the Mississippian marine faunas, all of Newfoundland and Nova Scotia are of European affinities, and North American marine faunas only begin to be present far to the west, well outside the confines of Greater Acadia. During the Middle and Late Ordovician, central Newfoundland and central New Brunswick as well as parts of northern Maine were chiefly of North American biogeographic character, with a varying admixture of Baltic Realm taxa. It is clear, however, that more biogeographic analysis of the marine fossils of these regions is required.

What utility do these biogeographic conclusions have for the geologist interested in Acadian tectonism? First, they certainly suggest that somewhere in southern New Brunswick and adjacent southeastern Maine there must be a very important tectonic boundary that juxtaposes an eastern Coastal Acadia block with European biogeographic affinities to a terrane with North American biogeographic affinities. In New Brunswick, this boundary must be located to the northwest of the Long Reach–Mascarene area and to the southeast of the isolated, most southeasterly known Eastern Americas Realm (Devonian) fossil locality. That isolated fossil locality is from the Hartin Formation located in a graben of Silurian-Devonian sedimentary and volcanic rocks (the Canterbury Basin) within the Cambro-Ordovician Miramichi Terrane west of Fredericton (Venugopal, 1979). The fauna of this locality (see Appendix) belong to the Eastern Americas Realm as contrasted with slightly earlier Devonian Old World Realm faunas well established in the Eastport Formation of Coastal Acadia in southeastern Maine and adjacent New Brunswick. The fossil locality is, unfortunately, separated from Coastal Acadia by the broad Fredericton Trough containing Acadian-deformed Silurian flysch intruded by plutons of the Pokiok Batholith that have recently yielded radiometric ages spanning the Silurian-Devonian systemic boundary (Bevier and Whalen, 1990). There are no clear candidates for a fault of Early-Middle Devonian age separating the two biogeographic units, and we lack sufficient fossil localities to narrow the gap at this time. No geological evidence is known to me for the claim by Neuman and others (1989, Fig. 2) that marine Devonian rocks are absent in the gap, implying that a nonmarine Early Devonian environment separated the known Eastern American Realm faunas of the Hartin locality from those of the Old World Realm, Rhenish-Bohemian Region type in Coastal Acadia. Such a hypothetical nonmarine barrier would make it very difficult to explain the presence of a few Old World Realm genera, such as *Rhenorensselaeria,* later Early Devonian in Gaspé, and of Eastern Americas Realm genera, such as *Meristella,* in rocks of the same age in Nova Scotia.

Thus, on the basis of the above biogeographic data, the

concept of a Eurafrican block (Coastal Acadia) finally becoming welded to a North American block during the Acadian orogeny is viable. As discussed below, this collision must have been completed by the mid-Middle Devonian, possibly during the latest Eifelian. Complicating this conclusion is the evidence favoring local uplift ranging in age from earlier Wenlock through Gedinnian in parts of the present North Atlantic region including Coastal Acadia (except for southwestern Nova Scotia and the Annapolis Valley region, where marine conditions persisted well into the Emsian, latest Early Devonian). The North Atlantic regions affected by such earlier uplift, additional to Coastal Acadia (except for southwestern Nova Scotia), include Scandinavia and Britain to the north of the Bristol Channel. Data adequate to answer this question for Newfoundland and eastern Greenland are missing. It is important to note, however, that the age of "Acadian orogeny" in the Fredericton Trough, to the northwest of Coastal Acadia, is unconstrained by paleontologic evidence because of the absence to date of any Devonian or even latest Silurian fossils. Although local uplift in northern Nova Scotia and the part of Coastal Acadia extending from southeastern New Brunswick through southeastern Maine and into eastern Massachusetts certainly resulted in nonmarine conditions by some time in the Gedinnian, the absence of post-Gedinnian, Early Devonian fossils precludes certainty about the lower limits for the local initiation of Acadian orogeny. Nevertheless the thick volcanic piles in such places as Vinalhaven are suggestive of a decent time interval being involved before the initiation of orogeny.

PALEONTOLOGIC LIMITS ON THE AGE OF THE ACADIAN OROGENY

Boucot (1968, Tables 6-1, 6-2) provided correlation charts for the Silurian through Middle Devonian of the Northern Appalachians; these need little modification. Boucot and St. Peter (1981) discussed brachiopods from Madawaska County, New Brunswick, that provided a local, upper age range for the Temiscouta Formation of Esopus or Schoharie age, that is, younger than the previous Oriskany limit.

The overall northern Appalachian paleontologic data indicate that the Acadian orogeny was initiated within a narrow Late Eifelian–Givetian time span. This is based on the fact that the Catskill Delta immediately to the west and northwest shows a marked upward shift from carbonate sedimentation of the Onondaga Group to siliciclastics of the Hamilton Group, including the Hamilton-age rocks and fossils of St. Helen's Island, Montreal (Boucot and others, 1986). At the same time there was a major biogeographic shift from purely Eastern Americas Realm faunas to a pronounced mixture of Eastern Americas Realm with Rhenish-Bohemian Region, Old World Realm forms. This shift was produced by a lowering of the global climatic gradient (Boucot, 1975, 1988) coinciding with a current circulation pattern alteration that permitted Old World Realm larvae access to New York and points south and west in central North America. Within the Northern Appalachians themselves the paleontologic

evidence shows the presence of Eifelian age units from the Lake Memphremagog area northeast to eastern Gaspé below, with overlying beds of Frasnian age (such as the Escuminac Formation). Farther to the southeast, the youngest beds below the unconformity are of Schoharie age in northern Maine with Givetian age beds (Mapleton Formation) overlying. Additionally in Coastal Acadia (southeastern Maine, New Brunswick, and Nova Scotia) the paleontologic constraints are provided by the Gedinnian age fossils of the lower parts of the Eastport Formation (in eastern Maine) and the Knoydart Formation (in Nova Scotia) below the Acadian unconformity with the overlying Upper Devonian of the Perry Formation and the basal Horton Group.

ACKNOWLEDGMENTS

I am very grateful to A. R. Palmer, Geological Society of America, Boulder, Colorado, and to Godfrey Nowlan, Geological Survey of Canada, Calgary for advice concerning certain Cambrian matters and to Palmer, D. C. Roy, and J. W. Skehan for constructively and patiently reviewing the manuscript. I am grateful to Reginald A. Wilson, New Brunswick Department of Natural Resources and Energy, Fredericton, for having revised the information concerning the regional significance of the Hartin Formation fossil locality. W. T. Dean, University of Wales, Cardiff, generously gave me a great deal of help in better understanding the complex relations of the Cambrian and Ordovician faunas of the region. I am very indebted to W. J. Sando, U.S. Geological Survey, Washington, D.C., for having provided a preprint of his forthcoming paper on global Mississippian marine biogeography and for having critically reviewed that part of the manuscript dealing with the Mississippian. M. G. Bassett, National Museum of Wales, Cardiff, very kindly provided a stimulating and critical review of the manuscript that was most helpful.

APPENDIX

The tectonically and biogeographically critical Hartin Formation fossil locality (Geological Survey of Canada Number 94984), discovered by D. V. Venugopal, yielded typical mid-to-lower Helderbergian brachiopods to Boucot. The previously unpublished assemblage consists of *Hedeina (Macropleura) macropleura, Atrypa "reticularis," "Gypidula" coeymanensis, Levenea* cf. *subcarinata, Platyorthis?* sp., *Nucleospira* sp., *Dicoelosia* sp., *Costistrophonella punctulifera, Leptaena "rhomboidalis," Megakozlowskiella?* sp., *Meristella?* sp., *Plicoplasia?* sp., leptocoelid?, leptostrophids, *Obturamentella?* sp., orthotetacid, *Dalejina?* sp., unidentified brachiopods, craniid, tetracorals, tabulates, trilobite, and bivalve.

REFERENCES CITED

Bell, W. A., 1929, Horton-Windsor District, Nova Scotia: Geological Survey of Canada Memoir 155, 268 p.

Beuf, S., Biju-Duval, B., de Charpal, D., Rognon, P., Gariel, D., and Bennacef, A., 1971, Les Grès du Paléozoique inférieur au Sahara: Publications de l'Institut Français du Petrole, Editions Technip, 464 p.

Bevier, M. L., and Whalen, J. B., 1990, Tectonic significance of Silurian magmatism in the Canadian Appalachians: Geology, v. 18, p. 411–414.

Boucot, A. J., 1968, Silurian and Devonian of the northern Appalachians, *in* Zen,

E., White, W. S., Hadley, J. B., and Thompson, J. B., Jr., eds., Studies of Appalachian geology: New York, Wiley Interscience, p. 83–94.

—— , 1975, Evolution and extinction rate controls: Amsterdam, Elsevier, 427 p.

—— , 1988, Devonian biogeography: An update, *in* McMillan, N. J., Embry, A. F., and Glass, D. J., eds., Devonian of the world, v. III, p. 211–227.

—— , 1990, Silurian biogeography, *in* McKerrow, M. S., and Scotese, C. R., eds., Palaeozoic palaeogeography and biogeography: Geological Society Memoir 12, p. 191–196.

Boucot, A. J., and Gray, J., 1986, Comment and reply *of* 'Oolitic ironstones and contrasting Ordovician and Jurassic paleogeography": Geology, v. 14, p. 634–635.

—— , 1987, The Tethyan concept during the Paleozoic, *in* McKenzie, K.G., ed., Shallow Tethys 2: Rotterdam, Balkema, p. 61–77.

Boucot, A. J., and St. Peter, C., 1981, Age and regional significance of brachiopods from the Temiscouata Formation of Madawaska County, New Brunswick: Maritime Sediments and Atlantic Geology, v. 17, p. 88–95.

Boucot, A. J., and eight others, 1974, Geology of the Arisaig Area, Antigonish County, Nova Scotia: Geological Society of America Special Paper 139, 191 p.

Boucot, A. J., Brett, C. E., Oliver, W. A., Jr., and Blodgett, R. B., 1986, The Devonian faunas of St. Helen's Island, Montreal, Quebec, Canada: Canadian Journal of Earth Science, v. 23, p. 2047–2056.

Chaloner, W. G., and Meyen, S. V., 1973, Carboniferous and Permian floras of the northern continents, *in* Hallam, A., ed., Atlas of palaeobiogeography: Amsterdam, Elsevier, p. 169–186.

Dean, W. T., 1976, Some aspects of Ordovician correlation and trilobite distribution in the Canadian Appalachians, *in* Bassett, M. G., The Ordovician System: Proceedings of a Palaeontological Association symposium: Cardiff, University of Wales Press and National Museum of Wales, p. 227–250.

—— , 1985, Relationships of Cambrian-Ordovician faunas in the Caledonide-Appalachian Region, with particular reference to trilobites, *in* Gayer, R. A., ed., The tectonic evolution of the Caledonide-Appalachian orogen: Braunschweig/Wiesbaden, Friedrich Vieweg & Sohn, p. 17–47.

Grabau, A. M., 1921, A textbook of geology: Part II, Historical geology: Boston, D.C. Heath, 976 p.

Jaanusson, V., 1979, Ordovician, *in* Robison, R. A., and Teichert, C., eds., Treatise on invertebrate paleontology: Part A, Introduction: Geological Society of America and University of Kansas Press, p. A136–166.

Kay, M., and Eldredge, N., 1968, Cambrian trilobites in central Newfoundland volcanic belt: Geological Magazine, v. 105, p. 372–377.

Landing, E., Nowlan, G. S., and Fletcher, T. P., 1980, A microfauna associated with Early Cambrian trilobites of the Callavia Zone, northern Antigonish Highlands, Nova Scotia: Canadian Journal of Earth Science, v. 17, p. 400–418.

Neuman, R. B., 1968, Paleogeographic implications of Ordovician shelly fossils in the Magog Belt of the Northern Appalachian Region, *in* Zen, E., White, W. S., Hadley, J. B., and Thompson, J. B., Jr., eds., Studies of Appalachian geology: Northern and Maritime: New York, Wiley Interscience, p. 35–48.

—— , 1984, Geology and paleobiology of islands in the Ordovician Iapetus Ocean: Review and implications: Geological Society of America Bulletin, v. 95, p. 118–1201.

Neuman, R. B., and Max, M. D., 1989, Penobscotian-Grampian-Finnmarkian orogenies as indicators of terrane linkages: Geological Society of America Special Paper 230, p. 31–45.

Neuman, R. B., Palmer, A. R., and Dutro, J. T., Jr., 1989, Paleontological contributions to Paleozoic paleogeographic reconstructions of the Appalachians: The Geology of North America, Vol. F-2, The Appalachian Orogen in the United States: Boulder, Colorado, Geological Society of America, p. 375–384.

Palmer, A. R., 1973, Cambrian trilobites, *in* Hallam, A., ed., Palaeobiogeography: Amsterdam, Elsevier, p. 3–12.

—— , 1974, Search for the Cambrian world: American Scientist, v. 62, p. 216–224.

Pickering, K. T., Bassett, M. G., and Siveter, D. J., 1988, Late Ordovician–early Silurian destruction of the Iapetus Ocean: Newfoundland, British Isles and Scandinavia—A discussion: Transactions of the Royal Society of Edinburgh, Earth Science, v. 79, p. 361–382.

Samson, S., Palmer, A. R., Robison, R. A., and Secord, D. T., Jr., 1990, Biogeographical significance of Cambrian trilobites from the Carolina Slate Belt: Geological Society of America Bulletin, v. 102, p. 1459–1470.

Sando, W. J., 1992, Global Mississippian coral zonation: Courier Forschungsinstitut Senckenberg (in press).

Schenk, P. E., 1972, Possible Late Ordovician glaciation of Nova Scotia: Canadian Journal of Earth Science, v. 9, p. 95–107.

Schuchert, C., 1915, Historical geology: New York, John Wiley and Sons, 415 p.

Sheehan, P. M., and Lespérance, P. J., 1981, Brachiopods from the Whitehead Formation (Late Ordovician–Early Silurian) of the Percé region, Quebec, Canada, *in* Field Meeting, Anticosti-Gaspé, Quebec, Volume II: Stratigraphy and paleontology, Lespérance, P. J., ed.: Montreal, International Union of Geological Sciences, I.U.G.S., Subcommission on Silurian Stratigraphy, Ordovician and Silurian Boundary Working Group, p. 247–256.

Skehan, J. W., Murray, D. P., Palmer, A. R. Smith, A. T., and Belt, E. S., 1978, Significance of fossiliferous Middle Cambrian rocks of Rhode Island to the history of the Avalonian microcontinent: Geology, v. 6, p. 694–698.

Skinner, R., 1974, Geology of Tetagouche Lakes, Bathurst, and Nepisiguit Falls map-areas, New Brunswick: Geological Survey of Canada Memoir 371, 133 p.

Venugopal, D. V., 1979, Geology of Debec Junction–Gibson Millstream–Temperance Vale–Meductic Region: New Brunswick Department of Natural Resources Mineral Resources Branch Map Report 79-5, 36 p.

Walcott, C. D., 1889, Stratigraphic position of the *Olenellus* fauna in North America and Europe: American Journal of Science, v. 30, p. 29–42.

—— , 1909, Evolution of Early Paleozoic faunas in relation to their environment: Journal of Geology, v. 17, p. 193–202.

Williams, A., 1973, Distribution of brachiopod assemblages in relation to Ordovician palaeogeography: Palaeontological Association Special Paper 12, p. 241–269.

Williams, H. S., 1912, Correlation of the Paleozoic faunas of the Eastport Quadrangle, Maine: Geological Society of America Bulletin, v. 23, p. 349–356.

Wright, A. D., 1968, A westward extension of the upper Ashgillian *Hirnantia* fauna: Lethaia, v. 1, p. 352–367.

—— , 1985, The Ordovician-Silurian boundary at Keisley, northern England: Geological Magazine, v. 122, p. 261–273.

MANUSCRIPT ACCEPTED BY THE SOCIETY JUNE 8, 1992

Geological Society of America
Special Paper 275
1993

The sequence of Acadian deformations in central New Hampshire

J. Dykstra Eusden, Jr.
Department of Geology, Bates College, Lewiston, Maine 04240
John B. Lyons
Department of Earth Sciences, Dartmouth College, Hanover, New Hampshire 03755

ABSTRACT

The Central Maine Terrane (CMT) includes the rocks that extend northeasterly from Connecticut to Maine and from the Monroe Fault on the west to the Campbell Hill–Nonesuch River Fault Zone on the east. A four-phase sequence of Acadian regional deformation is recognized for the CMT cover sequence. D_1, the earliest phase, is characterized by F_1 nappes that have east or west vergence; the sense of vergence switches at the Central New Hampshire anticlinorium (CNHA). D_1 is also characterized by early, rarely observed, low-angle and "blind" T_1 thrust faults. The CNHA (or "dorsal zone") is analogous to a "pop up" structure and is the likely root zone for both east- and west-verging Acadian D_1 thrust-nappes. D_2 is characterized by abundant F_2 tight to isoclinal, inclined to recumbent folds with northeast-trending axes and east-southeast vergence. Most of these folds face downward, a reflection of D_2 refolding the inverted limbs of D_1 structures, and these structures are identifiable chiefly in eastern New Hampshire. F_2 folds define a regional map-scale fold, the Lebanon antiformal syncline. During D_3 broad, open, upright to inclined F_3 folds with west- or northwest-trending axes were developed across the entire belt. F_3 map-scale syntaxial folds are well defined by the outcrop pattern of the metasedimentary rocks. D_4, the last phase of deformation, is characterized by F_4, tight to isoclinal, inclined folds with north-northeast–trending axes and east vergence and is restricted to the western part of the CMT. F_4 folds refolded the earlier structures and significantly modify the map pattern, tightening some of the earlier major structures in the CMT, for example the Kearsarge–Central Maine synclinorium. D_2 and D_4 are similarly oriented but spatially and temporally distinct. Deformation phases D_1 through D_4 are geographically restricted. This uneven distribution of structures is critical to correlations of deformation sequences across the orogen. Any local sequence of deformation in the CMT of central New Hampshire will commonly have only three of the four regional phases preserved in outcrop.

INTRODUCTION

This chapter focuses on the structure and stratigraphy of a group of rocks in central New Hampshire that occupy a region variously termed the Kearsarge–Central Maine synclinorium (KCMS) (Lyons and others, 1982), the Merrimack synclinorium (Billings, 1956), and the Merrimack Belt (Zen, 1983). The rocks in this structure have been interpreted as a large Silurian to early Devonian depositional basin that was multiply deformed and metamorphosed during the Acadian and perhaps the Alleghanian (?) orogenies.

Because of the variety of different names currently used to describe the major geologic features in this region, there has been much semantic confusion. Mapping over the last decade by geologists working in this belt has shown that there is not just one synclinorium, but several, with intervening anticlinoria (for example, see E. F. Duke, 1984; P. J. Thompson, 1985; Eusden, 1984, 1988a). It is clear that neither the Kearsarge–Central

Eusden, J. D., Jr., and Lyons, J. B., 1993, The sequence of Acadian deformations in central New Hampshire, *in* Roy, D. C., and Skehan, J. W., eds., The Acadian Orogeny: Recent Studies in New England, Maritime Canada, and the Autochthonous Foreland: Boulder, Colorado, Geological Society of America Special Paper 275.

Maine synclinorium nor the Merrimack synclinorium is an adequate descriptive term for the entire region.

For this chapter the term *Central Maine Terrane* (CMT) (Zen and others, 1986; Zen, 1989) will be adopted to include the rocks that extend northeasterly from Connecticut to Maine and from the Monroe fault on the west to the Campbell Hill–Nonesuch River fault zone (CHNRFZ) on the east (Fig. 1). The Bronson Hill anticlinorium, in Zen's (1989, Fig. 4) classification, is the western portion of the CMT. The KCMS is restricted to axial trace of the synclinorium defined by the distribution of the youngest unit, the Devonian Littleton Formation. The following major structures are recognized from west to east in the CMT of New Hampshire: the Bronson Hill anticlinorium (BHA), the Kearsarge–Central Maine synclinorium (KCMS), the Central New Hampshire anticlinorium (CNHA), and the Lebanon (Maine) Antiformal syncline (LAS) (Fig. 1) (Eusden and others, 1987; Eusden, 1988a, b; Lyons, 1988; Lyons and Bothner, 1989). In Zen's (1989) terminology the Brompton-Cameron Terrane lies west of the CMT and the Massabesic-Merrimack sequence to the east.

This reconciliation of regional nomenclature will probably not satisfy all geologists, some of whom argue that the belt should extend southeast beyond the CHNRFZ to include the formations of the Merrimack Group (Berwick, Eliot, and Kittery Formations) and hence prefer to keep the term *Merrimack Belt* or *synclinorium* as originally described by Billings (1956). However, the aforementioned Merrimack Group and Massabesic Gneiss Complex are part of a probable Precambrian exotic terrane quite different in origin from the rocks of the CMT (Lyons and others, 1982, 1986; Bothner and others, 1984; Gaudette and others, 1984; Naylor, 1985; Lyons and Bothner, 1989; Zen, 1989).

PREVIOUS STRUCTURAL MODELS

Billings (1956) interpreted the folds in the Siluro-Devonian metasedimentary rocks of the CMT as the product of the Devonian "Acadian Revolution." That summary work laid the foundation for later studies and also resulted in the recognition of the polydeformed nature of these rocks. However, Billings did not categorize the folded rocks into a regional sequence of deformations. In a now classic paper concerning the sequence of deformation along the gneiss domes of the BHA, Thompson and others (1968) proposed that early, west-verging nappes were later backfolded into east-facing structures and then domed to produce tight to isoclinal folds. This hypothesis was later modified and reaffirmed by Robinson (1979), Robinson and Hall (1980) and Hall and Robinson (1982). Adding to this scheme, Thompson (1985), Elbert (1986), Berry (1987), Robinson (1987), and Thompson and others (1987) recognized an early stage of thrusting that followed nappe-stage folding and preceded backfolding. Osberg and others (1989) outlined an eight-step sequence of faulting and folding events for the Acadian of New England that essentially follows the above regional models.

In central and eastern New Hampshire different structural models were developed. Lyons and his students (E. F. Duke, 1984; Englund, 1976; Nielson, 1981; Lyons, 1979; and Lyons and others, 1982) recognized three major Acadian folding events in central New Hampshire: (1) F_1, west-verging nappes; (2) F_2, broad open folds with west- or northwest-trending axes; and (3) F_3 isoclinal to open folds with northeast-trending axes. Farther east, Eusden and others (1984, 1987) also identified three major folding events: (1) F_1, east-verging nappes; (2) F_2, tight to isoclinal folds with northeast-trending axes; and (3) F_3, broad open warps with west-trending axes that define a major map pattern syntaxis.

A NEW STRUCTURAL MODEL

We believe that neither the nappe-thrust-backfold-dome nor the various F_1-F_2-F_3 scenarios can be uniquely applied to the cover sequence of the CMT. We propose here a revised sequence of Acadian deformations. Based on the extreme structural complexity in this belt and the potential for variations in structural style along and across strike, it remains unclear whether or not there is a regionally correlative sequence of structural events across the entire CMT.

Our model for the deformation of the CMT cover sequence has four phases, D_1 through D_4. This sequence of deformation dovetails the models proposed by Eusden and others (1987) and Lyons and others (1982) and accommodates the nappe-thrust-backfold-dome sequence of Thompson and others (1987) as well as the sequence proposed by Osberg and others (1989). D_1, the earliest phase, is characterized by F_1 nappes that have east or west vergence; the sense of vergence switches at the CNHA. D_1 is also characterized by early, T_1 thrust faults. D_2 is characterized by abundant F_2 tight to isoclinal, inclined to recumbnet folds with northeast-trending axes and east-southeast vergence. Most of these folds face downward, a reflection of D_2 refolding the inverted limbs of D_1 structures. D_2 folds are identifiable chiefly in eastern New Hampshire. During D_3, broad, open, upright to inclined F_3 folds with west- or northwest-trending axes were developed. D_4, the last phase of deformation, is characterized by F_4, tight to isoclinal, inclined folds with north-northeast–trending axes and east vergence. D_3 and D_4 structures are extremely well preserved in large, map-scale fold patterns throughout the CMT (Fig. 2).

An important aspect of this new model is that deformation phases D_1 through D_4 are geographically restricted. For instance,

Figure 1. Generalized geologic map of New Hampshire, modified from Lyons and others (1986). The Central Maine Terrane strikes northeast-southwest and extends northwest from the Campbell Hill–Hall Mountain–Nonesuch River fault zone to the Monroe fault. Metasedimentary rocks are shown in white. All gray shades and patterns are igneous rocks with the exception of the Piermont Allochthon and the Massabesic Gneiss Complex, which are metamorphic.

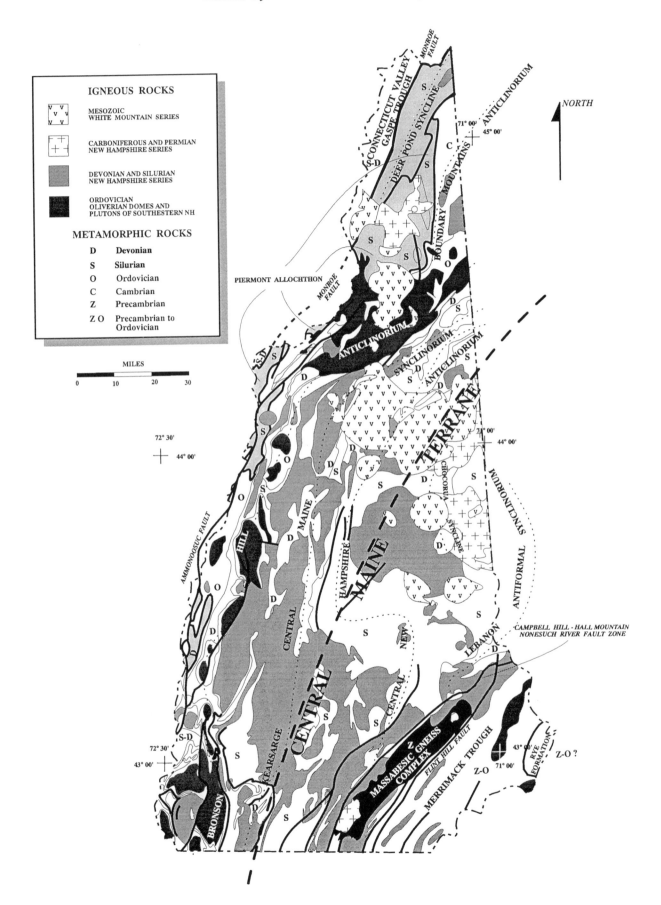

IGNEOUS ROCKS

MESOZOIC
WHITE MOUNTAIN SERIES

CARBONIFEROUS AND PERMIAN
NEW HAMPSHIRE SERIES

DEVONIAN AND SILURIAN
NEW HAMPSHIRE SERIES

ORDOVICIAN
OLIVERIAN DOMES AND
PLUTONS OF SOUTHESTERN NH

METAMORPHIC ROCKS

D   Devonian

S   Silurian

O   Ordovician

C   Cambrian

Z   Precambrian

Z O Precambrian to
     Ordovician

MILES

0   10   20   30

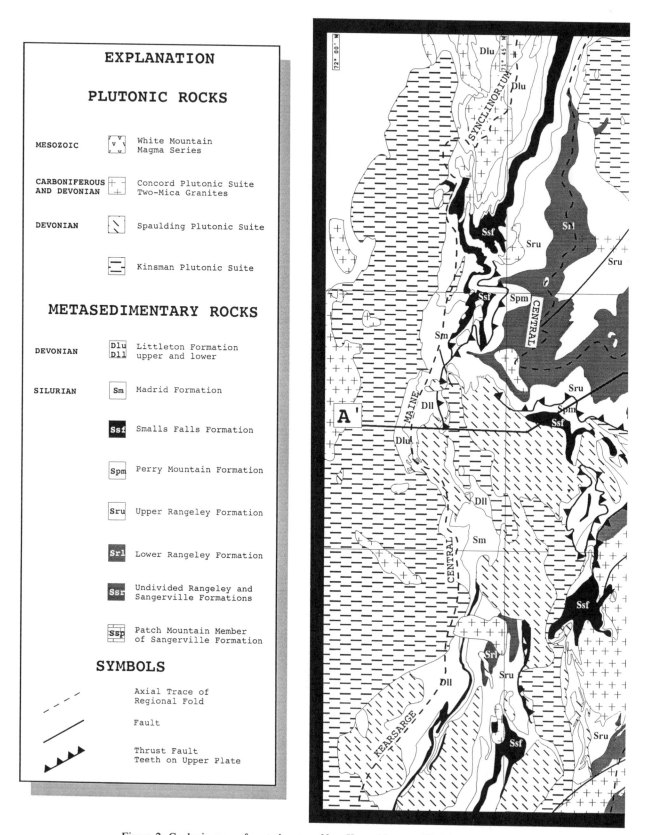

Figure 2. Geologic map of central-eastern New Hampshire, modified from Duke (1984), Englund (1976), Eusden (1988a), Eusden and others (1984), Lyons (1988), and Nielson (1981). The cross section along line A-A′ is shown in Figure 4.

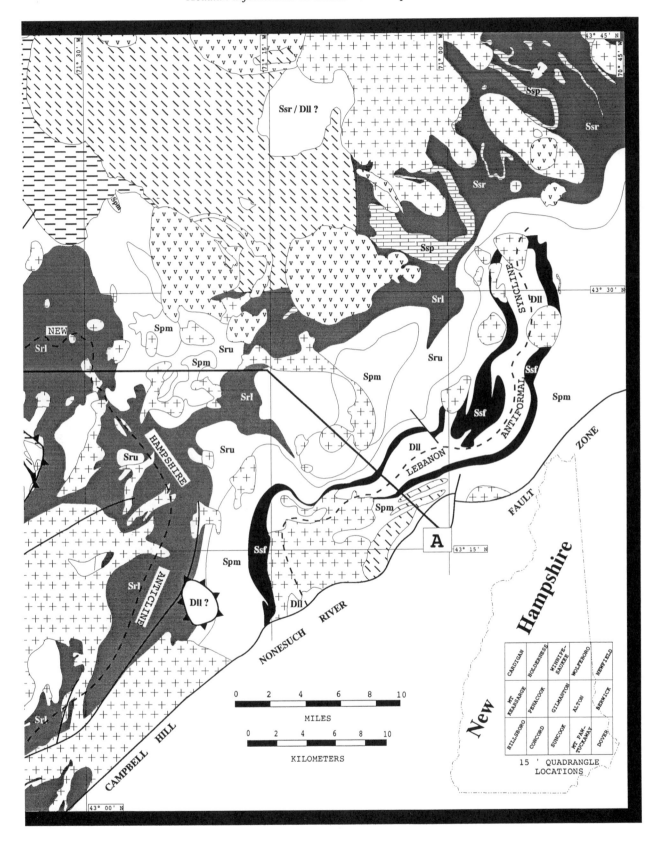

F_1 folds switch sense of vergence at the CNHA, D_2 structures are concentrated on the east side of the CNHA, D_3 structures appear ubiquitously throughout the region, and D_4 structures are chiefly identifiable west of the CNHA; their eastern limit may define the eastern limit of backfolding. This uneven distribution of structures is critical to correlations of deformation sequences across the orogen. Not all of the proposed phases of deformation are present in one particular area; only the partial sequence may be observed. The complexity and variation of Acadian deformation is far greater than previously imagined.

THE "DORSAL ZONE" OR "POP UP"

The CNHA acts as a "dorsal zone" or "pop up" (Eusden and others, 1987), structurally splitting the Siluro-Devonian cover of the CMT. West of the zone the early nappes verged only west (e.g., Thompson et al., 1968), but east of it they verged only east (e.g., Eusden and others, 1987); essentially they were mirror images of each other. Another critical observation is that there is no evidence for backfolding (defined as refolding of once west-verging structures into now east-facing structures) in most regions east of the CNHA. The importance of establishing the vergence of early structures is critical to understanding the Acadian tectonic transport and deformation mechanisms and has an obvious bearing on the interpretation of the deeper structure of the Northern Appalachians. The deep structure remains one of the major unsolved questions in tectonic models for the Acadian orogeny.

It has been well documented that west of the CNHA most east-facing D_1 structures were west-verging D_1 thrust-nappes that have been backfolded; perhaps the best example is the "Billings Fold" on Mt. Monadnock (Robinson, 1979; Robinson and Hall, 1980; Hall and Robinson, 1982; Thompson, 1985; Elbert, 1986; Robinson, 1987; Thompson and others, 1987; Berry, 1987). The backfolds are spatially near the BHA where granitic gneiss domes rose up through the cover rocks and backfolded D_1 thrust-nappes that had already been transported to the west. However, east of the BHA, away from the effects of the buttressing domes, the D_1 structures were not backfolded. West of the CNHA some of the D_1 west-verging structures still face west; for example, the nappes mapped in the Concord, New Hampshire, quadrangle by G. I. Duke (1984).

Because the CNHA marks the change from east- to west-verging D_1 structures, it qualifies as the probable but previously unrecognized root zone for many of the Acadian D_1 thrust-nappes. Thus the CNHA is more than a simple anticlinorium. It is marked not only by outcrops of the oldest formation in the CMT, the lower Rangeley Formation, but also by local "hot spots" of granulite facies metamorphism (Chamberlain and Lyons, 1983) and an unusual trend of soapstones interpreted as ultramafic slivers injected into thinned continental crust along the Concord Tectonic Zone of Lyons and others (1982). The "dorsal zone" with its east- and west-verging structures may have arisen during the Acadian orogeny by the collapse and tight closure of the subsiding trough of Silurian and Devonian age, first identified by

Boone and others (1970). The axis of this trough was not far from the axial trace of the present CNHA, as is revealed by the stratigraphy.

This model of the geologic structure across New Hampshire is analogous to that now recognized as "pop up" tectonics (Butler, 1982) in a number of mountain belts, for example, the Southern Irumbide Belt of Africa (Daly, 1986). Tectonic interpretations of the Variscan Belt in Europe (from which the term "dorsal zone" was taken) (Martin and Behr, 1983), the Caledonian Belt of Norway and East Greenland (Ziegler, 1985), and the Grampian/Caledonian of Scotland (Thomas, 1979) also show equivalent bilateral symmetry regarding vergence of early thrust-nappes. All of these belts have widths roughly the same as the CMT, are complexly refolded, and are similar in cross section to our model.

STRATIGRAPHY

Critical to this structural model is a revision of the stratigraphy for the cover sequence in central New Hampshire as originally set forth by Billings (1956). This revision was determined by mapping in detail small areas of the geology and also by reworking, often for the second or third time, the geology of the 15-minute quadrangles. Mapping along the axis of the CMT established the correlation of the unfossiliferous, high-grade metasedimentary rocks of central New Hampshire with the low-grade Silurian and Devonian, fossiliferous rocks of western Maine. Much of what was assigned by Billings (1956) to the Lower Devonian Littleton Formation is now interpreted as a thick section of Silurian turbidites correlative to the stratigraphy of the Rangeley, Maine, area (Moench and Boudette, 1970; Nielson, 1981; Moench, 1984; Hatch and others, 1983; Thompson, 1983, 1984, 1985; Chamberlain, 1984; G. I. Duke, 1984; E. F. Duke, 1984; Eusden and others, 1984, 1987; Eusden, 1988a; Lyons, 1988, 1989; Duke and others, 1988).

The Silurian and Devonian stratigraphy of central New Hampshire represents the middle or distal portion of the depositional basin. This trough was filled with nearly 3.5 km of Silurian clastics, carbonates, and volcanics derived from the west and then overlain by 2.5 km of easterly derived Devonian turbidites and volcanics. The axis of the basin or trough is probably coincident with the trace of the CNHA and is east of the "tectonic hinge" (Hatch and others, 1983) that hugs the BHA. The following section is a description of the stratigraphy in central New Hampshire from youngest to oldest. Figure 3 shows the stratigraphic correlations across the CMT.

Littleton Formation—Early Devonian: Gedinnian to Emsian

The Littleton Formation is a gray schist and quartzite and is subdivided into an upper and lower member. The upper member is a light gray, thinly bedded, well-graded metaturbidite. The lower member is a dark gray, massive metapelite. At the base of the formation in the Alton quadrangle (cf. Fig. 2) there is a thin,

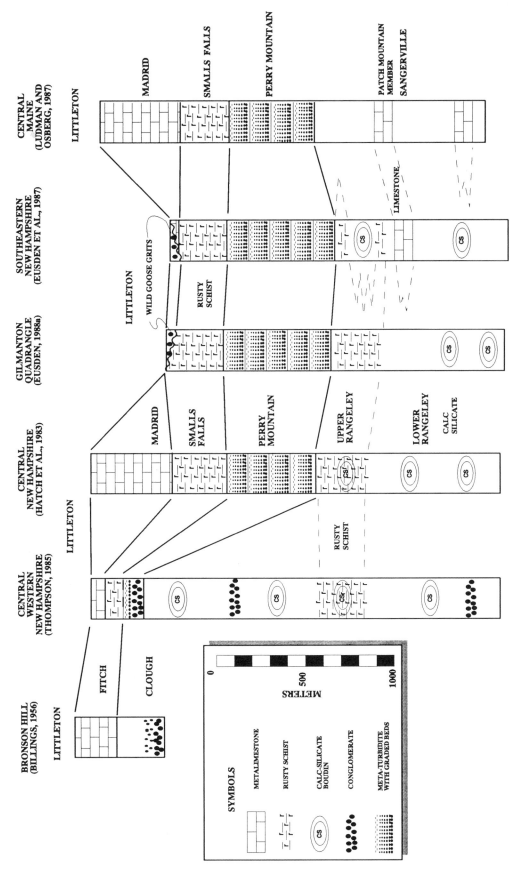

Figure 3. Stratigraphic correlations for the Silurian across the Central Maine Terrane. The Devonian Littleton Formation forms the top of the section and is not shown. The thickest section from central New Hampshire consisted of 3.5 km of Silurian overlain by 2.5 km of Devonian strata.

coarse-grained grit or conglomerate with gray and rusty-weathering lithic fragments of the Smalls Falls Formation. Eusden and others (1987) informally named these exposures the Wild Goose Grits, which presumably represent a basal Littleton transgressive deposit on a disconformity or shallow angular unconformity. The Littleton is at the top of the known Paleozoic section, and because of erosion only minimum estimates of the thickness can be made. The Littleton may have had a depositional thickness of 2.5 km prior to erosion. The Wild Goose Grits are thin—only about 100 to 300 m thick.

Madrid Formation—Late Silurian: Pridolian (?)

The Madrid Formation is a massive to foliated, purple, biotite, plagioclase, and quartz granofels with layers and lenses of calc-silicate. The Madrid thins from west to east across central New Hampshire. It is approximately 300 to 500 m thick in the west and in the east thins to a few tens of meters or is entirely absent.

Smalls Falls Formation—Early to Late Silurian: Ludlovian to Wenlockian

The Smalls Falls is a red-brown to brown-black, deeply rusty-weathering, sulfidic, graphitic schist with pyrrhotitic, calc-silicate–bearing granofels. The Smalls Falls is between 100 and 500 m thick. In places it is absent owing to nondeposition, erosion, or faulting. Clasts of Smalls Falls in the Wild Goose Grits suggest that the true thickness was probably greater and that erosion probably played a greater role in its thickness distribution than nondeposition.

Perry Mountain Formation—Early Silurian: Wenlockian (?)

The Perry Mountain Formation is a well-bedded gray schist and quartzite. The schist-quartzite pair makes up a turbidite couplet that is sharply bounded and rarely graded. The quartzites are light in color and lack micas. Infrequent garnet coticules as well as calc-silicate boudins are also found. The Perry Mountain is about 200 to 500 m thick.

Rangeley Formation—Early Silurian: Llandoverian

The Rangeley Formation is the most varied of all the formations in the stratigraphy. It is subdivided into upper and lower members. The upper Rangeley is a red-brown, rusty weathering, often graphitic, schist and sulfidic quartzite. In places it contains abundant calc-silicate boudins; in other regions there are none. The lower Rangeley has lentils of gray to slightly rusty red-weathering well-bedded schist and quartzite with some calc-silicate boudins. The matrix is characteristically dark gray thinly laminated schist. Calc-silicate boudins are highly variable in abundance but are generally considered the hallmark of the Rangeley Formation in this area.

Along the Maine–New Hampshire border, two or more rusty horizons are mapped within the upper Rangeley. In the lower Rangeley of the Alton and Newfield quadrangles a metalimestone unit is mapped.

The exposed Rangeley in central New Hampshire is between 1,000 and 1,500 m thick. The complete thickness is not well constrained as the bottom of the formation has not been observed in central New Hampshire. The upper member is not as thick as the lower. In places the upper Rangeley thins to 100 m or less. This may be because of either tectonic or depositional thinning. The thickness of the members varies considerably across strike. In the Alton quadrangle, for example, the upper Rangeley thins to only a few tens of meters. These thickness variations may represent facies variations within the Siluro-Devonian basin.

A correlation to the east with the Sangerville Formation of south-central Maine is quite likely for the lower Rangeley exposed in the Alton, Newfield, and Wolfeboro quadrangles, where thin sequences of metalimestones crop out. These are lithically identical to the ribbon limestone Patch Mountain member of the Sangerville. No limestones are found west of the Alton quadrangle, and it would seem likely that in this area a regional facies change occurs in the Rangeley Formation where, to the east, more Sangerville-like lithologies are found and, to the west, more typical Rangeley lithologies are found. This is in accord with a recent revision of the stratigraphy by Gilman (1988) in southwestern Maine.

STRUCTURAL GEOLOGY

The terminology used in this section follows that outlined by Turner and Weiss (1963). Penetrative surfaces—for example, axial-planar foliation and cleavage and bedding—are termed S-surfaces. Bedding is termed S_0, and succeeding planar fabrics S_1, S_2, S_3, and so forth. Successive folding events are termed F_1, F_2, F_3, and so on. Thrust faults are termed T_1, T_2, etc. Deformation phases are called D_1, D_2, D_3, and D_4. All folds are treated as cylindrical or near-cylindrical folds in this discussion, except for the F_4 folds, which may be sheath folds. The cross section related to this section is shown on Figure 4.

S_0 bedding

Bedding is common in all of the metasedimentary rocks. It is best preserved in the Littleton, Perry Mountain, and Lower Rangeley formations. In the Rangeley Formation elongate calc-silicate boudins are parallel to bedding and are assumed to have been continuous, now attenuated, calcareous, sandy beds. Alternatively they were clasts slumped from the paleocontinental shelf edge into the turbidite section. The boudins indicate that the section has undergone considerable thinning due to tectonic attenuation. This probably occurred during the first phase of deformation.

Except in the upper Littleton Formation, excellent tops from graded beds are infrequently found in the CMT. More often one

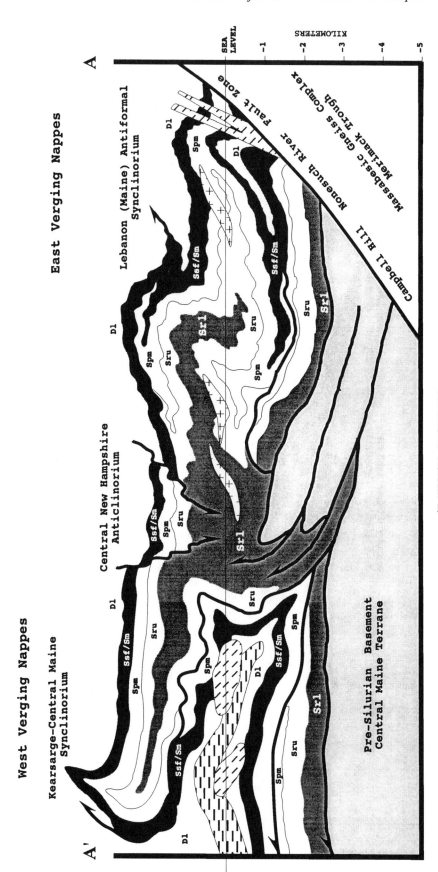

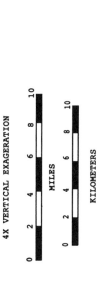

Figure 4. Geologic cross section through central-eastern New Hampshire. Figure 2 shows the location of the section as well as an explanation of patterns and symbols used here. The thin line indicating sea level is approximately the erosional level exposed today in central New Hampshire.

finds outcrops with unequivocal bedding and ambiguous topping sense. Most of the metasedimentary rocks in the eastern part of the CMT that have reliable topping indicators are inverted. Specifically, from the CHNRFZ west across the CNHA and to the axial trace of the KCMS proper, a distance of 75 km across strike, the section is predominately inverted. From the KCMS west to the BHA, about 30 km across strike, the section is largely upright. In the Alton quadrangle there is an exception to this where an F_1 axial trace closes on itself, and interior to the trace the stratigraphy is upright.

Periods of deformation

Four phases of deformation have been recognized in the central New Hampshire cover sequence of the CMT. Each phase is attributed to a stage in the deformational history. The phases are referred to as D_1, D_2, D_3, and D_4. The nomenclature used should in no way be construed to imply that each phase represents a discrete event. It is possible that they do, but we argue that the sequence of deformation was a continuum and not a series of separate pulses. The distribution and relative age of the major structures are shown on Figure 5.

D_1—first phase of deformation. D_1 is characterized by an S_1 axial planar schistosity, F_1 nappes, and T_1 thrust faults. S_1 is normally bed-parallel, that is, parallel to S_0. It is axial planar to the F_1 folds. This foliation is defined by an early stage of biotite and muscovite porphyroblasts. In F_1 hinge zones, S_1 is oblique to S_0. This fabric is the hallmark of the first phase of deformation: nappe-stage folding. It is common to see S_1 folded by later folds—clear evidence of polydeformation.

S_1 is found in all outcrops of metasedimentary rock in the central New Hampshire CMT. We often show S_1 on the same map symbol as S_0 because they have the same attitude. In extensively migmatized areas where S_0 is obliterated, only S_1 is found. In these migmatites S_1 is also defined by quartzo-feldspathic lenses, similar in appearance to augen. The lenses are asymmetric and commonly have tails that may be used as kinematic indicators. We have not done a study of the porphyroblast kinematics; however, the plane of shear, or shear envelope, is always parallel to S_1. The deformation seen in these quartzo-feldspathic lenses is attributed to nappe-stage folding and shear along the nappe limbs, not to discrete shear within or adjacent to mylonite zones.

F_1 folds are not common, and, when mapping 15-minute quadrangles, generally only 10 to 20 minor F_1 folds are discovered. All of the F_1 folds are isoclinal recumbent folds having a well-developed axial planar schistosity, S_1. The minor F_1 folds seen in the field range in amplitude from 2 to 10 m. The interlimb angles are 5° or less and are thus classified as isoclinal.

Evidence that early nappe-stage folding took place is incontrovertible, but vergence remains a major question. For many of the folds facing direction is indeterminate. However, in all the folds where topping direction could be unambiguously interpreted and the fold hinge observed, the F_1 folds face east to the east of the CNHA and west or east (if backfolded) to the west of the CNHA.

The cross section (Fig. 4) and the reconstruction of Acadian deformations (Fig. 6) show the development of this structural relationship. Based on the mapped extent of the folded Lower Rangeley Formation, the regional F_1 nappes have a minimum width of 10 to 20 km. Eusden and others (1987) thought that the east-verging Blue Hills Nappe in the Alton and Berwick quadrangles had rooted somewhere unspecified to the west. The CNHA is now regarded as the likely root zone for the Blue Hills Nappe. Based on map distances, that nappe must have had a minimum of 50 km of transport. A better approximation, considering the later shortening effects of refolding, would be closer to 100 km. Similar nappe transport distance is indicated in the west part of the CMT if the CNHA is the root zone of the classic Skitchewaug, Fall Mountain, and Cornish nappe sequences exposed along the BHA. Total shortening across the Central Maine Terrane must be approximately 200 km.

The CNHA is not precisely an anticline. As shown on Figures 4 and 6, it is more properly a root zone or region of divergence for the nappes, hence the term "dorsal zone." However, since the oldest unit in the stratigraphy crops out within, and thus defines, the CNHA, it is conceptually satisfying to show the structure as an anticline on the map. The fact that formations older (e.g., early Ordovician Quimby or latest Silurian Greenvale Cove Formations) than the Rangeley are not exposed in the CNHA strongly suggests a major décollement surface at or below the base of that formation.

On the cross section (Fig. 4), the east- and west-verging nappes are hypothesized to be floored by T_1 thrust faults that verge in the same direction as the nappes. In the Suncook, Concord, Penacook, and Mt. Kearsarge quadrangles several T_1 thrust faults have been mapped at the surface. They are based on marked stratigraphic discontinuities. Where the T_1 faults are shown on the cross sections and do not penetrate the surface, they are "blind thrusts" that are, of course, speculative. However, their existence is quite probable in light of fold propagation theory (see discussion by Suppe, 1985, on fault-related folds). According to this theory, incipient D_1 deformation (see Fig. 6) is thought to be controlled by fault-bend folds during the early, brittle phase of deformation.

The T_1 thrusts shown in the cross sections are floored by a basal décollement separating the lower Rangeley Formation from the older, pre-Silurian "basement" (Fig. 4). This basement is thought to be equivalent to the 1.5-Ga Gander Terrane of Boone and Boudette (1988) but may also include Cambrian and Ordovician formations exposed to the west along the Bronson Hill anticlinorium. The décollement is shown at about 2- to 3-km depth and is presumed to be a west-verging T_1 thrust. The depth to this structure is poorly constrained but is consistent with the interpretations of the Quebec–Western Maine and Ministère de l'Energie et les Resources, Quebec (MERQ) seismic profiles across the Maine part of the CMT, where flat reflectors at depths of 2 to 10 km are shown (Stewart and others, 1986; Unger and others, 1987). The vergence of this structure is equivocal because of its flat attitude, as shown in the migrated seismic sections. It is

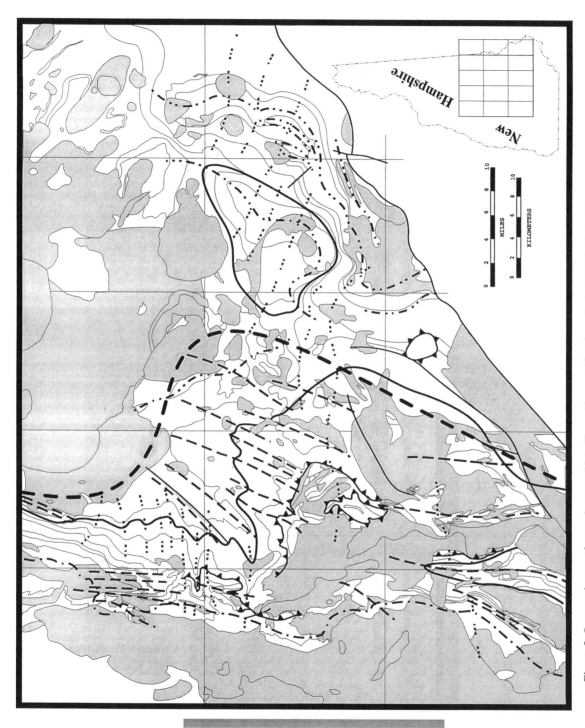

Figure 5. Structural map of central-eastern New Hampshire modified from Duke (1984), Englund (1976), Eusden (1988a), Eusden and others, 1984), Lyons (1988), and Nielson (1981). Plutons are shaded gray, and metasedimentary rocks are colored white. This figure covers the same area as shown in Figure 2.

EXPLANATION

F1 AXIAL TRACE

F2 AXIAL TRACE

F3 AXIAL TRACE

F4 AXIAL TRACE

F1/F4 COMPOSITE
AXIAL TRACE

EASTERN LIMIT OF
D4 – BACKFOLDING

TI THRUST FAULT

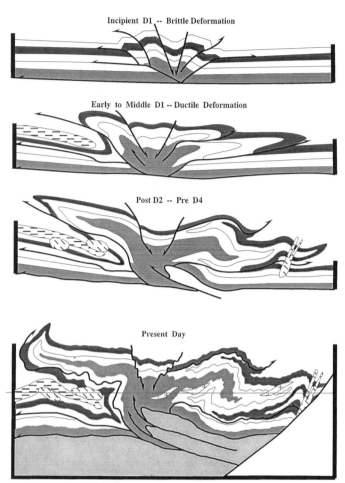

Figure 6. Schematic development of Acadian deformations in central-eastern New Hampshire. The symbols and patterns are the same as shown for Figure 4. The effects of D_3 cannot be shown in this sequence as F_3 axes are approximately parallel to the trend of the section. No attempt has been made to sequentially scale the horizontal axis throughout this sequence.

possible that an east-verging thrust rather than a west-verging thrust floors the CMT.

The Quebec–Western Maine and MERQ seismic reflection profiles in Maine show shallow east-dipping reflectors on the west edge of the CMT and shallow west-dipping reflectors on the east edge of the CMT. These are interpreted by us to be the nappe-propagating thrusts, T_1 thrusts, that root in the CNHA. This seismic geometry supports our hypothesis of oppositely verging thrust-nappes on either side on the CNHA. Though these reflectors are here construed to be early Acadian in age, there is no way of determining their true age from the seismic records.

The overall geometry that we present, matched, we believe, by the Maine seismic reflection profiles, is similar to the ECORS (ECORS Pyrenees team, 1988) seismic reflection profiles across the Pyrenees of Europe. The Pyrenees have oppositely verging thrust-nappes on either side of what is called an axial zone (ECORS Pyrenees team, 1988). This again points out the archi-

tectural similarity between our model for the CMT and other orogens.

D₂—second phase of deformation. D_2 is characterized by S_2 axial planar schistosity and F_2 folds that have east-southeast vergence. The second foliation, S_2, is axial planar to F_2 folds and is defined by a second generation of foliated muscovite and biotite porphyroblasts. This foliation is less well defined then S_1 and approaches or bridges the gray zone between foliation and cleavage. The distinction between the S_1 and S_2 fabrics is best seen in the hinges of F_2 folds. In these folds S_1 clearly wraps around the F_2 folds, whereas S_2 is axial planar to the fold. S_2 generally strikes north northeast and dips moderately to the west. Although this is the dominant orientation, there are places where S_2 varies from this attitude due to later folding. Eusden and others (1987) have used the variation of S_2 attitudes in the Alton and Berwick quadrangles to determine the orientation of F_3 fold axes.

F_2 folds are common in the eastern part of the study area, east of the CNHA. They are absent or unrecognizable west of the CNHA. This lopsided distribution demonstrates the across-strike variability of the deformation phases; D_2 effects are restricted to the east of the CNHA and D_4 effects, mentioned below, are concentrated to the west of the CNHA.

F_2 axes generally trend north northeast and are flat or shallowly plunging. F_2 folds in the CMT are characteristically downward facing. Not all F_2 folds have unequivocal topping sense, but those that do are invariably downward facing. The downward-facing orientation of the F_2 folds is a function of position on the inverted limb of a major F_1 nappe at the present erosion surface.

F_2 refolds the F_1 nappes coaxially. As a result the interference pattern of these two fold types is the "redundant type" or "type O," using the nomenclature of Ramsay and Huber (1987). The minor F_2 folds are generally asymmetric and gently to steeply inclined, that is, nearly recumbent to nearly upright. They range in wavelength from 0.5 to 4 m, have interlimb angles in the range of 0 to 30°, and as such are tight to isoclinal.

Most F_2 folds are either antiformal synclines or synformal anticlines and are parasitic to the regional Lebanon Antiformal Syncline (LAS). Along the length of the LAS the S_2 axial surface changes dip continually from vertical to moderate to subhorizontal and then back to vertical. It is common to find domains of steep-dipping S_2 and/or S_1 surfaces that lead on strike into domains of shallow-dipping or flat surfaces throughout the CMT; this is a function of later refolding.

D₃—third phase of deformation. D_3 is characterized by broad, open, west-northwest–trending folds. The axial surface to F_3 folds, S_3, is rarely seen in the CMT. This is because F_3 folds are characteristically broad open warps of the bedding and foliation. The warps only rarely have an associated axial planar foliation. In places (e.g., Penacook quadrangle) F_3 folds are expressed as crenulations, where a distinct crenulation cleavage, S_3, can also be measured.

F_3 folds in the CMT are characterized by warps and minor crenulations of the bedding and foliation and are found throughout the study area. They fold F_2 axial traces in the east and are folded by D_4 structures to the west. The F_3 axes trend west northwest and plunge shallowly in the same direction.

F_3 map-scale folds are quite evident from the geometry of the folded metasedimentary rocks. Figures 2 and 5 show such folds, roughly about west-trending axes. The F_3 axial traces shown on these figures bisect these regional bends in the stratigraphy. In the past the folds have been referred to as "cross folds" and are generally thought to be broad, open warps with shallowly plunging axes (Chamberlain, 1985; E. F. Duke, 1984; Eusden and others, 1987).

D_4—fourth phase of deformation. D_4 is characterized by north- to northeast-striking, moderately west dipping S_4 axial plane cleavage and F_4 folds. S_4 is expressed as a distinct crenulation cleavage or, in areas of tight structure, as a pervasive schistosity along the regional structural grain (Englund, 1976). It is clearly seen both meso- and microscopically as crenulation of S_1 and/or S_2 foliations. The average S_4 orientation is similar to the average S_2 surface.

F_4 folds are abundant in the western portion of the study area. They are unrecognizable or rare east of the CNHA. F_4 axes trend north northeast and plunge shallowly both to the north and south.

Based on detailed mapping of the Mt. Kearsarge summit (Lyons, 1988), it is clear that portions of the KCMS are a D_4/D_2 composite antiformal syncline. The style of refolding is similar to the coaxial refolding by D_2 of the D_1 thrust-nappes along the LAS.

The spatial distribution of D_4 effects, concentrated west of the CNHA, would suggest a relationship to the backfold-dome stage of the deformation surrounding the BHA. It is possible that the eastern limit of the F_4 folds marks the eastern extent of the backfolding caused by the rise of the BHA gneiss domes. We have shown the hypothetical limit of backfolding on Figure 5, based on the eastern extent of D_4 structures.

GEOMETRY OF MAP-SCALE FOLDS

As shown on Figure 2, particularly by the mapped distribution of the Smalls Falls Formation, the syntaxial folds in the CMT present an unusual geometry. West of the CNHA the pattern of the fold shows counterclockwise rotation; east of the CNHA the pattern is clockwise. It is assumed that these map-scale structures are of similar age. There are at least two possibilities that may explain the map pattern geometry.

The first argues that the complex map pattern is a result of superposed folding. Evidence for superposed folding comes from the minor symmetry of the sinistral and dextral folds on either side of the CNHA. The broad, open, shallow-plunging F_3 folds refolded by F_4 produced a complex refolded form. Horizontal or subhorizontal slices through this composite structure would give a map pattern similar to that seen in the CMT of New Hampshire. The belts of Smalls Falls that show apparent clockwise and counterclockwise motion may then be merely a function of the high

order of symmetry produced by these intersecting fold patterns and not related to opposing shear regimes working in the same belt of rocks.

The angle between the F_3 and F_4 fold axes is near 90°, and the angle between S_3 and S_4 is between 80 and 90°. Therefore, the type of interference is "Dome-Crescent-Mushroom" pattern as outlined by Ramsay and Huber (1987).

A second possible mechanism to produce such a map pattern is one involving extrusion and sheath folding. After D_2 folding the section was shortened in a northeast direction by a unidirectional squeeze centered over the CNHA such that the resulting ductile flow pattern would be symmetrically divided. F_3 folds would form like "rumples in a carpet." The sinistral and dextral patterns outlined by the Smalls Falls would then be F_3 folds and a result of regional extrusion caused by constriction along the edge of the CMT trough. The shear sense would be equivalent to that observed in salt diapirs, which show reverse drag sense.

This model would involve the development of a zone of intense strain on the D_2 structures while D_3 developed. The region affected by such strain could be between the Bronson Hill Anticlinorium and Campbell Hill–Nonesuch River Fault or perhaps between the large Early Devonian syntectonic Kinsman Suite plutons (the Winnipesaukee quartz diorite and Cardigan Pluton). It appears on Figure 2 that the two belts of Smalls Falls Formation deflect around the large Winnipesaukee sheetlike pluton. The bounding areas, whatever they might be, would act as rigid blocklike massifs sandwiching the interior metamorphic rocks and extruding them to the northeast. Perhaps the impingement of the Massabesic-Merrimack terrane supplied the necessary force for this deformation.

It is difficult to choose among these two intriguing possibilities: superposed F_3/F_4 folding and F_3 extrusion/sheath folding. We have normally interpreted the folds as cylindrical, and such structures would not fit a sheath fold model. In the Gilmanton quadrangle several outcrops showing complex interference patterns were observed that resembled calculated two dimensional dome-crescent-mushroom interference patterns. However, these fold patterns often look like quasi-sheath or eyed-folds that could be produced during a regional stage of sheath folding. Neither model is as yet proven.

TIMING OF DEFORMATION, METAMORPHISM, AND PLUTONISM

The abundant sheetlike (Nielson and others, 1976) Acadian plutons in this part of the CMT are synkinematic (Kinsman suite), late kinematic (Spaulding and early Concord suites), and postkinematic (remaining Concord suites). The Kinsman sheets were thrust along with the T_1/F_1 thrust-nappes.

Acadian regional metamorphism is characterized by early andalusites overprinted by widespread lower sillimanite zone metamorphism punctuated by hot spots of upper sillimanite zone and granulite facies metamorphism (Chamberlain and Lyons,

cal Society of London Special Publication 8, p. 205–211.

Thompson, J. B., Robinson, P., Clifford, T. N., and Trask, N. J., 1968, Nappes and gneiss domes in west-central New England, *in* Zen, E-an, and others, eds., Studies of Appalachian geology northern and maritime: New York, Interscience Publishers, p. 203–218.

Thompson, P. J., 1983, Silurian-Devonian stratigraphy, Monadnock quadrangle, New Hampshire: Geological Society of America Abstracts with Programs, v. 15, p. 186.

—— , 1984, Stratigraphy and structure of Monadnock quadrangle, New Hampshire: Refolded folds and associated fault zones: Geological Society of America Abstracts with Programs, v. 16, p. 67.

—— , 1985, Stratigraphy, structure and metamorphism in the Monadnock quadrangle, New Hampshire: University of Massachusetts Department of Geology and Geography Contribution 58, 191 p.

Thompson, P. J., Elbert, D. C., and Robinson, P., 1987, Thrust nappes superimposed on fold nappes: A major component of early Acadian tectonics in the central Connecticut Valley region, New England: Geological Society of America Abstracts with Programs, v. 19, p. 868.

Turner, F. J., and Weiss, L. E., 1963, Structural analysis of metamorphic tectonites: New York, McGraw-Hill, 545 p.

Unger, J. D., Stewart, D. B., and Phillips, J. D., 1987, Interpretation of migrated seismic reflection profiles across the northern Appalachians in Maine: Geophysical Journal of the Royal Astronomical Society, v. 89, p. 171–176.

Zen-E-an, ed., 1983, Bedrock geologic map of Massachusetts: U.S. Geological Survey and Commonwealth of Massachusetts, Department of Public Works, scale 1:125,000.

—— , 1989, Tectonostratigraphic terranes in the northern Appalachians: Field trip guidebook T359 (28th International Geological Congress): American Geophysical Union, Washington, D.C., 68 p.

Zen, E-an, Stewart, D. B., and Fyffe, L. R., 1986, Paleozoic tectonostratigraphic terranes and their boundaries in the mainland Northern Appalachians: Geological Society of America Abstracts with Programs, v. 18, p. 800.

Ziegler, P. A., 1985, Late Caledonide framework of western and central Europe, *in* Gee, D. G., and Sturt, B. A., eds., The Caledonian orogen—Scandinavia and related areas: New York, John Wiley and Sons, p. 3–18.

Manuscript Accepted by the Society June 8, 1992

Geological Society of America
Special Paper 275
1993

Nature of the Acadian orogeny in eastern Maine

Allan Ludman, John T. Hopeck*, and Pamela Chase Brock
Department of Geology, Queens College of the City University of New York, Flushing, New York 11367

ABSTRACT

New insight into the nature of the Acadian orogeny in eastern Maine has been gained by combining detailed field studies in six lithotectonic belts with geochemical data from the igneous rocks of the region. Revised stratigraphies and deformation histories of the tracts reveal their sedimentological and structural evolution from Ordovician through Early Devonian times, and variations in the isotope geochemistry of the igneous rocks permit delineation of the basement blocks beneath the supracrustal belts. Combined, these results yield a model for plate interactions that followed Taconian deformation and culminated in the Acadian orogeny.

Large basins (e.g., Aroostook-Matapedia, Central Maine) formed immediately after the Taconic orogeny on the recently accreted eastern margin of ancestral North America. These filled with thick clastic sequences derived from post-Taconian highlands during Late Ordovician through at least Middle Silurian times and characteristically preserve complex facies patterns at their margins. At the same time, sedimentation continued in the Fredericton Trough, inferred to be the only remaining oceanic crust in the region. This ocean basin separated the composite North American terrane from an equally complex Avalonian continent. Closing of this basin resulted in the Acadian orogeny.

The onset of the Acadian suturing of Avalon to North America is indicated by a change from local basin filling to a more homogeneous blanket of sandstones whose deposition appears to have begun in the east (Flume Ridge Formation) and migrated westward. Collision of basement blocks led first to westward thrusting of parts of the Avalonian continent over the Fredericton belt. Later Acadian thrusting caused by final collision between Avalon and ancestral North America transported supracrustal Miramichi belt strata eastward over the Fredericton belt and parts of the Fredericton belt eastward over the western edge of the Avalonian allochthon. Acadian thrusting has displaced the original boundaries between supracrustal belts in southeastern Maine so that they no longer coincide with boundaries between the basement blocks that originally lay beneath them.

INTRODUCTION

The correspondence of particular tectonic lithofacies in the Northern Appalachians with specific environments in the plate tectonic model has greatly improved our understanding of the Taconic orogeny and continental accretion during Cambrian through Middle Ordovician times. Tectonic indicators such as ophiolites, melanges, and blueschists are absent from the record of the Late Silurian–Early Devonian Acadian orogeny, however, suggesting that a mechanism other than plate collision following subduction of oceanic crust was the cause of that event. The recognition of exotic terranes in the northwestern American Cordillera (Coney and others, 1980; Monger and Irving, 1980) has led to models for the Acadian orogeny that invoke accretion of analogous exotic blocks by transcurrent faulting (Williams and

---

*Present address: Department of Environmental Protection, Augusta, Maine 04333.

Ludman, A., Hopeck, J. T., and Brock, P. C., 1993, Nature of the Acadian orogeny in eastern Maine, *in* Roy, D. C., and Skehan, J. W., eds., The Acadian Orogeny: Recent Studies in New England, Maritime Canada, and the Autochthonous Foreland: Boulder, Colorado, Geological Society of America Special Paper 275.

Hatcher, 1983; Zen, 1983). Recent discussions of the Acadian event in the Northern Appalachians have become a debate between proponents of accretion of unrelated terranes by transcurrent faulting and those who envisage subduction-related tectonism acting on terranes that originally lay within a single ocean basin (Bradley, 1983; Ludman, 1981, 1986).

East-central and southeastern Maine play a vital role in our attempts to resolve the Acadian controversy. Early, largely reconnaissance mapping in the region (e.g, Bastin and Williams, 1914; Larrabee and others, 1965) described belts composed of volcanic and sedimentary rocks of very different lithologies and ages. Tectonic modelers have interpreted the volcanic-dominated tracts as island arcs separated by marginal seas or ocean basins (McKerrow and Ziegler, 1971; Bradley, 1983) and more recently as exotic terranes separated by major transcurrent fault sutures (e.g., Zen, 1983). Our detailed mapping has led to substantial revisions of the stratigraphic and structural histories of southeastern Maine (Ludman, 1988, 1990b; Ludman and Hill, 1990; Hopeck, 1989) and to a new view of its accretion history.

This chapter presents a model of pre-Acadian and Acadian events based on our interpretation of the region's supracrustal strata and deformation history, coupled with indirect analysis of the underlying basement blocks based on isotopic and geochemical data presented by Brock (1989, 1990), unpublished data). The model synthesizes 20 person-years of mapping in eastern Maine and draws on local seismic reflection data (Ludman and others, 1990; Costain and others, 1989, Luetgert and others, 1987) as well as geochemical information. It is not feasible to present our entire data base in this summary; interested readers are referred to the references cited below for details of stratigraphy, structure, and isotope geochemistry.

BACKGROUND

Eastern Maine consists of several northeast-trending lithotectonic belts (Fig. 1). Understanding the tectonic evolution of the region requires a detailed description of each belt, determination of the original relationships between belts, and establishment of the time and manner of their juxtaposition. Until recently, however, very little was known of the region between Calais and Houlton, an area that spans the boundaries of six of these belts. Detailed work in Aroostook County to the north (Pavlides, 1968, 1971, 1972, 1974) and northwest (Neuman, 1967; Ekren and Frischknecht, 1967) and in coastal Maine to the south (Gates, 1975; Gates and Moench, 1981) described very different strata whose tectonic relationships were blurred by imprecise definition of the intervening area. Reconnaissance and some detailed mapping of this poorly known region by Larrabee and others (1965; see also Larrabee, 1963, 1964a, 1964b; Larrabee and Spencer, 1963; Amos, 1963) helped delineate major boundaries, but the stratigraphies and deformation histories of the individual belts have only been established in the past few years (Fyffe, 1982; Ludman, 1985, 1987, 1990a, 1990b; Hopeck, 1988, 1989; Hopeck and others, 1988, 1989; Sayres, 1986; Sayres and Ludman, 1985).

Several new faults have been mapped, and most of the lithotectonic belts shown in Figure 1 are now known to be fault bounded (Ludman and Hopeck, 1988). Each is thus a candidate for an exotic terrane as described by Coney and others (1980). Evidence for the timing of juxtaposition of these belts demonstrates that geographic relationships comparable to those of today had been achieved by the end of the Acadian orogeny (Pollock and others, 1988; Ludman, 1981, 1986). This conclusion invalidates early exotic terrane models for the region that called upon post-Carboniferous accretion by transcurrent faulting (e.g., Kent and Opdyke, 1978).

The complex surface geology revealed by these studies represents only the uppermost part of the architecture of the Northern Appalachians. Various methods have been used to identify the basement blocks whose interaction led to the deposition and deformation of the supracrustal cover rocks. Osberg (1978) described several basement blocks based on ages of rocks and rock type. Other workers have used the major and trace element chemistry and isotopic signatures of plutonic rocks to postulate basement blocks often modeled as terranes (Loiselle and Ayuso, 1979; Andrew and others, 1983; Ayuso, 1986; Brock, unpub-

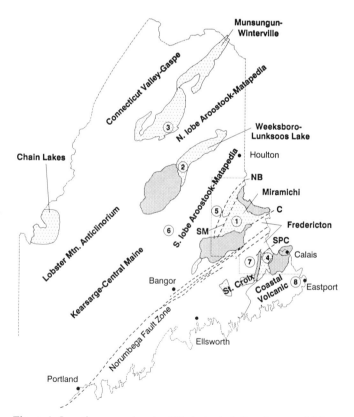

Figure 1. Location map showing lithotectonic belts of eastern Maine. Dotted line shows area detailed in Figure 2. Light pattern highlights dominantly pre-Silurian belts. Dark pattern indicates plutons discussed in text. Numbers show locations of stratigraphic columns given in Figures 3 and 4. Faults are shown in boldface: NB—North Bancroft; SM—Stetson Mountain; C—Codyville; SPC—South Princeton/Crawford.

lished data). Geophysical methods have also been applied, including seismic reflection (Stewart and others, 1985; Unger and others, 1987; Luetgert and others, 1987; Costain and others, 1989; Ludman and others, 1990) and gravity modeling (Hodge and others, 1982; Coblentz, 1988).

Supracrustal and basement block boundaries rarely coincide, probably because of the allochthonous nature of large portions of the rock exposed at the surface. The most fruitful regional analysis must therefore combine geological, geophysical, and geochemical data (see, for example, Osberg, 1978; Roy, 1980; Zen, 1983). In this chapter we will present a tectonic model for eastern Maine based on the first detailed geologic mapping carried out in a region encompassing several supracrustal belts and on a compilation of isotope and major-element geochemistry from plutonic rocks of eastern Maine and adjacent New England and Maritime Canada. We will construct our model in three steps: (1) a brief description of each lithotectonic belt and discussion of the original relationships of those belts will be used to constrain the nature and timing of the Acadian orogeny; (2) a similarly brief summary of isotopic and geochemical data will identify the basement blocks whose interaction caused the orogeny; and (3) an explanation of the processes that led to the Acadian orogeny and to the current relationships between supracrustal belts and deep-crustal blocks.

SUPRACRUSTAL BELTS

The supracrustal sedimentary and volcanic rocks record the timing and nature of accretion in the region. These rocks have long been divided into the subparallel tracts shown in Figure 1. Figure 2 shows a more detailed view of our study area, and Figure 3 summarizes the stratigraphies of the six belts that it comprises. Two tracts, the Miramichi and St. Croix belts, are composed mostly of pre-Silurian strata. Rocks of similar age in these belts belong to very different lithofacies, and correlation between them is highly problematical. The Kearsarge–Central Maine, Aroostook-Matapedia, Fredericton, and Coastal Volcanic belts consist for the most part of Silurian and Early Devonian rocks, but major differences in lithofacies also exist here, and correlation among these tracts has been debated for many years.

We will briefly describe the stratigraphy of the six belts in order to demonstrate the differences between Acadian and pre-Acadian accretion. Rocks of Caradocian and older age exhibit significant differences in sedimentation, volcanism, and deformation history from those that are Ashgillian and younger. Accordingly, our discussion of the supracrustal rocks is divided into two sections, the first dealing with Cambrian through Caradocian strata and the second covering those of Ashgillian, Silurian, and Devonian age. Unfortunately, fossil age control is available for only a few units; age assignments of the others are based on inferred equivalence or interfingering with fossiliferous strata and by correlation with well-dated, lithologically similar units. Italicized names are informal at this time but are being proposed for formal status (see Ludman, 1990a, for detailed descriptions of these units).

Cambrian through Caradocian strata

Rocks of Cambrian(?) through Caradocian age comprise the entire St. Croix belt and most of the Miramichi belt (Figs. 1, 2). In addition, a large sliver of anomalously high-grade pelitic rocks interpreted as being of pre-Ashgillian age is exposed along the Norumbega Fault Zone in the Kellyland and Waite quadrangles (Op in Fig. 2).

Miramichi belt. Rocks of the Miramichi belt are separated by an unconformity into two packages (the *Baskahegan Lake Formation* and *Kossuth Group*), both of which are pre-Ashgillian. The oldest rocks in the Miramichi belt in Maine, the *Baskahegan Lake Formation,* are a thick sequence of chalky weathering, quartzofeldspathic turbiditic wackes with minor interbedded pelites and even rarer granule conglomerates. The *Baskahegan Lake Formation* is generally thick bedded and exhibits well-developed Bouma sequence features. Most beds are green or greenish-gray, but the lower part of the formation contains dark red and maroon varieties. The *Baskahegan Lake Formation* is more complexly deformed than the overlying formations of the *Kossuth Group,* and an unconformity is inferred to separate them. The lower part of the *Kossuth Group* contains black shales, thin-bedded turbidites, and mafic volcanic rocks assigned to the *Bowers Mountain Formation*; these pass upward, apparently conformably, into the felsic volcanic rocks of the *Stetson Mountain Formation.* Graptolites indicate that the *Stetson Mountain* (and possibly the upper part of the *Bowers Mountain*) *Formation* is Middle Caradocian (Larrabee and others, 1965; see also Ludman, 1990a). An early Ordovician age [Arenigian(?)-Llandeilan] is inferred for the lower part of the *Bowers Mountain Formation,* and a Late Cambrian through earliest Ordovician age for the *Baskahegan Lake Formation.*

St. Croix belt. The Cookson Group of the St. Croix belt also contains rocks of Late Cambrian(?) through Middle Caradocian age, but the lithologies and sequence are very different from those of the Miramichi belt (Ludman, 1987; modified in Ludman, 1990b). The oldest rocks, time-equivalents of the *Baskahegan Lake Formation,* are coticule rocks of the lower Calais Formation. Pillow basalts and Tremadocian black shales (upper Calais Formation) and euxinic Arenigian-Llandeilan(?) turbidites (Woodland Formation) are superficially similar to the *Bowers Mountain* rocks, although their sequence is not identical. The Middle Caradocian Kendall Mountain Formation of the St. Croix belt consists of thick orthoquartzite beds intercalated with sparse black shales, not at all similar to the coeval *Stetson Mountain* volcanics of the Miramichi suite.

Regional correlation. Strong similarities in both lithology and sequence indicate that the Miramichi belt was closely related to the Weeksboro–Lunksoos Lake and Munsungun-Winterville belts to the northwest during the Early Paleozoic (Fig. 4). The basal units of these tracts are the *Oldhamia*-bearing quartzofeldspathic turbidites of the Grand Pitch (Weeksboro–Lunksoos Lake; Neuman, 1967) and Chase Brook (Munsungun-Winterville; Hall, 1970) formations. Both units are very similar to the

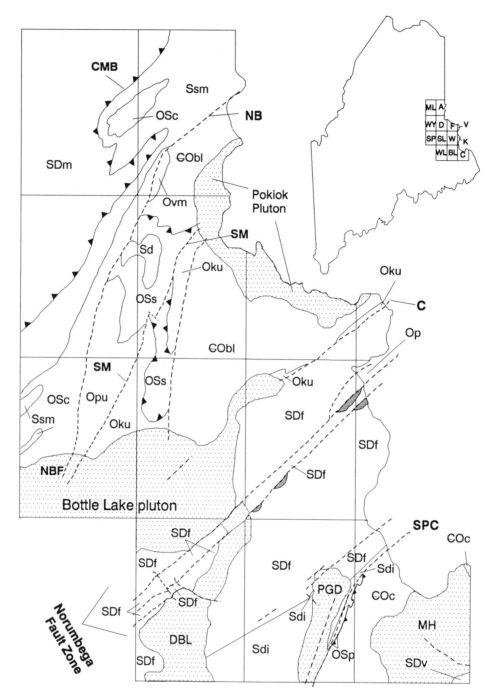

Figure 2. Simplified geologic map of central-eastern Maine. Dark pattern indicates post-Acadian red beds. Light pattern highlights plutons: DBL—Deblois pluton; PGD—Pocomoonshine gabbro-diorite; MH—Moosehorn complex.

15′ quadrangles shown in inset: ML—Mattawamkeag Lake; A—Amity; WY—Wytopitlock; D—Danforth; F—Forest; V—Vanceboro; SP—Springfield; SL—Scraggly Lake; W—Waite; K—Kellyland; WL—Wabassus Lake; BL—Big Lake; C—Calais.

Stratigraphic units (italics indicate informal names now in review): €Obl—*Baskahegan Lake Formation;* €Oc—Cookson Group, undivided; *Oku—Kossuth Group,* undivided; Op—unnamed pelitic rocks; *Opu-Prentiss Group, undivided*; OSs—*Sam Rowe Ridge Formation;* Ovm—unnamed mafic volcanic rocks; OSc—Carys Mills Formation; OSp—Pocomoonshine Lake Formation; Sd—Daggett Ridge Formation; SDf—Flume Ridge Formation; Sdi—Digdeguash Formation; Ssm—Smyrna Mills Formation; SDm—Madrid Formation; SDv—Coastal Volcanic Belt, undivided.

Faults as in Figure 1, plus: CMB—Central Maine Boundary Fault.

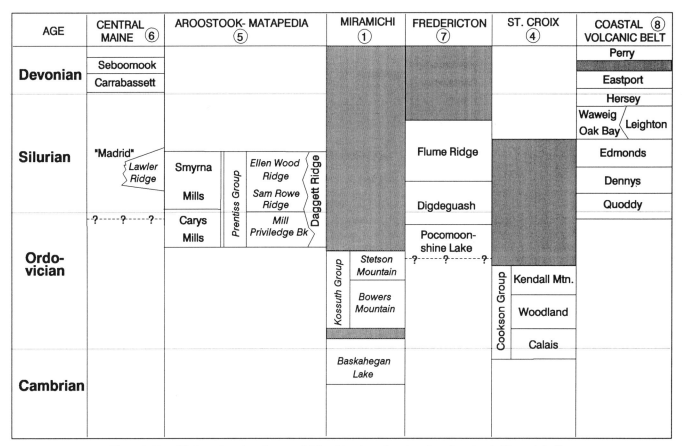

Figure 3. Simplified stratigraphy of eastern Maine. Numbers show location of section in Figure 1. Italics indicate informally designated units.

Baskahegan Lake Formation, except that they contain melange (Osberg and others, 1985); none has been recognized in the Miramichi belt. The melange suggests correlation of the two northern belts with the Hurricane Mountain melange of northwestern Maine (Fig. 1; Boone and Boudette, 1989). Tremadocian through Middle Caradocian black shales, turbidites, cherts, and bimodal volcanic units unconformably overlie the basal turbidites in the Weeksboro–Lunksoos Lake and Munsungun-Winterville belts; these are comparable to, if not one-to-one correlatives of, rocks in the Miramichi belt (Fig. 4). The unconformity in the two northern pre-Silurian belts marks the Late Cambrian–earliest Ordovician Penobscot orogeny of Neuman (1967), and the unconformity separating the *Baskahegan Lake Formation* from the *Kossuth Group* is also interpreted as reflecting the Penobscot event.

Thick orthoquartzites of the Kendall Mountain Formation were deposited in the St. Croix belt at the same time that widespread volcanism was taking place in the three more northerly tracts, suggesting a very different tectonic setting. Van Staal and others (1990) have used volcanic geochemistry, blueschist metamorphism, and structural evolution to postulate a subduction environment for the Miramichi belt. We believe that the St. Croix rocks are more closely related to the shallow-water Cambro-

Ordovician continental shelf strata of the Saint John Group of southern New Brunswick (Pickerill and Tanoli, 1985) than to the Miramichi and its companion tracts to the northwest. The Saint John Group is widely accepted as representing the continental shelf of the Avalonian continent (e.g., Rast and Stringer, 1974); we suggest that the Cookson Group was deposited on the west-facing continental slope of Avalon (Ludman, 1987).

Comparison of the rare earth compositions of basalts from the St. Croix and Miramichi belts supports this interpretation (Fig. 5). Calais Formation pillow basalts of the former exhibit the relatively flat patterns (when compared with chondritic abundances) that are typical of within-plate volcanism, and these are repeated throughout the St. Croix belt in coastal Maine and New Brunswick (Fyffe and others, 1988). In contrast, massive basalts of the *Bowers Mountain Formation* display the much greater enrichment of light rare earth elements characteristic of island arc environments.

Op. The unnamed pelites exposed along the Norumbega Fault Zone are problematic because they are not similar to strata in either of the flanking pre-Silurian tracts. They are intensely polydeformed and have experienced a biotite-grade regional metamorphism not recorded in either the Miramichi or St. Croix belts

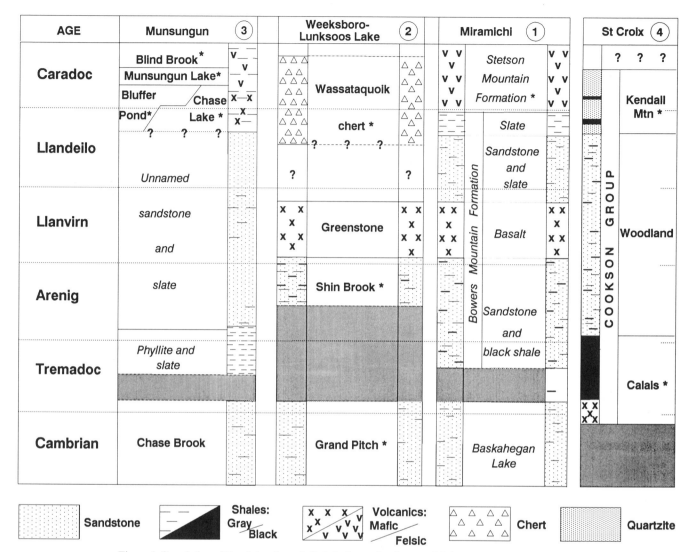

Figure 4. Correlation of Cambrian through Caradocian rocks of eastern Maine. Numbers show locations of sections in Figure 1. Heavy shading indicates inferred unconformities. Asterisks indicate faunal age control.

or in any of the Fredericton belt strata against which they are now juxtaposed (Ludman, 1990a). Their present position appears to be caused by uplift during late-stage motion on the Norumbega faults, and these rocks may represent an otherwise completely buried basement block described below.

Cambrian through Caradocian accretion. The similarities in Cambrian through Caradocian sedimentation, volcanism, and deformation histories of the Miramichi, Weeksboro–Lunksoos Lake, and Munsungun-Winterville belts suggest that they experienced the same tectonic regime during the Early Paleozoic. We agree with Boone and Boudette (1989) and Pollock and others (1988) that the Penobscot orogeny was caused by assembly of a composite terrane within the Iapetus Ocean, that this terrane included the three tracts named here as well as the Chain Lakes Massif of western Maine (Fig. 1), and that the Taconic

orogeny welded this composite terrane onto ancestral North America.

We suggest that the Miramichi belt formed the eastern edge of North America immediately after the Taconic orogeny. The Avalonian continent, represented in our study area by the St. Croix belt and by the basement upon which the Coastal Volcanic belt was erupted, was separated from Mid-Ordovician North America by a sedimentary basin named the Fredericton Trough by McKerrow and Ziegler (1971). (To avoid confusion, the term Fredericton Trough will be used in this paper to refer to this depositional basin; the term Fredericton belt will be used for the rocks deposited in that basin that now form a supracrustal belt in eastern Maine and adjacent New Brunswick.) Events that occurred between the end of Taconic deformation and onset of the Acadian orogeny are recorded in the younger cover rocks.

a)Miramichi belt

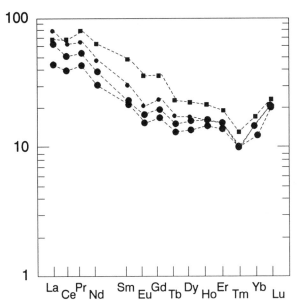

b)St. Croix belt

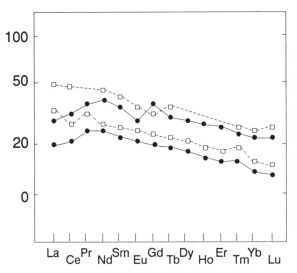

Figure 5. Comparison of chondrite-normalized rare earth element patterns for volcanic rocks of the Miramichi and St. Croix belts. Data for (a) from Ludman and Hopeck, unpublished data; solid squares represent unnamed basalts from west flank of Miramichi belt, circles from *Bowers Mountain Formation.* Data for (b) from Fyffe and others, 1988. Open squares represent samples from Penobscot Formation, mid-coastal Maine; circles indicate basalts of the Calais Formation, Calais, Maine, and St. Stephen, New Brunswick.

Ashgillian through Early Devonian rocks

Large basins formed within the composite Grenville-Munsungun-Weeksboro-Miramichi terrane immediately after the Taconic orogeny and received thick deposits of Late Ordovician through Silurian age (locally extending into the Early Devonian)

that characterize the Connecticut Valley–Gaspé, Kearsarge–Central Maine, and Aroostook-Matapedia belts. Acadian and post-Acadian thrust, normal, and strike-slip faulting have modified the original facies patterns, but relationships at the margins of these basins show that the pre-Silurian belts mentioned above (e.g., Miramichi and Weeksboro–Lunksoos Lake) served as sources of sediment from at least the Late Ordovician through the Middle Silurian. The nature of the Fredericton Trough during this time period is critical to any tectonic model; unfortunately, faulting at both margins of the Fredericton belt clouds its original relationships to the adjacent tracts, particularly the St. Croix belt. Post-Taconic sedimentary linkage of the Fredericton and Miramichi belts is indicated at the western margin of the Fredericton belt in New Brunswick by coarse clastic rocks containing detritus recognizable as having been derived from the Miramichi volcanic suite. Figure 6 summarizes our interpretation of the correlation of rocks deposited in eastern Maine during the interval between the Taconic and Acadian orogenies.

The southern lobe of the Aroostook-Matapedia belt (see Fig. 1) is the best-understood basin in southeastern Maine because of the detailed work of Neuman (1967) at its northwest margin (adjacent to the Weeksboro–Lunksoos Lake belt) and our recent mapping along its southeastern contact with the Miramichi belt (Hopeck, 1988, 1989; Ludman, 1988, 1990a). We will describe the evolution of this basin as an example of events following the Taconic orogeny.

Southern lobe of the Aroostook-Matapedia belt. The southern marginal facies of the Aroostook-Matapedia belt consists of the *Prentiss Group,* a complex assemblage of conglomerates, lithic and quartzose wackes, and pelitic rocks that lies tectonically and/or unconformably above the Caradocian volcanic rocks of the *Kossuth Group* along the western margin of the Miramichi belt (Figs. 2, 3, 6). Several formations are included in the *Prentiss Group*; these grade into one another laterally as well as vertically, revealing complex facies relationships. The two lowest units also interfinger with rocks that make up most of the Aroostook-Matapedia belt, confirming sedimentary linkage of the Miramichi and Aroostook-Matapedia belts and indicating an Ashgillian age for the inception of *Prentiss Group* sedimentation.

The oldest rocks in this post-Caradocian suite are the pyritiferous and locally calcareous or carbonaceous lithic and quartzose wackes and black shales of the *Mill Priviledge Brook Formation.* Coarse conglomerates containing clasts up to 30 cm in diameter occur locally within the formation, apparently as debris flows. The entire Cambrian(?) through Caradocian sequence of the Miramichi belt is represented in clasts in *Mill Priviledge Brook* wackes and conglomerates, leaving no doubt as to their source area. The basal *Mill Priviledge Brook* rocks pass rapidly upward into interbedded feldspathic and lithic wackes and burrowed, noncarbonaceous gray pelites of the *Sam Rowe Ridge Formation*; these in turn grade into similar but slightly finer grained and green strata of the *Ellen Wood Ridge Formation.* Stratified conglomerates are common in these formations near their contact with the pre-Ashgillian Miramichi suites. Many of their clasts are identical

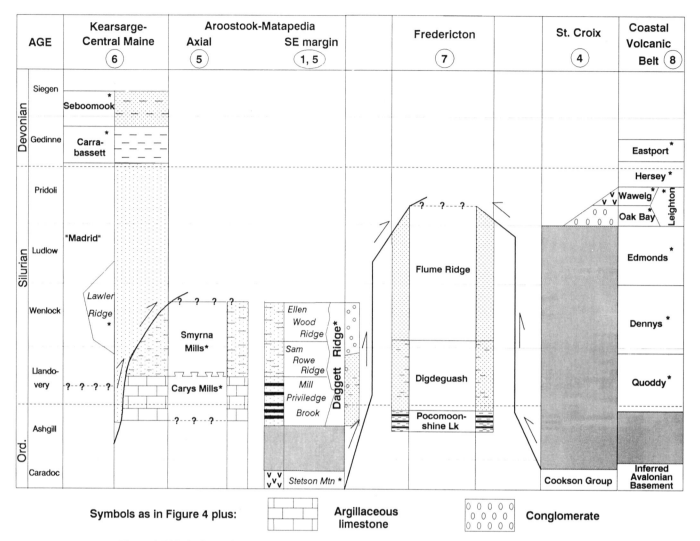

Figure 6. Lithologic relationships among post-Caradocian rocks of southeastern Maine. Numbers indicate locations of sections as shown in Figure 1. Asterisks indicate faunal age control. Heavy lines and arrows show interpreted tectonic relationships between belts.

to those of the *Mill Priviledge Brook* conglomerates, but some are composed entirely of intraformational fragments. A few channel deposits have been identified, but most of the conglomerates lie in layers parallel to bedding. Disruption of underlying beds and a few exposures exhibiting mixing of conglomerate and substrate indicate that some of the conglomerates are slump deposits or debris flows. Others appear to be related to disrupted zones caused by motion along faults at the basin margin.

All three of these formations pass eastward, and locally upward, into coarse conglomerates and sandstones assigned to the Daggett Ridge Formation by Larrabee and others (1965). This formation contains the coarsest clasts of any in eastern Maine, some reaching diameters of 3 m, generally set in a slightly to highly calcareous matrix. The clast population is dominated by lithic fragments of the *Baskahegan Lake* and *Stetson Mountain*

formations. An unnamed limestone conglomerate containing clasts of dolomitic limestone up to 25 cm in diameter crops out patchily in the Danforth quadrangle near the unconformity with the Caradocian volcanic rocks.

The apparently gradational lateral transition from thin bedded, fine-grained, generally noncalcareous *Sam Rowe Ridge* or *Ellen Wood Ridge* wackes with sparse conglomeratic interbeds to massive, coarse, calcareous Daggett Ridge conglomerates is interpreted as a transition from a distal facies to one more proximal to the source area within the Miramichi belt. We cannot unambiguously assign the Daggett Ridge conglomerates to a shallow-water facies, but rugose corals (some broken but many intact), abundant crinoid columnals, and brachiopod fragments have been collected from Daggett Ridge conglomerates and coral fragments from the limestone conglomerate. Some of the rugose corals appear to have lived among the boulders; the others may have been transported, but all were certainly close to emergent parts of the Miramichi belt. These fossils indicate a Silurian age for the two conglomerates: Specimens of *Tryplasma* and *Squameofavosites* restrict the limestone conglomerate to Middle Silurian times (Larrabee and others, 1965, p. 13), whereas a single orthoid brachiopod suggests a Silurian age for the Daggett Ridge (Larrabee and others, 1965, p. 15).

Pelitic rocks in the *Mill Priviledge Brook–Sam Rowe Ridge–Ellen Wood Ridge* sequence display an upward transition from black, carbonaceous types to medium gray noncarbonaceous varieties and finally into green shales rich in detrital chlorite. Coarse conglomerates of the Daggett Ridge Formation exhibit the same transition; this, along with the interfingering relationships described above, suggests a Silurian age for the three formations.

The axial portion of the southern lobe of the Aroostook-Matapedia belt underlies most of the Amity, Wytopitlock, and Mattawamkeag and large parts of the Danforth and Springfield quadrangles. It comprises two formations that can be traced continuously to their type localities near Houlton, where they were named the Carys Mills and Smyrna Mills formations (Pavlides, 1971, 1972; see Fig. 1). The Carys Mills Formation in our study area consists of thinly layered argillaceous micritic limestones and noncalcareous shales with very minor amounts of pyritiferous calcareous wacke and calcareous conglomerate. Bedding has been obliterated tectonically in most exposures and replaced by transposed layering that had for years prevented recognition of these rocks as equivalent to the Carys Mills Formation of the Houlton area (Osberg and others, 1985; Doyle and Hussey, 1967). Where bedding is preserved, it occurs as 5- to 25-cm-thick layers of buff or gray, variably dolomitic and calcitic micrite with thinner (2 to 5 cm) pelite beds identical to many exposures in the Houlton area.

The conglomerates, wackes, and limestones of the Carys Mills Formation interfinger with the *Mill Priviledge Brook Formation* to the east. Clasts in the Carys Mills conglomerates appear to be of the same Cambro-Ordovician lithologies found in the *Mill Priviledge Brook* but are enclosed in a more calcareous matrix. Similar wackes and fine-grained conglomerates were reported in the Carys Mills Formation in the Houlton area (Pavlides, 1965, 1971), but coarser strata comparable to the Daggett Ridge Formation were not described there.

The Carys Mills Formation passes conformably upward into thin-bedded pelites and fine-grained sandstones of the Smyrna Mills Formation. The upward transition from gray to green, chlorite-rich strata described for the *Sam Rowe Ridge, Ellen Wood Ridge,* and Daggett Ridge formations is also recognized in the Smyrna Mills Formation in the western part of our study area. Reconnaissance in the type area of the Smyrna Mills Formation near Houulton suggests that this change may also be present there. Lenses and horizons of argillaceous limestone and dolostone similar to the dominant lithology of the Carys Mills Formation occur throughout the Smyrna Mills and are now also

known from a few localities in the *Sam Rowe Ridge Formation.* Thicker-bedded sections of the Smyrna Mills are virtually identical to the *Sam Rowe Ridge* and *Ellen Wood Ridge* strata.

The southern lobe of the Aroostook-Matapedia belt is in fault contact to the northwest with the largely pre-Silurian Weeksboro–Lunksoos Lake belt (Fig. 1). Despite the tectonic complications, facies relationships comparable to those of the southeastern margin described above are recognizable. Pebble and cobble conglomerates containing clasts derived, for the most part, from the underlying Grand Pitch Formation grade basinward into thick-bedded sandstones of the Allsbury Formation and then into thinly intercalated pelites and sandstones of the Frenchville and Allsbury formations (Neuman, 1967; Osberg and others, 1985). A proximal conglomerate and wacke sequence contains stromatoporoid reefs, reef detritus, and calcarenite, indicating shallow water conditions in late Silurian time for at least that part of the eroded Weeksboro–Lunksoos Lake belt.

The similarities in stratigraphic sequence and facies change recognized on both flanks of the southern lobe of the Aroostook-Matapedia belt suggest the basin model shown in Figure 7. Analogous relationships were mapped by Hall (1970) at the boundary between the Munsungun-Winterville belt and the northern lobe of the Aroostook-Matapedia tract (location 3 in Fig. 1). This suggests that similar basin formation and filling occurred at several different sites during Late Ordovician through Silurian times upon the composite terrane accreted to North America during the Taconic orogeny. Roy's description of the northwest margin of the Connecticut Valley-Gaspe belt depicts comparable events there as well, although with slightly different timing (Roy, 1989).

Kearsarge–Central Maine belt. Early and Middle Silurian deposits in the Kearsarge–Central Maine belt in western and central Maine also preserve original facies patterns. A complex basin margin facies clearly derived from the adjacent Boundary Mountain antiform passes eastward into intermediate and distal facies described in detail by Ludman (1976), Moench and others (1988), Pankiwskyj and others (1976), and Osberg (1988). The symmetrical facies pattern of a two-sided basin exhibited by the Aroostook-Matapedia belt is not preserved, however. The southeastern contact of the Kearsarge–Central Maine belt in our area is a thrust fault (the Central Maine Boundary Fault) that transported the central Maine suite eastward over Aroostook-Matapedia rocks in the Mattawamkeag, Wytopitlock, and Mattawamkeag Lake quadrangles (Fig. 2).

The complex facies relationships typical of Late Ordovician through Middle Silurian rocks of central Maine were replaced in Ludlow times in the Kearsarge–Central Maine belt by a much more homogeneous pattern. Thick sequences of sandstones (Madrid Formation, *sensu stricto*) and then slates (Carrabassett Formation) blanketed the earlier deposits and were followed by turbidites of the Early Devonian Seboomook Group. Only the Silurian-Devonian part of the Kearsarge–Central Maine sequence is preserved in our study area, where the thick sandstones have been assigned to the Madrid Formation (*sensu lato*; see Osberg and others, 1985) and the pelites to the Carrabassett Formation

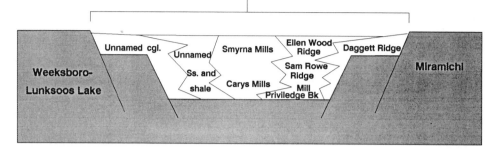

Figure 7. Schematic diagram showing original relationships of units discussed in text within the two-sided southern lobe of the Aroostook-Matapedia belt. Line of section is between localities 2 and 5 in Figure 1.

(Figs. 2, 3). Roy (1981) mapped Madrid-like sandstones just west of our study area as the *Lawler Ridge Formation*; his discovery of Llandoverian (?) graptolites in black shales just below these sandstones suggests that the change to a homogeneous sedimentation style may have begun earlier in east-central Maine than in central and western parts of the state.

Fredericton belt. The Fredericton belt is nearly 50 km wide in eastern Maine and contains rocks very similar to those present in the Kearsarge–Central Maine belt in west-central Maine. Fossils are absent from the Fredericton belt in Maine, however, so that the ages of the units are uncertain; indeed, there is debate about whether some units shown as part of the belt in Figures 2 and 3 more properly belong in the adjacent St. Croix belt (see Fyffe, 1989; Fyffe and Riva, 1990; Ludman, 1991). It is generally agreed, however, that the Fredericton belt in Maine consists for the most part (ca. 85%) of the Flume Ridge Formation, a generally calcareous quartzofeldspathic wacke with subordinate pelitic interbeds that greatly resembles parts of the Madrid and *Lawler Ridge* formations of the Kearsarge–Central Maine suite.

We interpret granule conglomerates and thick-bedded turbidites of the Digdeguash Formation as underlying the Flume Ridge Formation and euxinic sandstone/shale interbeds of the Pocomoonshine Lake Formation as being the oldest rocks of the Fredericton belt. Coarse Digdeguash conglomerates described by Westerman (1978) near the eastern boundary of the Fredericton belt suggest proximity to the original basin margin, and the formation apparently fines westward toward the center of the Fredericton belt. The Digdeguash strata are nearly identical to an early to mid-Silurian intermediate facies (Sangerville Formation) of the Kearsarge–Central Maine belt, and the transition from poorly sorted Digdeguash lithic wackes to nearly homogeneous quartzofeldspathic wackes of the Flume Ridge Formation mirrors the change from pre-Madrid sedimentation to homogeneous Madrid deposition in that belt. Digdeguash-like proximal and intermediate facies are not present in Maine along the northwestern margin of the Fredericton belt, but comparable rocks are

found at the Miramichi/Fredericton boundary in New Brunswick (L. R. Fyffe, personal communication).

The Fredericton belt contains the same rock types, similar stratigraphic sequence, and the same type of change from a locally heterogeneous to regionally homogeneous sedimentation style exhibited by the Kearsarge–Central Maine belt. However, we believe that it alone represents material deposited upon oceanic crust and will discuss evidence for this interpretation below.

Coastal volcanic belt. There is little debate that the Coastal volcanic belt records the Silurian–Early Devonian history of the Avalonian continental margin. This belt is composed largely of lava flows, ashfalls, and ashflows of mostly bimodal (rhyolite/basalt) composition intercalated with shallow-water sandstones, siltstones, mudstones, and volcanogenic breccias (Fig. 3; see Gates, 1975; Gates and Moench, 1981). The sedimentary rocks contain a rich faunal assemblage of Acado-Baltic affinity, indicating that the belt belongs with the Avalonian rocks it erupted on rather than with Early Paleozoic North America.

We interpret the Oak Bay and Waweig formations as a basin marginal facies of the Coastal volcanic belt. These units lie unconformably upon the Cookson Group of the St. Croix belt at Cookson Island in Oak Bay, New Brunswick. The Waweig contains a fauna identical in age, Acado-Baltic affinity, and depositional environment to that of the Leighton Formation of the Coastal volcanic belt. This confirms juxtaposition of the St. Croix and Avalon rocks by upper Silurian times at the very latest.

Despite extensive geochemical examination of the Coastal volcanic belt from New Brunswick to mid-coastal Maine, the tectonic setting of these volcanic rocks remains uncertain. Gates and Moench (1981) proposed either an intraplate extensional setting or a subduction-related (transtensional) environment for the classic section at Eastport, Maine. Van Waggoner and others (1988) suggested a nonarc setting for equivalent rocks of the Mascarene peninsula in New Brunswick. Studies of the correlative Castine Formation by Pinette and Osberg (1989) in the Penobscot Bay area south of Ellsworth led to further ambiguous results in which either an ensialic rift or Andean subduction setting is possible. We favor an Andean-type continental margin and will discuss our reasoning below.

Constraints on Acadian tectonism based on the supracrustal belts

Stratigraphic, paleontologic, and paleomagnetic data compiled from the supracrustal belts provide constraints on the timing and deformation styles of the Acadian orogeny but do not resolve the problems concerning the mechanism involved. Sedimentary overlap sequences, geographic and geologic distribution of deformation features, and radiometric dating of plutons that seal contacts between many of the lithotectonic belts of eastern and southeastern Maine permit development of a tectonic linkage timetable for the region (Fig. 8). Facies relationships discussed above demonstrate sedimentary linkages during late Ordovician and Silurian times for all of the tracts west of and including the Miramichi belt. Common Penobscot and Taconic deformation histories in the pre-Silurian belts of east-central and northeastern Maine support a Cambro-Ordovician linkage of these belts.

Similarly, sedimentary cover rocks of the Coastal volcanic belt prove linkage of the St. Croix belt with the main part of the Avalon belt by late Silurian times at latest. The only pre-Ashgillian linkage that cannot be firmly demonstrated is between the St. Croix and Miramichi belts, two pre-Silurian tracts containing widely different lithofacies separated by the Fredericton belt. The Fredericton belt may thus represent deposits in an ocean basin that separated North America/Miramichi and St. Croix/Avalon continents, but seismic reflection studies of central (Stewart and others, 1985; Unger and others, 1987) and east-central Maine (Costain and others, 1989; Doll and others, 1989a, b; Ludman and others, 1990) show that no ocean crust is present today beneath the Fredericton belt. It may have been subducted during Silurian through Early Devonian times, although, as indicated above, a clear subduction signature is not present in the appropriate volcanic rocks of the region.

It is also possible that the Fredericton Trough of McKerrow and Ziegler (1971) was an intraplate basin similar to our view of the Kearsarge–Central Maine and Aroostook-Matapedia belts. Indeed, one of us had suggested such a model previously (Ludman, 1981). Two lines of evidence suggest, however, that whatever the Fredericton Trough was, it was a very wide feature.

Studies of faunal provinciality show that the Fredericton Trough was a barrier to communication between fauna of Old World (Acado-Baltic) and North American affinities as late as Early Devonian times (see Berry and Osberg, 1989, for a summary of these data; see also McKerrow and Cocks, 1976; Pickerill, 1976). Such a barrier need not have been a broad ocean basin, but paleomagnetic data suggest that this was indeed the case. The paleolatitude separation between Avalon and North America decreased from Cambrian through Early Devonian times, but at least 20° of latitude difference remained in the Early

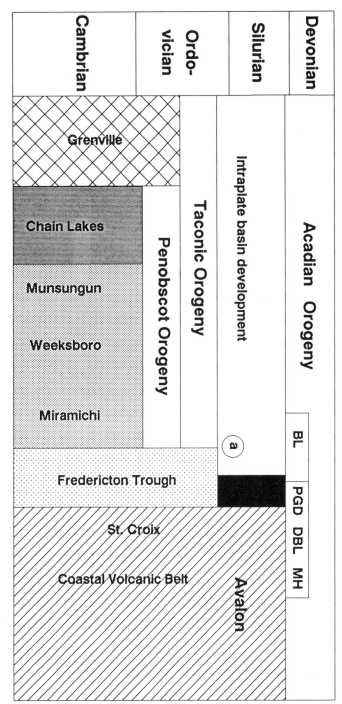

Figure 8. Accretion timetable for eastern and southeastern Maine. Horizontal bars crossing boundaries between supracrustal belts (shaded vertical columns) indicate geologic events shared by the supracrustal belts at times shown on the left. In most cases these document linkage of the belts. a—unnamed conglomerates in Fredericton belt in New Brunswick described in text. BL—Bottle Lake pluton; PGD—Pocomoonshine gabbro-diorite; DBL—Deblois pluton; MH—Moosehorn complex.

Silurian (Van der Voo, 1988). There was effectively no paleolatitude difference between the two plates by the end of the Acadian orogeny in mid-Devonian time.

The stratigraphic, deformational, paleontologic, and paleomagnetic data taken together indicate that the Acadian orogeny marked the suturing of Avalon to North America. The critical question that remains is how that suturing took place: by transcurrent fault juxtaposition of the two continents, by subduction of ocean crust that had once separated them, or by some combination of these mechanisms. Information from the supracrustal stratigraphy does not permit distinction between these tectonic mechanisms. The answer, in our opinion, lies in delineation and tectonic identification of the basement blocks (i.e., plates) whose interaction during the Early Paleozoic led to the Acadian orogeny.

BASEMENT BLOCKS

Detailed mapping at the contacts between belts demonstrates that Acadian thrusting has transported portions of the supracrustal section both eastward and westward over adjacent belts (Ludman and Hopeck, 1988; Ludman, 1990b). The potential for large-scale detachment requires that basement relationships be elucidated in as much detail as possible in order to reconstruct plate motions leading to the Acadian orogeny. Models identifying as few as two (Coblentz, 1988) to as many as five (Brock, 1990, unpublished data) blocks have been proposed from gravity, seismic, lithologic, and geochemical data.

Definition of basement blocks

The basement configuration used here is the one proposed by Brock (1989, 1990, unpublished data). It is based on a compilation of Sr, Pb, O, and Nd data from more than 130 igneous bodies of Silurian to Carboniferous age in New England and adjacent Canada. Geographic trends were found in both the absolute isotopic values and in the breadth of variation of isotopic compositions. Based on these trends, the Northern Appalachians have been divided into five major basement blocks. Relationships of these blocks to the classic Appalachian zones of Williams (1978, 1979) are shown in Figure 9a; Figure 9b shows their relationships to the supracrustal belts in eastern Maine. Table 1 summarizes the isotopic characteristics of the five basement blocks.

Igneous rocks in three of the five blocks (Grenville, Central, and Avalon) display a broad range of isotopic compositions, including many values characteristic of the upper continental crust (e.g., $^{87}Sr/^{86}Sr_i > .710$, $\delta^{18}O > 11^0/_{00}$. Some plutons in these belts contain Middle Proterozoic xenocrystic zircons or isotope values that yield Middle Proterozoic Sm/Nd model ages. These three belts are therefore interpreted as being underlain by ancient (Middle Proterozoic) continental basement and derivative sedimentary cover.

In contrast, the two intervening blocks (Boundary Mountain and Sebago) contain plutons that exhibit only a narrow range of isotopic compositions, a range that excludes typical upper crust values (e.g., all $^{87}Sr/^{86}Sr_i < .706$, most $\delta^{18}O \leqslant 10^0/_{00}$). The Boundary Mountain and Sebago blocks do contain a wide range of bulk igneous chemical compositions, including granitic rocks, but isotopic variations in these blocks do not correlate simply with silica content (e.g., Ayuso and others, 1988). Plutons in these blocks do not show isotopic evidence of derivation from Middle Proterozoic source materials. Because fusion of ancient upper continental crust did not contribute significantly to granite magmatism in these blocks, it is inferred that such a crust was not present. Accordingly, the Boundary Mountain and Sebago blocks are interpreted as being underlain by Late Proterozoic to Early Paleozoic crusts of island arc origin.

A broad spectrum of geological and geophysical evidence supports the basement block identification proposed here. $^{40}Ar/^{39}Ar$ studies of the post-Acadian cooling history of New England have delineated discontinuities in crustal uplift that correspond to the margins of the Boundary Mountain block in New Hampshire (Harrison and others, 1989), the Central/Sebago boundary in New Hampshire (Lux and West, 1989), and the Avalon/Sebago boundary in Maine (West and Lux, 1989). Hodge and others (1982) concluded that the density structure of the crust beneath Maine is consistent with the presence of three major basement blocks. These correspond to our Boundary Mountain, Central, and a composite of the Sebago and Avalon blocks. According to the geophysical model of Coblentz (1988), shallow-level dense rocks of eastern Maine are confined to our isotopically defined Avalon block. Seismic studies also suggest a discontinuity between the Sebago and Avalon blocks; steps in crustal thickness occur on both the northwestern and southeastern boundaries of the Sebago block (Unger and others, 1987).

Comparison of basement blocks and supracrustal belts

It is clear from Figure 9b that the supracrustal belts usually do not coincide with the isotopically defined basement blocks. For example, the Chain Lakes Massif contains material of ancient (Middle Proterozoic) origin but lies within the more primitive Boundary Mountain block. The Miramichi belt in New Brunswick belongs entirely to the Central block and is intruded by many plutons with distinct upper-crust continental signatures (e.g., Bevier, 1987), but in eastern Maine the Miramichi rocks lie atop primitive Sebago block basement. The Fredericton belt lies upon parts of the Sebago and Avalon basements.

These discrepancies suggest that there are major detachments between the supracrustal belts and the basement blocks beneath them. West-directed thrusting of the Chain Lakes Massif during the Taconic orogeny has been suggested by Boone and Boudette (1989) and could explain the first discrepancy cited above. East-directed thrusting of the Miramichi belt over the Fredericton belt and both east- and west-directed thrusting involving the St. Croix and Fredericton belts during the Acadian orogeny have been proposed by Ludman (1990b) and Ludman and Hopeck (1988). This thrusting is supported by seismic reflection studies within our study area (Doll and others, 1989a, b;

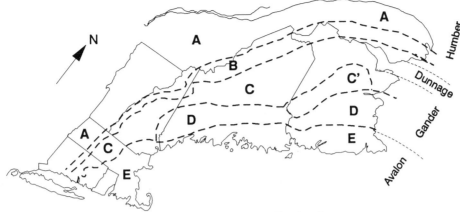

a) Isotopically defined basement blocks in New England

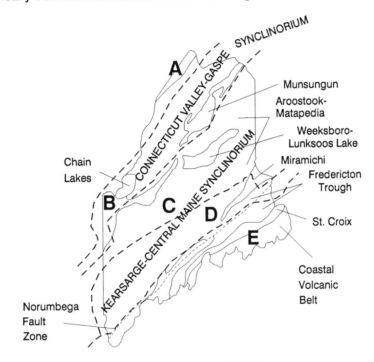

b) Basement blocks and supracrustal terranes in Maine

Figure 9. Basement blocks in the Northern Appalachians defined by isotope geochemistry (after Brock, unpublished data). Appalachian lithotectonic zones shown in (a) after Williams, 1978. Basement blocks: A—Grenville; B—Boundary Mountain; C—Central; D—Sebago; E—Avalon. (b) Relationships among supracrustal terranes and proposed basement blocks in Maine.

Ludman and others, 1990). It appears that the discrepancies between supracrustal and basement zonation of the Northern Appalachians in Maine can be explained by tectonic transport of the supracrustal section.

THE ACADIAN OROGENY IN EASTERN MAINE

Figure 10 is a cartoon illustrating our concept of the tectonic history of eastern Maine from the end of the Taconic orogeny through Acadian deformation. In addition to the stratigraphic, isotopic, faunal, and paleomagnetic information discussed above,

the model draws upon radiometric dating of plutons that seal terrane boundaries, structural data for vergence direction during Acadian deformation, and tectonic affinities of the eruptive and plutonic igneous rocks. It is not possible to describe all of these details here. See Ludman (1990b), Ludman and Hill (1990), Hopeck (1988, 1989), and Ludman and Hopeck (1988) for information concerning eastern Maine deformation history and Brock (unpublished, manuscript b) for a larger-scale view of the Acadian orogeny throughout New England and adjacent Canada incorporating plutonic, structural, and metamorphic data.

TABLE 1. CHARACTERISTICS OF NORTHERN APPALACHIAN BASEMENT ZONES*†

	Grenville	Boundary Mountain	Central	Sebago	Avalon
$^{87}Sr/^{88}Sr$ Initial ratio	0.704–0.7165	0.7045–0.7051 (Ordovician)	0.705–0.7161	0.7031–0.7056	0.703–0.719
$^{206}/^{204}Pb$	18.02–19.04	17.96–18.44	18.21–19.13	18.25–18.44	18.00–19.21
$^{207}Pb/^{204}Pb$	15.52–15.64	15.51–15.58	15.57–15.71	15.56–15.58	15.52–15.69
$^{208}Pb/^{204}Pb$ (feldspar in Dev. plutons)	37.80–29.10	37.80–38.37	38.01–38.63	38.06–38.29	38.05–38.58
$\delta^{18}O$ (whole rock)	....	6.8–8.2	7.2–13.3	7.4–11.1	5.8–12.6
Inferred age of basement§ (b.y.)	0.9–1.4	<0.750	1.4–1.8	<0.750	>1.8

*After Brock (unpublished data).
†Data from Silurian to Carboniferous igneous rocks unless otherwise indicated.
§Based on xenocrystic zircon, Nd and Pb model ages.

The tectonic setting at the end of the Taconic orogeny is shown in Figure 10a. The Cambro-Ordovician Penobscot orogeny assembled a composite terrane consisting of the Boundary Mountain and Central basement blocks, along with their supracrustal Chain Lakes, Munsungun-Winterville, Weeksboro–Lunksoos Lake, and Miramichi cover rocks. Melange described by Pollock (1989) appears to record the suturing of these basement blocks in the supracrustal section. By the end of the Taconic orogeny, this composite terrane had been accreted to ancestral North America.

By Late Ordovician times, "North America" consisted of the Grenville, Boundary Mountain, and Central basement blocks, along with all supracrustal belts inboard of the Fredericton Trough. Extension within these accreted blocks followed closely upon Taconic convergence and formed the Aroostook-Matapedia, Central Maine, and Connecticut Valley–Gaspé depositional basins. Roy (1989) suggested that extension in the Connecticut Valley–Gaspé belt may have begun earlier and might have been contemporaneous with late Taconic deformation. The isotopic data confirm an ensialic setting for these basins. A possible modern analogue for this environment has been reported in Indonesia by Charlton (1991). He suggested that such extension is a "virtually unavoidable consequence" of suturing an island arc onto a continent by subduction; detachment of subducting oceanic lithosphere from unsubductable continental lithosphere causes the extension.

Ocean crust continued to separate ancestral North America from the Sebago block and from the Avalon continent, but by Early Silurian time the ocean basin separating Avalon from the Sebago block began to close (Fig. 10b). Although geochemical data for the Coastal Volcanic and Piscataquis volcanic belts do not conclusively identify their tectonic settings, they do not rule out an Andean-type continental margin. Indeed, these volcanic belts have many aspects of lithology and geochemistry in common with early Tertiary volcanic rocks of the Patagonian Andes (40 to 43°S), and data from both regions plot in similar positions on trace-element and tectonic discrimination diagrams. Accordingly, we suggest a subduction environment for both volcanic tracts and thus subduction as the mechanism for closing the Fredericton Trough.

In the absence of a clear tectonic signature, the geographic distribution of post-Taconic, pre-Acadian volcanic rocks in the Northern Appalachians provides some support for this interpretation (Figure 11). Brock (unpublished data) compiled data for well-dated Silurian and Early Devonian volcanic rocks of the Coastal and Piscataquis volcanic belts. Each of the units was assigned to one of four tectonic settings: arc, indeterminate, transitional, and intraplate. An arc setting was restricted to calc-alkaline, mafic to intermediate rocks with enriched light rare earth elements; high-alumina basalts; and largely calc-alkaline, unimodal, andesitic volcanic assemblages. Tholeiites with depleted light rare earth elements, peralkaline, alkaline, and alkalic rocks were assigned to the intraplate category. Mildly alkaline basaltic suites were considered to be "transitional." Tectonic assignments were made in as conservative a manner as possible. Thus, all other rock types, including bimodal, calc-alkaline volcanic suites, peraluminous rhyolites, and normal tholeiites, were classified as "indeterminate." Some rocks were assigned to this category because the available chemical data are insufficient for tectonic assignment.

Figure 11 shows the distribution of these post-Taconic, pre-Acadian volcanic rocks along with their inferred tectonic settings. This episode of volcanism evidently occurred in linear belts that are disposed about the Sebago block with mirror symmetry. Rocks of mostly "indeterminate" affinity lie adjacent to the Sebago block on both the North American and Avalon blocks, with

a)After Taconic orogeny (Late Caradocian)

b)Early Silurian

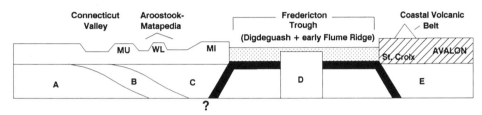

c)Early Acadian (Pridoli-Gedinne)

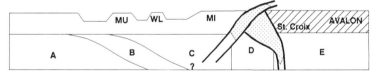

d)Late Acadian

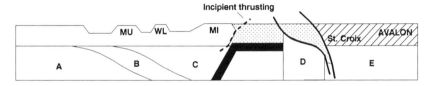

Figure 10. Diagram showing sequence of Acadian tectonism in eastern and southeastern Maine. Basement blocks named as in Figure 9, supracrustal belts as in Figure 1, plus: MU—Munsungun-Winterville; CL—Chain Lakes Massif; MI-Miramichi belt. See text for explanation.

some arc rocks in northeastern Massachusetts. Belts of intraplate activity occur on North America and Avalon farther away from the Sebago block. On the North American block, a region of transitional rocks separates the indeterminate and intraplate suites, whereas the transition is more abrupt on Avalon. These parallel volcanic suites are most readily explained by a change from arc to back-arc igneous activity and thus appear to support a subduction model. The symmetrically paired suites on both margins of the Sebago block further imply two subduction zones of opposite polarity, as shown on Figure 11: One subduction zone dipped eastward beneath Avalon, the other westward beneath North America. This model is remarkably similar to that proposed by Bradley (1983) based on stratigraphic analysis and is in accord with vergence directions of Acadian deformation in eastern Maine (Ludman, 1988).

The initiation of subduction is recorded by the Early Silurian volcanicity of the Coastal Volcanic belt. Debris eroded from Avalon (and from the Sebago block?) filled the eastern part of the Fredericton Trough while detritus shed from the Central block (Miramichi belt) entered the trough from the west. Westward subduction beneath the Central block in Maine may have begun concurrently but was most probably initiated slightly later in the Piscataquis volcanic belt. Deep-water deposition in the basins on the North American margin continued from Early Silurian through Early Devonian (Siegenian) time.

Recent dating of plutons in southeastern Maine by Jurinski (1989; see also Jurinski and Sinha, 1989) suggests that the earliest Acadian deformation occurred in Late Silurian times (Fig. 10c), possibly as the result of collision between the Avalon and Sebago blocks. Collision of these basement blocks brought an end to

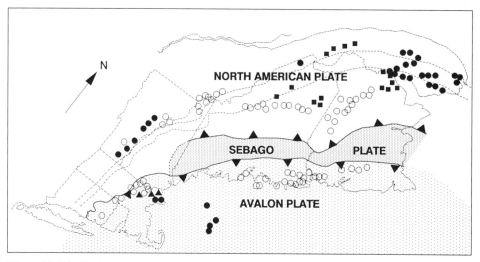

Figure 11. Distribution of post-Caradocian, pre-Acadian volcanic rock types in the Northern Appalachians. Inferred tectonic affinities: solid circles = intraplate; solid squares = transitional; open circles = indeterminate; triangles = volcanic arc.

subduction beneath Avalon and led to thrusting of the St. Croix belt westward over the sediments of the Fredericton belt and possibly over the Sebago basement block as well. As subduction beneath ancestral North America continued (Fig. 10c), supracrustal rocks of the Miramichi belt were thrust eastward over the Fredericton strata (Ludman and Hopeck, 1988).

Subduction beneath the margin of North America continued (Fig. 10d), as did eastward transport of supracrustal rocks. The thickened Avalon/Sebago block could have been the source of the Madrid Formation sandstones and younger sediments that blanketed central and western Maine. By the end of the Acadian orogeny the Sebago-Avalon composite block was sutured to North America, and strata of the Fredericton belt were thrust *eastward* over the St. Croix belt during late Acadian deformation.

ACKNOWLEDGMENTS

Our (AL, JH) fieldwork in Maine has been carried out under the auspices of the Maine Geological Survey under the direction of State Geologist Walter Anderson and Robert Marvinney, senior physical geologist. Laboratory and additional field support has been provided by National Science Foundation Grant EAR 870693 and PSC-CUNY Grants 13601, 663194, 664161, 666320, and 667212 to Ludman.

We are indebted to many colleagues for their information, advice, and arguments and for sharing their field and laboratory data with us. We thank Gary Boone, Cahit Coruh, John Costain, William Doll, Les Fyffe, John Hogan, Arthur M. Hussey, Joseph Jurinski, Marc Loiselle, Malcolm MacLeod, Steven McCutcheon, Philip Osberg, Nicholas Rast, Krishna Sinha, and David B. Stewart for their roles in shaping the thoughts presented here.

We are grateful to Les Fyffe and Marc Loiselle for thoughtful and most helpful comments on an earlier version of this paper that have significantly improved the final version. Finally, very special thanks are extended to Patrick W.G. Brock for his counsel during all stages of this study.

REFERENCES CITED

Amos, D. H., 1963, Petrography and age of plutonic rocks, extreme southeastern Maine: Geological Society of America Bulletin, v. 74, p. 169–194.

Andrew, A. S., Loiselle, M. C., and Wones, D. R., 1983, Granitic plutonism as an indicator of microplates in the Paleozoic of central and eastern Maine: Earth and Planetary Sciences Letters, v. 66, p. 151–165.

Ayuso, R. A., 1986, Lead isotopic evidence for distinct sources of granite and for distinct basements in the Northern Appalachians: Geology, v. 14, p. 322–325.

Ayuso, R. A., Horan, M. F., and Criss, R. E., 1988, Pb and O isotopic geochemistry of granitic plutons in northern Maine: American Journal of Science, v. 288-A, p. 421–460.

Bastin, E. S., and Williams, H. S., 1914, Description of the Eastport quadrangle, Maine: U.S. Geological Survey Atlas, Folio 192, 15 p.

Berry, H. N. IV, and Osberg, P. H., 1989, A stratigraphic synthesis of eastern Maine and western New Brunswick, *in* Tucker, R. D., and Marvinney, R. G., eds., Studies in Maine geology, Vol. 2: Structure and stratigraphy: Augusta, Maine Geological Survey, p. 1–32.

Bevier, M. L., 1987, Pb isotopic ratios of Paleozoic granitoids from the Miramichi terrane, New Brunswick, and implications for the nature and age of the basement rocks, *in* Radiogenic age and isotopic studies: Report 1: Geological Survey of Canada Paper 87-2, p. 43–50.

Boone, G. M., and Boudette, E. L., 1989, Accretion of the Boundary Mountains terrane within the northern Appalachian orthotectonic zone, *in* Horton, J. W., and Rast, N., eds., Melanges and olistostromes of the U.S. Appalachians: Geological Society of America Special Paper 228, p. 17–42.

Bradley, D. C., 1983, Tectonics of the Acadian orogeny in New England and adjacent Canada: Journal of Geology, v. 91, p. 381–400.

Brock, P. C., 1989, Isotopically-defined crustal zonation of New England and adjoining Canada, and its implications for Acadian tectonics: Geological Society of America Abstracts with Programs, v. 21, p. 7.

——, 1990, Basement blocks of the north-central Appalachians defined by Sr, Pb, O, and Nd isotopic compositions of igneous rocks: Geological Society of America Abstracts with Programs, v. 22, p. 6.

Charlton, T. R., 1991, Postcollision extension in arc-continent collision zones, eastern Indonesia: Geology, v. 19, p. 28–31.

Coblentz, D. D., 1988, Crustal modeling in Maine through the simultaneous inversion of gravity and magnetic data [M.A. thesis]: Chestnut Hill, Massachusetts, Boston College, 82 p.

Coney, P. J., Jones, D. L., and Monger, J.W.H., 1980, Cordilleran suspect terranes: Nature, v. 288, p. 329–333.

Costain, J., Domoracki, W. J., and Coruh, C., 1989, Processing and preliminary interpretation of Bottle Lake seismic reflection data: Maine Geological Survey Open File Report, 28 p.

Doll, W. E., and six others, 1989a, Results of a seismic reflection and gravity study of the Bottle Lake Complex, Maine: EOS Transactions of the American Geophysical Union, v. 70, p. 401.

Doll, W. E., Costain, J. K., Domoracki, W. J., Coruh, C., Ludman, A., and Hopeck, J., 1989b, Interpretation of seismic reflection lines crossing the Norumbega Fault and Bottle Lake Plutonic Complex, eastern Maine: Geological Society of America Abstracts with Programs, v. 21, p. 320.

Doyle, R. G., and Hussey, A. M. III, 1967, Preliminary geologic map of Maine: Augusta, Maine Geological Survey, scale 1:500,000.

Ekren, E. B., and Frischknecht, F. C., 1967, Geological-geophysical investigation of bedrock in the Island Falls quadrangle, Aroostook and Penobscot counties, Maine: U.S. Geological Survey Professional Paper 527, 36 p.

Fyffe, L. R., 1982, Taconian and Acadian structural trends in central and northern New Brunswick, *in* St. Julien, P., and Beland, J., eds., Major structural zones and faults of the Northern Appalachians: Geological Association of Canada Special Paper 24, p. 117–130.

——, 1989, Bedrock geology of the Moores Mills area, Charlotte County, New Brunswick, *in* Abbott, S. A., ed., Project summaries for 1989: New Brunswick Department of Natural Resources, Minerals and Energy Division Information Circular 89-2, p. 28–39.

Fyffe, L. R., and Riva, J., 1990, Revised stratigraphy of the Cookson Group of southwestern New Brunswick and adjacent Maine: Atlantic Geology, v. 26, p. 271–275.

Fyffe, L. R., Stewart, D. B., and Ludman, A., 1988, Tectonic significance of black pelites and basalts in the St. Croix terrane, coastal Maine and New Brunswick: Maritime Sediments and Atlantic Geology, v. 24, p. 281–288.

Gates, O., 1975, Geologic map and cross-sections of the Eastport quadrangle, Maine: Maine Geological Survey Geologic Map Series 3, 19 p.

Gates, O., and Moench, R. H., 1981, Bimodal Silurian and Lower Devonian volcanic rock assemblages in the Machias-Eastport area, Maine: U.S. Geological Survey Professional Paper 1184, 32 p.

Hall, B. A., 1970, Stratigraphy of the southern end of the Munsungun anticlinorium, Maine: Maine Geological Survey Bulletin 22, 63 p.

Harrison, T. M., Spear, F. S., and Heizler, M. T., 1989, Geochronologic studies in central New England II: Post-Acadian hinged and differential uplift: Geology, v. 17, p. 185–189.

Hodge, D. S., Abbey, D. A., Harbin, M. A., Patterson, J. L., Ring, M. J., and Sweeney, J. F., 1982, Gravity studies of subsurface mass distributions of granitic rocks in Maine and New Hampshire: American Journal of Science, v. 282, p. 1289–1324.

Hopeck, J., 1988, Tectonic fabrics of the Passadumkeag River pluton and contact aureole: Maine Geological Survey Open File Report, 21 p.

——, 1989, Preliminary bedrock geology of parts of the Wytopitlock and Springfield 15-minute quadrangles, Maine: Maine Geological Survey Open File Report, 18 p.

Hopeck, J., Ludman, A., and Hon, R., 1988, Ages of tectonic fabrics in the Bottle Lake Complex, eastern Maine: Geological Society of America Abstracts with Programs, v. 20, p. 28.

Hopeck, J., Ludman, A., and Brock, P. C., 1989, Acadian evolution of the Miramichi anticlinorium and Aroostook-Matapedia belt, eastern Maine: Geological Society of America Abstracts with Programs, v. 21, p. 23.

Jurinski, J., 1990, Petrogenesis of the Moosehorn Igneous Complex, Maine [M.A. thesis]: Blacksburg, Virginia Polytechnic Institute and State University, 125 p.

Jurinski, J., and Sinha, A. K., 1989, Igneous complexes within the coastal Maine magmatic province: Evidence for a Silurian extensional environment: Geological Society of America Abstracts with Programs, v. 21, p. 25.

Kent, D. V., and Opdyke, N. D., 1978, Paleomagnetism of the Devonian Catskill redbeds: Evidence for motion of the coastal New England–Canadian Maritime region relative to cratonic North America: Journal of Geophysical Research, v. 83, p. 4441–4450.

Larrabee, D. M., 1963, Geologic map and section of Kellyland and Vanceboro quadrangles, Maine: U.S. Geological Survey Mineral Investigations Field Studies Map MF-269, scale 1:62,500.

——, 1964a, Bedrock geologic map of the Big Lake quadrangle, Washington County, Maine: U.S. Geological Survey GQ-358, scale 1:62,500.

——, 1964b, Reconnaissance bedrock geology of the Wabassus Lake quadrangle, Washington County, Maine: U.S. Geological Survey Mineral Investigation Field Studies Map MF-282, scale 1:62,500.

Larrabee, D. M., and Spencer, C. W., 1963, Bedrock geology of the Danforth quadrangle, Maine: U.S. Geological Survey GQ-221, scale 1:62,500.

Larrabee, D. M., Spencer, C. W., and Swift, D.J.P., 1965, Bedrock geology of the Grand Lake area, Aroostook, Hancock, Penobscot, and Washington counties, Maine: U.S. Geological Survey Bulletin 1202-E, 38 p.

Loiselle, M. C., and Ayuso, R. A., 1979, Geochemical characteristics of granitoids across the Merrimack Synclinorium, eastern and central Maine, *in* Wones, D. R., ed., The Caledonides in the USA: International Geological Correlation Program Project 27, Proceedings, Blacksburg, Virginia, p. 117–121.

Ludman, A., 1976, A fossil-based stratigraphy in the Merrimack Synclinorium, central Maine, *in* Page, L., ed., Contributions to New England stratigraphy: Geological Society of America Memoir 148, p. 65–78.

——, 1981, Significance of transcurrent faulting in eastern Maine and location of the suture between Avalonia and North America: American Journal of Science, v. 281, p. 463–483.

——, 1985, Pre-Silurian rocks of eastern and southeastern Maine: Maine Geological Survey Open File Report OF85-78, 29 p.

——, 1986, Timing of terrane accretion in eastern and east-central Maine: Geology, v. 14, p. 411–414.

——, 1987, Pre-Silurian stratigraphy and tectonic significance of the St. Croix Belt, southeastern Maine: Canadian Journal of Earth Sciences, v. 24, p. 2459–2469.

——, 1988, Revised bedrock geology of central-eastern and southeastern Maine: Maine Geological Survey Open File Report, 36 p.

——, 1990a, Revised bedrock geology of the Danforth, Scraggly Lake, Forest, Waite, Vanceboro, and Kellyland quadrangles, Maine: Maine Geological Survey Open File Report OF90-42, 20 p.

——, 1990b, Bedrock geology of the Big Lake quadrangle, Maine: Maine Geological Survey Open File Report OF90-28, 22 p.

——, 1991, Revised stratigraphy of the Cookson Group in southeastern Maine and southwestern New Brunswick: An alternate view: Atlantic Geology, v. 27, p. 49–57.

Ludman, A., and Hill, M., 1990, Bedrock geology of the Calais quadrangle, Maine: Maine Geological Survey Open File Report OF90-27, 32 p.

Ludman, A., and Hopeck, J., 1988, Classification and significance of faults in eastern Maine: Geological Society of Maine Field Trip Guidebook, 9 p.

Ludman, A., Hopeck, J., Costain, J. K., Domoracki, W. J., Coruh, C., and Doll, W. E., 1990, Seismic reflection evidence for the NW limit of Avalon in east-central Maine: Geological Society of America Abstracts with Programs, v. 22, p. 32.

Luetgert, J. H., Mann, C. E., and Klemperer, S. L., 1987, Wide-angle deep crustal reflections in the northern Appalachians: Geophysical Journal of the Royal Astronomical Society, v. 89, p. 183–188.

Lux, D. R., and West, D. P., Jr., 1989, Late Paleozoic exhumation of northern New England: Implications from $^{40}Ar/^{39}Ar$ thermochronology: Geological Society of America Abstracts with Programs, v. 21, p. A141.

McKerrow, W. S., and Cocks, L.R.M., 1976, Progressive faunal migration across the Iapetus Ocean: Nature, v. 263, p. 304–305.

McKerrow, W. S., and Ziegler, A. M., 1971, The Lower Silurian paleogeography of New Brunswick and adjacent areas: Journal of Geology, v. 79, p. 635–646.

Moench, R. M., and six others, 1988, Geologic map of western interior Maine: U.S. Geological Survey Miscellaneous Investigations map MI-1692, 21 p.

Monger, J.W.H., and Irving, E., 1980, Northward displacement of north-central British Columbia: Nature, v. 285, p. 289–294.

Neuman, R. B., 1967, Bedrock geology of the Shin Pond and Stacyville quadrangles, Penobscot County, Maine: U.S. Geological Survey Professional Paper 524-I, 37 p.

Osberg, P. H., 1978, Synthesis of the geology of the northeastern Appalachians, U.S.A.: Geological Survey of Canada Paper 78-13, p. 137–147.

—— , 1988, Geologic relations within the shale-wacke sequence in south-central Maine, *in* Tucker, R. D., and Marvinney, R. G., eds., Studies in Maine geology, Vol. 1: Structure and stratigraphy: Augusta, Maine Geological Survey, p. 51–73.

Osberg, P. H., Hussey, A. M. III, and Boone, G. M., 1985, Bedrock geologic map of Maine: Maine Geological Survey, scale 1:500,000.

Pankiwskyj, K. A., Ludman, A., Griffin, J. R., and Berry, W.B.N., 1976, Stratigraphic relationships on the southeast limb of the Merrimack Synclinorium in central and west-central Maine, *in* Brownlow, A., and Lyons, P., eds., Studies in New England geology: Geological Society of America Memoir 146, p. 263–280.

Pavlides, L., 1965, Geology of the Bridgewater quadrangle: U.S. Geological Survey Bulletin 1206, 72 p.

—— , 1968, Stratigraphic and facies relationships of the Carys Mills Formation of Silurian and Ordovician age, northeast Maine: U.S. Geological Survey Bulletin 1264, 44 p.

—— , 1971, Geologic map of the Houlton quadrangle, Aroostook County, Maine: U.S. Geological Survey GQ-920, scale 1:62,500.

—— , 1972, Geologic map of the Smyrna Mills quadrangle, Aroostook County, Maine: U.S. Geological Survey GQ-1024, scale 1:62,500.

—— , 1974, General bedrock geology of northeastern Maine, *in* Osberg, P. H., ed., New England Intercollegiate Geologic Conference guidebook to geology of east-central and north-central Maine: Orono, Maine, p. 61–85.

Pickerill, R. K., 1976, Significance of a new fossil locality containing a *Salopina* community in the Waweig Formation (Silurian–Uppermost Ludlow/Pridoli) of southwest New Brunswick: Canadian Journal of Earth Sciences, v. 13, p. 1328–1331.

Pickerill, R. K., and Tanoli, S. K., 1985, Revised lithostratigraphy of the Cambro-Ordovician Saint John Group, southern New Brunswick—A preliminary report, *in* Current research, part B: Geological Survey of Canada Paper 85-1B, p. 441–449.

Pollock, S. G., 1989, Melanges and olistostromes associated with ophiolitic metabasalts and their significance in Cambro-Ordovician forearc accretion in the Northern Appalachians, *in* Melanges and olistostromes in the U.S. Appalachians: Geological Society of America Special Paper 228, p. 43–64.

Pollock, S. G., Roy, D. C., Ludman, A., Hopeck, J., Hall, B. A., and Repetski, J., 1988, Terrane accretion in the Northern Appalachians of the northeastern United States and Canada: Geological Society of America Abstracts with Programs, v. 20, p. 124.

Rast, N., and Stringer, P., 1974, Recent advances and the interpretation of geological structure of New Brunswick: Geoscience Canada, v. 1, p. 15–25.

Roy, D. C., 1980, Tectonics and sedimentation in northeastern Maine and adjacent New Brunswick, *in* Roy, D. C., and Naylor, R. S., eds., New England Intercollegiate Geological Conference guidebook to the geology of northeastern Maine and neighboring New Brunswick: Chestnut Hill, Massachusetts, p. 142–168.

—— , 1981, Reconnaissance bedrock geology of the Sherman, Mattawamkeag, and Millinocket 15′ quadrangles, Maine: Maine Geological Survey Open File Report 81-46, 18 p.

—— , 1987, Geologic map of the Caribou and northern Presque Isle 15-minute quadrangles, Maine: Maine Geological Survey Open File Report 87-2, 44 p.

—— , 1989, The Depot Mountain Formation: Transition from syn- to post-Taconian Basin along the Baie Verte–Brompton line in northwestern Maine, *in* Tucker, R.D., and Marvinney, R. G., eds., Studies in Maine geology, Vol. 2: Stratigraphy: Augusta, Maine Geological Survey, p. 85–99.

Sayres, M., 1986, Stratigraphy, polydeformation, and tectonic setting of Ordovician volcanic rocks in the Danforth area, eastern Maine [M.A. thesis]: Flushing, New York, Queens College, 135 p.

Sayres, M., and Ludman, A., 1985, Stratigraphy and polydeformation of Tetagouche (Ordovician) volcanic rocks of the Miramichi Anticlinorium in the Danforth quadrangle, eastern Maine: Geological Society of America Abstracts with Programs, v. 17, p. 62.

Stewart, D. B., Unger, J. D., and Phillips, J. D., 1985, Quebec–Western Maine seismic reflection profile 3B across the coastal volcanic belt, Maine: Geological Society of America Abstracts with Programs, v. 17, p. 64.

Unger, J. D., Stewart, D. B., and Phillips, J. D., 1987, Interpretation of migrated seismic reflection profiles across the Northern Appalachians in Maine: Geophysical Journal of the Royal Astronomical Society, v. 89, p. 171–176.

Van der Voo, R., 1988, Paleozoic paleogeography of North America, Gondwana, and intervening displaced terranes: Comparisons of paleomagnetism with paleoclimatology and biogeographical patterns: Geological Society of America Bulletin, v. 100, p. 311–324.

Van Staal, C., Ravenhurst, C. E., Winchester, J. A., Riddick, J. C., and Langton, J. P., 1990, Post-Taconic blueschist suture in the Northern Appalachians of northern New Brunswick, Canada: Geology, v. 18, p. 1073–1077.

Van Wagoner, N. A., McNeil, W., and Fay, V., 1988, Early Devonian bimodal volcanic rocks of southwestern New Brunswick: Petrology, stratigraphy, and depositional setting: Maritime Sediments and Atlantic Geology, v. 24, p. 310–319.

West, D. P., Jr., and Lux, D. R., 1989, Evidence for Late Paleozoic–Early Mesozoic extensional tectonism along the Norumbega fault zone, southwestern Maine: Geological Society of America Abstracts with Programs, v. 21, p. A141.

Westerman, D. S., 1978, Bedrock geology of the Wesley quadrangle, *in* Ludman, A., ed., New England Intercollegiate Geologic Conference guidebook for fieldtrips in southeastern Maine and southwestern New Brunswick: Queens College Geological Bulletin 6, p. 120–132.

Williams, H., 1978, Tectonic lithofacies map of the Appalachian orogen: Memorial University of Newfoundland Map 1, scale 1:1,000,000.

—— , 1979, Appalachian orogen in Canada: Canadian Journal of Earth Sciences, v. 16, p. 792–807.

Williams, H., and Hatcher, R. D., 1983, Appalachian suspect terranes, *in* Hatcher, R. D., Williams, H., and Zietz, I., eds., Contributions to the tectonics and geophysics of mountain chains: Geological Society of America Memoir 158, p. 33–53.

Zen, E-an, 1983, Exotic terranes in the New England Appalachians—Limits, candidates, and ages: A speculative essay, *in* Hatcher, R. D., Williams, H., and Zietz, I., eds., Contributions to the tectonics and geophysics of mountain chains: Geological Society of America Memoir 158, p. 55–81.

MANUSCRIPT ACCEPTED BY THE SOCIETY JUNE 8, 1992

Geological Society of America
Special Paper 275
1993

Acadian deformations in the southwestern Quebec Appalachians

Pierre A. Cousineau
Sciences de la Terre, Université du Québec, Chicoutimi, Québec G7H 2B1, Canada
Alain Tremblay
INRS-Géoressources, Centre Géoscientifique de Québec, 2700 Einstein, CP 7500 Sainte-Foy, Québec G1V 4C7, Canada

ABSTRACT

Regional deformation encountered in Ordovician rocks of the Ascot Complex and the Magog Group of the Dunnage zone is synchronous with the development of the La Guadeloupe fault. Structural analysis indicates that along most of the length of this fault, Silurian and Devonian rocks of the Saint-Francis Group were thrust over the Dunnage zone and its post-Ordovician cover sequence. Structural fabrics of these units are thus related to the Acadian Orogeny, and pre-Acadian deformations are of much less importance.

Major Acadian faults in the northeastern part of the Québec Appalachians are dextral strike-slip faults, whereas they are largely thrust faults in its southwestern parts. The northern part of the Beauce area lies within the transition zone between these two fault regimes. Transected folds on both sides of the La Guadeloupe fault suggest a dextral oblique-slip tectonic transport. Northwest of the Magog Group, Upper Silurian rocks of the Cranbourne Formation unconformably overlie the Baie Verte–Brompton Line. The Cranbourne Formation is folded and cut by the same regional cleavage as the rocks below the unconformity. This demonstrates that the Acadian Orogeny is the major regional phase of deformation of all these units.

INTRODUCTION

Historically, regional deformation of Cambro-Ordovician rocks of the southwestern Québec Appalachians was interpreted as the result of the Taconian Orogeny (St-Julien and Hubert, 1975; Doolan and others, 1982). These models proposed that tectonic accretion and deformation occurred during the Ordovician, following an island arc collision with the continental margin of Laurentia (Osberg, 1978; Robinson and Hall, 1979; Rowley and Kidd, 1981; Stanley and Ratcliffe, 1985). The extent to which the Acadian Orogeny affected these terranes is not specified in models.

This study concerns the timing of regional folding and faulting within a segment of the southwestern Québec Appalachians, including the Dunnage zone and some parts of adjacent units. The focus is on the structural style of two specific areas of the Dunnage zone of southwestern Québec (Fig. 1): the Sherbrooke and Beauce areas. We will show that the development of the regional cleavage and associated folds and faults can be assigned to the Acadian rather than the Taconian Orogeny.

In the Québec Appalachians, the Acadian tectonics vary along the orogen. In its northeastern part, most Acadian faults have a dextral strike-slip movement (Stockmal and others, 1987; Malo and Béland, 1989; Malo and Bourque, this volume), whereas reverse faulting prevails in its southwestern part (St-Julien and others, 1983). The onset of a change from purely dip-slip to transcurrent movement is first recognized east of the Chaudière River along the La Guadeloupe fault.

REGIONAL GEOLOGICAL SETTING

Cambro-Ordovician rocks of the Québec Appalachians were grouped into a continental and an oceanic domain by St-Julien and Hubert (1975). Later, the rocks were respectively assigned to the Humber and Dunnage zones of Williams's (1979) classification (St-Julien and others, 1983). In southern Quebec, the Dunnage zone is made up of four stratigraphic units interpreted as the remnant of a forearc region (Cousineau and St-Julien, 1985): ophiolite complexes, the Saint-Daniel Melange, clastics of the Magog Group, and the volcanic-rich Ascot Com-

Cousineau, P. A., and Tremblay, A., 1993, Acadian deformations in the southwestern Quebec Appalachians, *in* Roy, D. C., and Skehan, J. W., eds., The Acadian Orogeny: Recent Studies in New England, Maritime Canada, and the Autochthonous Foreland: Boulder, Colorado, Geological Society of America Special Paper 275.

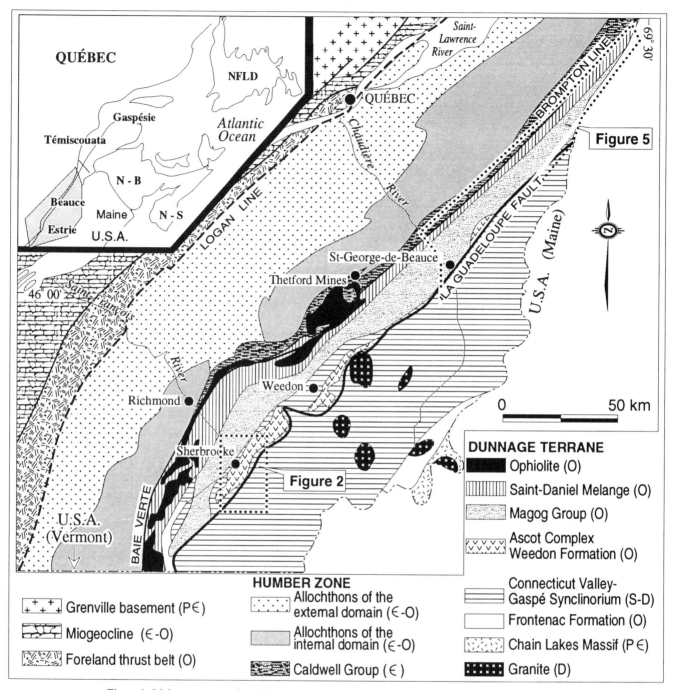

Figure 1. Major tectonostratigraphic units of the southwestern Québec Appalachians. Modified after St-Julien and others (1983) and Laurent (1975).

plex (Fig. 1). The Dunnage zone forms a narrow band between the Baie Verte–Brompton line (BBL) to the northwest and the Silurian and Devonian rocks of the Connecticut Valley–Gaspé Synclinorium (CVGS) to the southeast.

Southwestern Québec ophiolites are interpreted as segments of oceanic crust and mantle accreted to the passive margin of Laurentia in Taconian time (Laurent, 1975, 1977). The Saint-

Daniel Melange represents a relict accretionary prism formed as the result of a Taconian southeast-dipping subduction zone (Cousineau, 1992). The Magog Group represents a relict forearc basin that evolved between the Ascot Complex and the Saint-Daniel Melange (St-Julien and Hubert, 1975; Cousineau and St-Julien, 1985; Cousineau, 1990). The Magog Group is made up of four formations that form the basement of the Saint-Victor Synclinorium. The Ascot Complex is interpreted as a Taconian-related volcanic arc that collided with Laurentia (St-Julien and Hubert, 1975).

Fossils recovered from the uppermost unit of the Magog Group yield a Caradocian age (St-Julien and Hubert, 1975). Based on a correlation with other fossiliferous units, the base of the Magog Group would be Late Arenigian to Early Llanvirnian in age (Cousineau, 1990). Fossils have not been found in other units of the Dunnage zone, and precise radiometric age determinations are still to be done; Dunning and others (1986), however, have obtained a 478-Ma (Lower Ordovician) U/Pb zircon age from a plagiogranite of the Thetford Mines ophiolites. Ages of both the Saint-Daniel Melange and Ascot Complex are actually based on correlation with rocks of presumed equivalent ages in the Magog Group. These units are inferred to be Early to Late Ordovician in age (St-Julien and Hubert, 1975; Cousineau, 1992).

The Dunnage zone is fault bounded (Fig. 1). The BBL defines its northwestern contact with the Humber Zone (Williams and St-Julien, 1982). Along the BBL, the ophiolite complexes and the Saint-Daniel Melange are thrust over the Caldwell Group, a feldspathic sandstone-shale unit of the Humber Zone (St-Julien and others, 1983).

The La Guadeloupe fault marks the contact between the Silurian and Devonian rocks of the CVGS (Saint-Francis Group) with the Dunnage zone. Structural analysis of various sites along the La Guadeloupe fault demonstrates that it is a northwestward-directed, high-angle reverse fault (Tremblay and others, 1989b; Tremblay and St-Julien, 1990; Labbé and St-Julien, 1989; Cousineau, 1990). On the 2001 Ministère Energie et Ressources du Québec (MERQ) deep seismic reflection profile, a major low-angle ($<30°$) reflector is interpreted as the extension at depth of the La Guadeloupe fault (St-Julien and others, 1983; Spencer and others, 1989). Therefore, the La Guadeloupe fault has been interpreted as an unusual young-over-old thrust fault.

THE SHERBROOKE AREA

In the Sherbrooke area, the Ascot Complex is subdivided into three metavolcanic-volcaniclastic assemblages separated by Saint-Daniel–type phyllites: the Sherbrooke, Eustis, and Stoke domains (Tremblay and St-Julien, 1990; Fig. 2). The Sherbrooke domain is made up of mafic volcanic rocks overlain by an approximately equal amount of felsic volcanics. Mafic volcanics are pillowed and massive metabasalts and foliated equivalents. Felsic volcanics are pyroclastic breccias, crystal tuffs, and quartz-sericite schists. The Eustis domain is dominated by quartzo-feldspathic, chlorite-sericite schists originating from metamorphosed, conglomeratic volcaniclastic sandstones. A minor amount of chlorite- and quartz-sericite schists is present. In the Stoke domain, felsic volcanics clearly predominate over mafic lavas. The former are fine-grained rhyolitic tuffs and flows and foliated equivalents, and the latter are pillowed and foliated metabasalts. A syn-volcanic (?), granitic to tonalitic intrusive body occurs within the extrusive

sequence of the Stoke domain. Each lithotectonic domain of the Ascot Complex is characterized by a distinctive lithological association along with a characteristic geochemical signature (Tremblay and others, 1989a). The domains are in fault contact with adjacent melange-type metapelites. The Ascot Complex itself is in fault contact with the Ordovician Magog Group to the northwest and with the post-Ordovician Saint-Francis Group to the southeast.

Three episodes of deformation are recognized in the Ascot Complex (Tremblay and St-Julien, 1990). The first two produced penetrative S1 and S2 schistosities, but the D2 phase is responsible for the regional structural style. It transposes all older structures (S0 and S1) into parallelism with S2, which forms a composite S1–2 foliation. The latest D3 phase is associated with a northwest-dipping crenulation cleavage and related, northeasterly plunging, open folds. These F3 folds affect all earlier structures and are associated with slight variations in the trend of D2-related fabrics (Fig. 3, folds and lineations, stretching lineation).

Along the La Guadeloupe fault, F1 and F2 fold axes of the Ascot Complex have been rotated into a unimodal clustering close to a D2-related stretching lineation (Fig. 3). Northeasterly striking, anastomosing D2 shear zones are widespread in the Eustis domain and are associated with mylonites and mylonitic schists (Fig. 4). These shear zones dip moderately to steeply to the southeast. The mylonitic foliation is subparallel to the regional S2 foliation. All kinematic indicators found within the mylonites consistently indicate a northwestward-directed sense of shear (Tremblay and others, 1989b) along a southeast-plunging tectonic vector (i.e., stretching lineation). Shear zones of the Eustis domain are interpreted as subsidiary fault zones of the La Guadeloupe fault (Tremblay and St-Julien, 1990).

Southeast of the La Guadeloupe fault, sedimentary rocks of the Saint-Francis Group have been deformed twice. These deformational events are correlated with the D2 and D3 phases of the Ascot Complex. As in the latter, high strain zones are developed along the La Guadeloupe fault in the Saint-Francis Group. These zones are characterized by a mylonitic foliation with down-dip stretching lineations (Fig. 3). This foliation has been interpreted as evidence that the La Guadeloupe fault is an Acadian structure (Tremblay and others, 1989a). The quantification of displacement along the La Guadeloupe fault is actually unknown, although Labbé and St-Julien (1989), based on stratigraphic relationships in the Saint-Victor synclinorium, estimated this displacement to be at least 10 km. However, there are no unambiguous stratigraphic markers on both sides of the fault that could be used to verify this hypothesis.

In the Magog Group, the regional deformation is coeval with the D2 phase observed in the Ascot Complex. The intensity of the deformation decreases away from the Ascot-Magog contact (Fig. 3) just as the D2 deformation in the Saint-Francis Group decreased away from the Ascot–Saint Francis contact. Based on vergence of folds, the contact between the Magog Group and the Ascot Complex is interpreted as a northwestward

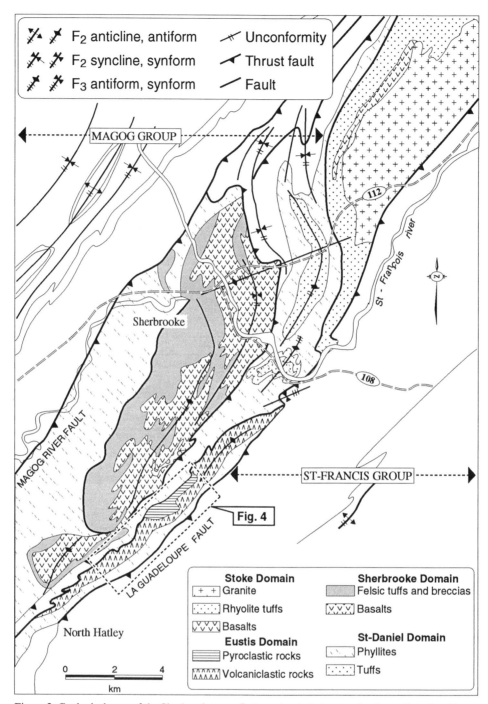

Figure 2. Geological map of the Sherbrooke area. Patterned units belong to the Ascot Complex. From Tremblay and St-Julien (1990).

thrust fault. Field relationships permit the interpretation of the Ascot Complex as a tectonic sliver within the La Guadeloupe fault system, which is consistent with interpretations of the MERQ 2001 deep seismic profile (St-Julien and others, 1983).

In summary, structural analysis of the Sherbrooke area indicates that the regional deformation is Acadian both in Ordovician and post-Ordovician units and that the contact between the Con-

necticut Valley–Gaspé Synclinorium and the Dunnage zone is a major reverse fault of Acadian age.

THE BEAUCE AREA

Along most of its length, the La Guadeloupe fault brings rocks of the Saint-Francis Group directly upon rocks of the Magog Group. The Ascot Complex and correlatives form a dis-

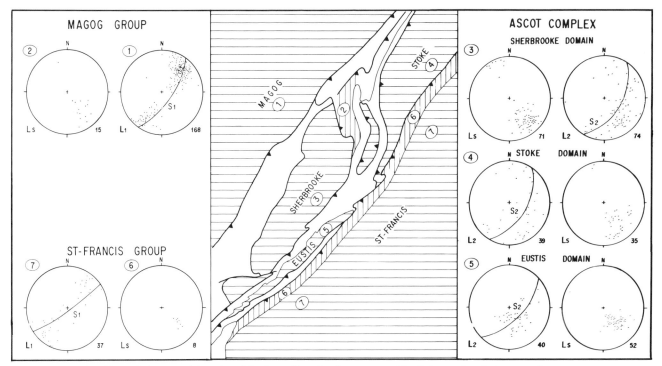

Figure 3. Structural domains of the Ascot Complex (same area as Fig. 2). Lower-hemisphere, equal-area projections of linear elements from the Sherbrooke area. The central map limits the numbered structural domains. Planes appearing on projections represent the mean attitude of the regional foliation in each domain. Linear fabrics are stretching lineations (Ls), L1 lineations, and F1 fold hinges in the Magog and Saint-Francis groups and L2 lineations and F2 fold hinges in the Ascot Complex. The number of data points appears at the lower left or the lower right of each projection.

continuous belt, also recognized at depth by airborne magnetic and gravimetric methods (Relevés géophysiques Inc., 1978) as well as by seismic reflection and refraction methods (Spencer and others, 1989). The Ascot volcanic belt extends for approximately 150 km along strike, that is, from the Sherbrooke area up to a point located about 10 km northeast of the Chaudière River (Fig. 1). At the latter locality, a small sliver of Ascot-like rocks crops out along a 5-km-long sinistral bend of the La Guadeloupe fault. Northeastward from this Ascot Complex sliver (Fig. 5), the fault has a linear trend, and investigations by airborne magnetic and gravimetric surveys indicate an absence of volcanic rocks at depth along the La Guadeloupe fault (Relevés géophysiques Inc., 1978).

In the Beauce area, the Saint-Victor Synclinorium and the CVGS are not affected by a polyphase deformation along the La Guadeloupe fault, as they are in the Sherbrooke area. Only one regional cleavage is associated with the regional folding (Figs. 6 and 10). In this area, the Magog Group is subdivided into two structural superdomains, each of which contains smaller domains separated by inferred and/or observed faults.

First structural superdomain

The first superdomain (Figs. 5 and 6) is located southwest of the small sliver of the Ascot Complex. It is made up of the Morisset and Lac Fortin structural domains.

Because of a gradual change in attitude of S1 and L1, there is no homogeneous data base from which it is possible to make stereographic projections representative of the entire first super-domain. Thus, areas lying between the La Guadeloupe fault to the southeast and the Cumberland and related faults to the northwest were put into the same structural domain, as was the Cumberland thrust slice with which those areas share similar structural features. These form the Morisset structural domain (MA-6, Figs. 5 and 6) and comprise the Saint-Victor and Beauceville formations. Along both the Cumberland and La Guadeloupe faults, younger rocks are thrust upon older ones and the orientation of the regional cleavage is deflected to become parallel to fault traces. The Cumberland fault is truncated by the La Guadeloupe fault and is interpreted as a La Guadeloupe-related subsidiary thrust fault.

In domain MA-6 (Fig. 6), several structural subareas are present, all of which are separated by faults. The orientations of structural elements vary accordingly to the orientation of the nearest fault (Figs. 6 and 7) and may differ from one fault block to the next.

Like the Morisset domain, the Lac Fortin structural domain (Figs. 5 and 6) comprises the Saint-Victor and Beauceville formations. Its southwestern extension is actually unknown, but the Magog Group rocks of the adjacent Thetford Mines–Saint-Victor area (St-Julien, 1987) are believed to be part of it.

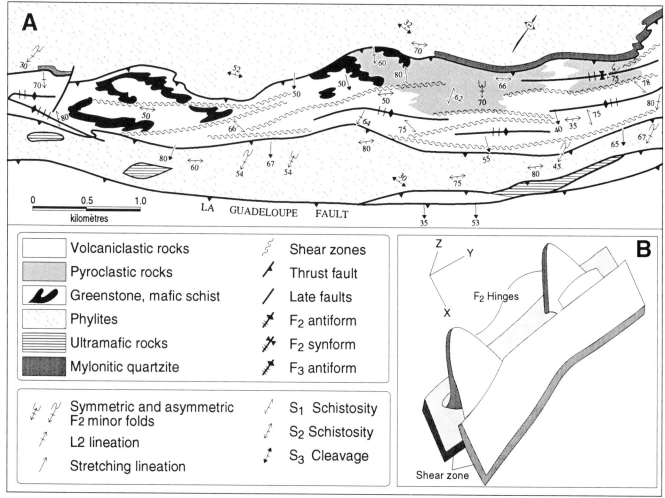

Figure 4. A, Detailed geological map of the Eustis domain. See Figure 2 for location. B, Three-dimensional interpretation of the structural pattern shown in A.

Even in the Lac Fortin domain, the gradual changes in attitude of S1 and L1 toward the La Guadeloupe fault persist and are indicated by the fact that the calculated B axes of folds do not lie on the regional S1 cleavage stereonets. However, by splitting the Lac Fortin domain into subareas, MA-4 and MA-5 (Fig. 6), the calculated B axis of the folds lies on the regional cleavage in MA-4 that is the most distant of the fault but not in MA-5 (Fig. 6). Again, by splitting of MA-5 into two smaller subareas (Fig. 8), only the B axis for the farthest subarea (relative to the La Guadeloupe fault) lies on the regional cleavage.

From this, we deduced that the deflection of the regional S1 cleavage, the tightening of folds, and the steepening of fold plunges increase toward the La Guadeloupe and related faults. As also indicated by field observations, away from these faults (domain MA-5, Fig. 6) folds are open and plunge gently (10 to 20°)

to the southwest, and the cleavage is subvertical. Along the La Guadeloupe fault, folds are tight and plunge 60 to 70° to the southeast, and the cleavage dips moderately to southeast (domain MA-6, Fig. 6). Similar structural relationships are present in the Saint-Francis Group on the southeast side of the La Guadeloupe fault (domain CSD-3, Fig. 9). The structural relationship found in these areas is similar to those observed in the Sherbrooke area.

Second structural superdomain

The second superdomain is located northeast of the sliver of the Ascot Complex. It is subdivided into Lac Lanigan and Lac à la Raquette structural domains (Figs. 5 and 6), each of these representing a distinct thrust slice of the Saint-Victor Synclinorium. The Saint-Victor Formation is the only Magog Group unit

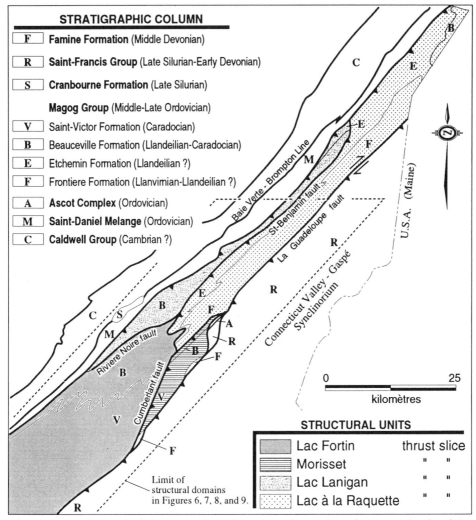

Figure 5. Geological map of the Beauce Area. Patterned units are the major structural units of the Saint-Victor synclinorium (Magog Group). Letters refer to the stratigraphic units. Structural data presented in Figures 6, 7, 8, and 9 come from the area within the dotted box.

that does not crop out within this superdomain. Contrasting with the first superdomain, folds and plunges here do not tighten or steepen toward the La Guadeloupe fault (Fig. 6).

The Lac Lanigan domain forms a 50-km-long thrust slice located 5 km from the La Guadeloupe fault. As a result of southwestern shallow plunging of folds, the Beauceville Formation crops out almost continuously in this domain. Fold trends are similar to those in the adjacent Lac Fortin domain (Fig. 6); these domains are separated by the Rivière Noire fault. The Lac à la Raquette domain forms a 75-km-long thrust slice bounded by the La Guadeloupe and Cumberland faults to the southeast and by the Saint-Benjamin fault to the northwest. The trace of the latter fault is folded (Fig. 5), although the regional cleavage is

neither folded nor reoriented near either the Saint-Benjamin or the Rivière Noire faults. Consequently, the cleavage is interpreted as younger than these faults.

The Lac à la Raquette thrust slice is a doubly plunging structure. As in other thrust slices of the Saint-Victor Synclinorium, fold plunges are generally to the southwest except in its northeastern extremity, where they become steeper and northeasterly plunging. However, even in the southwest part of the Lac à la Raquette structural domain, folds are typically tight, bedding-cleavage intersections plunge gently to steeply to the southwest or to the northeast, and the cleavage dips steeply (Fig. 6). The B axes of the S0 stereonets do not lie on the regional cleavage. Moreover, there is always a clockwise rotation of the B axis with

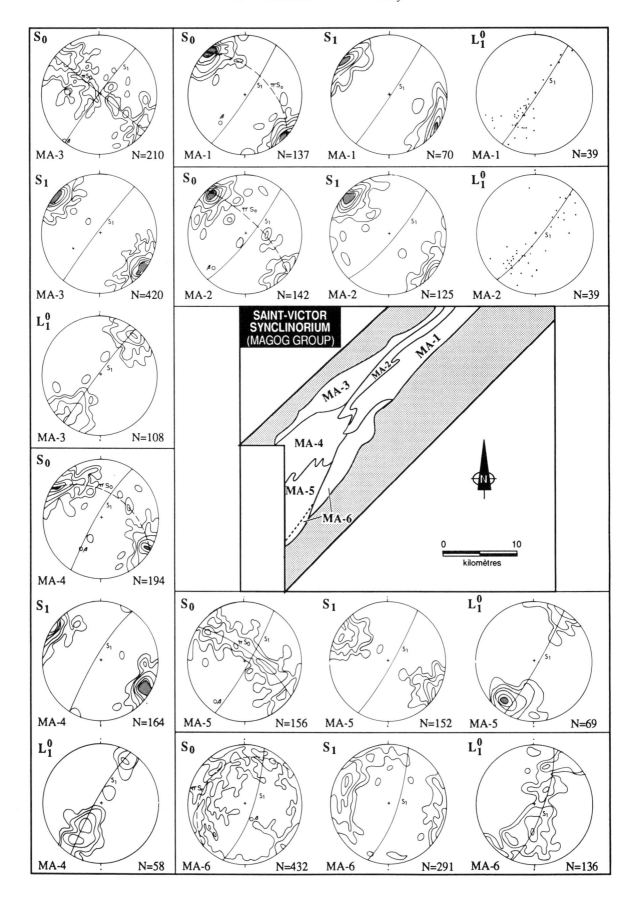

respect to the cleavage in all subareas that can be made from the Lac à la Raquette domain (see MA-1 and MA-2 in Fig. 6). This observation is also verified southeast of the La Guadeloupe fault, in the Saint-Francis Group lying northeast of the Ascot sliver (Fig. 9).

In the Lac à la Raquette domain, the clockwise rotation of B axes with respect to the regional cleavage represents a meaningful structural observation that is here interpreted as an indication of transected folds. However, it is impossible to observe these transected folds in the field because of a small transection angle (<10°) as well as limited fold hinges exposed in the area.

THE LA GUADELOUPE FAULT: KINEMATICS AND AGE CONSTRAINTS

The Acadian tectonic regime in the Québec Appalachians is viewed as resulting in strike-slip faults in the Gaspé Peninsula and thrust faults in the Eastern Townships (Tremblay and St-Julien, 1990; Malo and Bourque, this volume; St-Julien and others, 1983). This change of tectonic regimes is attributed to the original geometry of North America as plate convergence occurred during the Taconian and Acadian orogenies (Stockmal and others, 1987). Therefore, segments of the orogen lying between the Gaspésie and Eastern Townships regions—that is, the Témiscouata region, northern Maine, and the northeastern part of the Beauce—are thought to represent transition zones between thrusting and strike-slip faulting.

Near the Chaudière River, the La Guadeloupe fault is exposed along Route 204 at St-Georges-de-Beauce. At this locality, the Eifelian (early Middle Devonian) Famine Formation (Boucot and Drapeau, 1968; Oliver, 1971) occurs within a small area (500 m × 2 km) and unconformably overlies the Caradocian Saint-Victor Formation. The Famine Formation is also sheared and thrusted upon by the Emsian (latest Early Devonian) Compton Formation of the Saint-Francis Group (Kelly, 1975). Therefore, the La Guadeloupe fault is there a classical young-over-old thrust. Rocks of Eifelian age are uncommon in the Québec Appalachians, although they are found in the Sherbrooke (Mountain House Wharf Formation) and Témiscouata (Touladi Formation) areas. Consequently, the apparent young-over-old relationship

along the La Guadeloupe fault is believed to be the result of erosion rather than an original structural phenomenon.

At St-Georges-de-Beauce, the base of the Famine Formation is a polymictic clast-supported conglomerate. It is directly overlain by a floatstone and a dark gray, shaly, fine-grained limestone. These rocks are cut by the same regional cleavage as are those of the Saint-Victor Formation. Toward the La Guadeloupe fault,

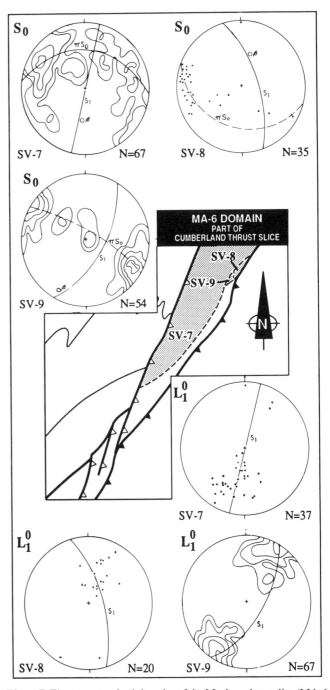

Figure 6. Structural domains of the Magog Group. Lower-hemisphere, equal-area projections of structural elements from part of the Beauce area. The map limits the numbered structural domains. Planes shown in cyclographic projection represent the mean attitude of the regional foliation in each domain. Pi-circles shown in cyclographic projection represent the mean attitude of the regional folds in each domain. Density contours are by 2, 5, 10, 15, and >20 percent. Linear fabrics are L1 lineations (or F1 fold hinges). The number of data points appears at the lower right of each projection. This legend also applies for stereonets on Figures 7, 8, and 9.

Figure 7. Three structural subdomains of the Morisset thrust slice (MA-6 in Fig. 6) showing the variations in the attitude of the structural elements along the La Guadeloupe and related faults. Compare with Figures 6, 8, and 9.

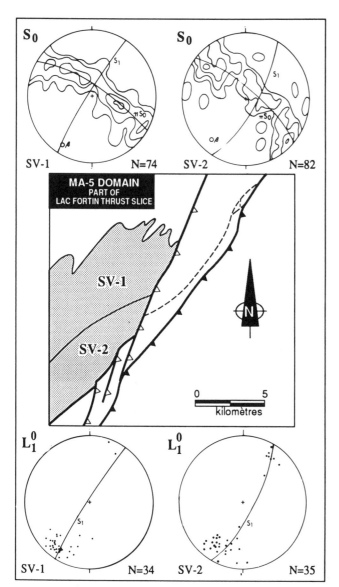

Figure 8. Two structural subdomains of the Lac Fortin thrust slice (MA-5 in Fig. 6) showing the progressive changes in the attitude of the structural elements toward the La Guadeloupe fault (from SV-1 to SV-2). Compare with Figures 6, 7, and 9.

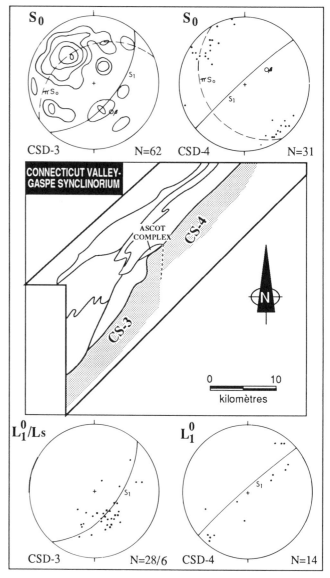

Figure 9. Two structural domains of the Connecticut Valley–Gaspé Synclinorium (Saint-Francis Group) showing the differences in the attitudes of the structural elements that exist along the La Guadeloupe fault on either side of the sliver of the Ascot Complex. Linear fabrics are stretching lineations (Ls; appear as squares) and L1 lineations or F1 fold hinges (appear as dots). Compare with Figures 6, 7, and 8.

the fine-grained limestone is transformed first into a banded calc-mylonite (Fig. 10) and then into a finely laminated ultramylonite. Along the main fault, a 2-m-wide band of sericite-rich schist and mylonite crops out. S-C fabrics indicate an east-over-west movement along a southeast-plunging tectonic vector, as suggested by stretching lineations (Fig. 11). The sericite-rich rocks are structurally overlain by sheared rocks of the Saint-Francis Group. The shear intensity decreases rapidly southeastward. Stretching lineation in rocks of the Saint-Francis Group are only present in rocks located within 100 m of the fault, although folds continue to plunge parallel to the stretching lineation (60° SSE) over a greater distance (domain CSD-3, Fig. 9). Parallelism of

fold axes and stretching lineations is explained by the development of sheath-folds (Quinquis and others, 1978; Cobbold and Quinquis, 1980).

Similar shear zones and mylonites within the Famine Formation also occur 15 km farther to the northeast. These show the same structural geometry as those just described.

The La Guadeloupe fault does not crop out northeast of the Ascot Complex sliver. Thus, kinematic indicators have not been observed. The nature of this segment of the La Guadeloupe fault can only be deduced from the regional structural analysis of rocks

Figure 10. Mylonitic rock of the Saint-Francis Group exposed along the Famine River where the La Guadeloupe fault crops out. Millimeter-thick bands of white recrystallized calcite alternate with darker bands of carbonaceous shale.

Figure 11. Stretching lineation (parallel to pencil) contained within the regional cleavage plane of the sericite schist along the La Guadeloupe fault (Route 204 at St-Georges-de-Beauce).

lying on both sides of the fault zone. By comparing the two structural domains located on the southwestern and northeastern sides of the sliver of Ascot Complex in the Magog Group (MA-6 and MA-1, Fig. 6), as well as those located in the CVGS (CSD-3 and CSD-4, Fig. 9), major differences can be noted. Southwest of the sliver of Ascot Complex, the S1 cleavage dips moderately to the southeast, and the B axes and L1 lineation also plunge moderately to the southeast. Northeast of the sliver of Ascot Complex, the trace of the fault is displaced 5 km to the northwest, and the S1 cleavage becomes subvertical and apparently transects the folds on both sides of the La Guadeloupe fault. The folds no longer plunge moderately to the southeast. To explain these structural variations, a kinematic change along the La Guadeloupe fault is invoked.

The clockwise rotation of the B axes of the inferred transected folds is used to indicate dextral strike-slip movement (Blewett and Pickering, 1988; Murphy, 1985; Sanderson and others, 1980) along this segment of the fault (Fig. 12). In the Gaspé Peninsula similar clockwise fold axes rotations exist (Malo and Béland, 1989; Malo and Bourque, this volume) and are associated with well-documented Acadian strike-slip faults. The nature of the La Guadeloupe fault equivalent in northern Maine is complex (Roy, 1989), but in the Timiscouata area it has been interpreted as a dextral strike-slip fault (David and others, 1985; David and Gariépy, 1990).

In the Beauce area, the importance of the strike-slip component of the fault cannot be evaluated. Since it is totally eliminated over a short distance (<30 km), it must not be very great. Hence, the northeast part of the La Guadeloupe fault is probably a dextral oblique-slip fault. Because there is no evidence of a strike-slip movement along the southwest segment of the La Guadeloupe fault, it must be eliminated by some means. This can be achieved by generating a series of small curved faults at the tip of the fault (Nicolas, 1984, p. 115) as in some transpression zones. In the Beauce area, these minor faults could be the Cumberland and other smaller faults (Fig. 12). These faults form where the Ascot Complex appears because it behaves as a buttress to the dextral oblique-slip movement.

The formation of small faults of this type has modified pre-Acadian structures of the Magog Group. Hence, the Cumberland fault cuts the pre-Acadian Saint-Benjamin fault, causing the Beauceville Formation to be emplaced against rocks of the stratigraphically lower Frontière Formation (Fig. 12).

In the Sherbrooke area, the regional deformation was contemporaneous or slightly preceded the development of the La Guadeloupe. The youngest rocks involved in the deformation are Pridolian (Late Silurian) rocks of the Lac Lambton Formation (Saint-Francis Group). In the Beauce area, the regional deformation is also coeval with the development of the La Guadeloupe fault, but the youngest sheared rocks are the Eifelian (lower Middle Devonian) rocks of the Famine Formation.

In the Weedon area (Fig. 1), Pridolian rocks (Boucot and Drapeau, 1968) of the Lac Aylmer Formation are imbricated within the La Guadeloupe fault (St-Julien and Hubert, 1975; Labbé and St-Julien, 1989). Moreover, the fault zone is cut by the undeformed Aylmer Pluton, which has superimposed a metamorphic aureole on fault rocks (Labbé, 1992). The Aylmer Pluton has been dated by Pb/Pb and U/Pb methods on monazites and yields a U/Pb age of 375 ± 3 Ma (Simonetti and Doig, 1990), indicating that the La Guadeloupe fault is younger than Eifelian but older than 375 Ma and therefore related to the Acadian Orogeny.

DISCUSSION: TACONIAN VERSUS ACADIAN DEFORMATION

In the Magog Group, the regional cleavage and folds are attributed to the Acadian Orogeny and were produced at the same time as the La Guadeloupe and related faults. Pre-Acadian

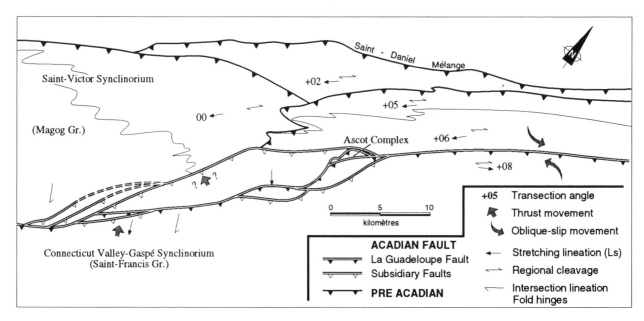

Figure 12. Schematic map of a segment of the Beauce area summarizing the structural data and the movements along the La Guadeloupe fault.

deformation is limited to younger thrust faulting, such as the Saint-Benjamin and Rivière Noire faults. Pre-Acadian thrusts were accompanied by bedding warping but no cleavage development at a regional scale. Thus, no pre-Acadian cleavage is found in the Magog Group, a feature also shared by folded rocks of the Mictaw Group, a correlative unit in Gaspésie (De Broucker, 1986). A similar situation exists in the Sherbrooke area, where the D2 and D3 phases were produced by the Acadian Orogeny. The D1 phase of deformation has been attributed to a pre-Acadian tectonic event that could be related to the Taconian Orogeny (Tremblay and St-Julien, 1990).

The Taconian Orogeny is interpreted to be the product of a continent–volcanic arc collision (Fig. 13) during the Middle Ordovician (Rowley and Kidd, 1981; Stanley and Ratcliffe, 1985). Why then is there so little Taconian deformation in the Magog Group? This can be explained by a better understanding of the paleogeography during the Taconian Orogeny. In modern forearcs, deformation is most intense in the accretionary prism (Cowan and Shilling, 1978; Cowan, 1985; Bally and Oldow, 1986; Moore and Lundberg, 1986), and deformation produced by normal compression is much less intense in the volcanic arc and back-arc basin. Albeit, these regions can be deformed early in their evolution by arc-splitting (Bally and Oldow, 1986; Karig and others, 1986; Hamilton, 1979, 1988). Forearc basins are less deformed (Lucas and Moore, 1986; Westbrooke, 1982) as a result of a decreasing intensity of deformations from the trench toward the arc–back-arc area (e.g., the Barbados ridge; Torrini and others, 1985). The relative absence of deformations should persist even during the collision with another volcanic arc or with a passive margin—the arc basement acting as a buttress and the accretionary prism, made up of soft, water-laden sediments, acting as a buffer.

In the southwest Québec Appalachians, the basement of the Magog Group was part of an accretionary-tectonic wedge consisting of accreted crustal blocks such as ophiolites, basement rocks of the Chain Lakes Massif (Cousineau, 1991), and Ascot volcanics that could have acted as buttresses. The Saint-Daniel argillites were present between these blocks and behaved as a buffer, leaving most of the blocks and adjacent-overlying forearc basin deposits almost undeformed.

In the Beauce area, the Cranbourne Formation of Pridolian age (Boucot and Drapeau, 1968) unconformably overlies the BBL (Fig. 5), thus indicating that it is not an Acadian suture but a younger, most probably Taconian suture. Below the unconformity, the Saint-Daniel Melange and the Caldwell Group (Humber zone) are both polydeformed but exhibit only one regional cleavage. As in the Cranbourne Formation, the cleavage strikes northeast-southwest with a steep dip (Cousineau, 1990, 1992). It follows that even though the tectonic collage of the Dunnage and Humber zones is interpreted as a Taconian event, rocks on both sides of the BBL were also deformed during the Acadian Orogeny. The intensity of the Acadian deformation in the Saint-Daniel Melange is such that the identification of younger deformation is difficult. In the Caldwell Group, the older deformation phase is Taconian related (St-Julien, 1987; St-Julien and Hubert, 1975).

Northwest of the BBL, the extent of Acadian deformation is unknown. Structures found in the external nappes of the Humber zone were formed entirely during the Taconian Orogeny (St-Julien and Hubert, 1975; Vallières, 1984). Hence, the limit of Acadian deformations must lie within the internal nappes of the

Humber zone where they are superimposed on Taconian deformations.

In the southwestern Québec Appalachians, the nature of the Acadian tectonics is just coming to be understood. Before the onset of the Acadian Orogeny, subsidence must have occurred where the CVGS now lies. This was probably associated with major faulting (Bradley, 1983, Naylor, 1989). However, available data allow only the broad elaboration of an Acadian tectonic model in the studied area.

The Acadian Orogeny is attributed to the collision of the Laurentian margin and accreted Taconian terranes with a large continental mass, i.e., Avalonia (e.g., Osberg, 1978; Bradley, 1983). In southern Québec, some Taconian faults were reactivated during the Acadian and new ones were formed (Bernard, 1987; St-Julien and others, 1983). Acadian faults are generally interpreted as southeast-dipping, high-angle reverse faults. In the Gaspé peninsula, strike-slip faults were activated. Major faults, like the La Guadeloupe fault, extend from southern Québec up to the Gaspé Peninsula and become oblique-slip faults. Plutonism occurred within the CVGS in Late Devonian, and metamorphic aureoles were formed. The structural pattern of the Dunnage zone as it now exists in southwest Québec was then developed (Fig. 13).

CONCLUSIONS

Historically, Cambrian and Ordovician rocks of the southwestern Québec Appalachians were assumed to have been deformed mostly during the Taconian Orogeny. The models proposed assumed that these rocks had been slightly affected by the Acadian Orogeny. Detailed work in the Ascot Complex and in the Magog Group proves the opposite.

Silurian and Devonian rocks of the Saint-Francis Group are transported northwestward upon Ordovician rocks of the Dunnage zone along a major Acadian fault: the La Guadeloupe fault. The intensity of the regional deformation decreases away from that fault in both the Saint-Francis Group and Dunnage rocks. The La Guadeloupe fault is contemporaneous with the development of the regional cleavage and related folds in all these units. Therefore, the deformational event affecting most, if not all, rocks of the Dunnage zone occurred during the Acadian Orogeny.

In the northeastern part of the Beauce area, structural analysis suggests the existence of transected folds on both sides of the La Guadeloupe fault. The angle of transection is consistent with dextral strike-slip movement. In this area, the La Guadeloupe fault is thus interpreted as a dextral oblique-slip fault. The Acadian tectonic regime is mostly related with dextral strike-slip faults in Gaspésie and reverse faults in Estrie-Beauce. The northeastern part of the Beauce area is thus interpreted as a part of the transition zone between strike-slip and dip-slip fault movements.

The extent of cratonward Acadian deformations is unknown. Upper Silurian rocks unconformably overlying the BBL indicate that it is a Taconian suture. Because the underlying Cambro-Ordovician rocks are affected by the same regional deformation as the Upper Silurian rocks, the Acadian Orogeny is

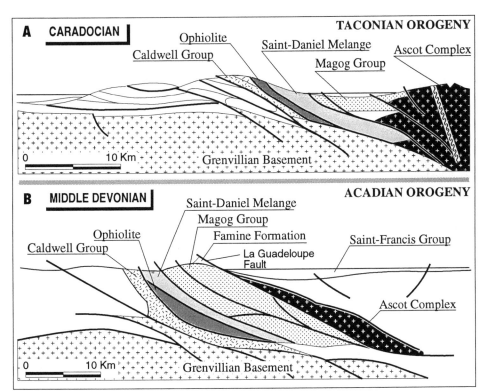

Figure 13. A, schematic structural evolution of a portion of southwestern Québec during the Taconian Orogeny; B, schematic structural evolution of the Beauce area during the Acadian Orogeny.

interpreted as being responsible for the actual structural pattern of a large segment of southwestern Québec Appalachians.

ACKNOWLEDGMENTS

This contribution is a portion of the doctoral studies of both authors at the Université Laval. The Ministère de l'Energie et des Ressources du Québec supported most of the field expanses. Comments by E. W. Sawyer, D. C. Bradley, D. C. Roy, and an anonymous reviewer on earlier versions were greatly appreciated. We thank Claude Dallaire for his excellent artwork.

REFERENCES CITED

Bally, A. C., and Oldow, F. J., 1986, Plate tectonics, structural styles and the evolution of sedimentary basins: Geological Association of Canada Cordilleran Section, Short Course 7, 237 p.

Bernard, D., 1987, Etude structurale et stratigraphique du synclinorium de Connecticut Valley–Gaspé dans le sud-est du Québec [M.Sc. thesis]: Québec, Université Laval, 30 p.

Blewett, R. S., and Pickering, K. T., 1988, Sinistral shear during Acadian deformation in north-central Newfoundland, based on transected cleavage: Journal of Structural Geology, v. 10, p. 125–128.

Boucot, A. J., and Drapeau, G., 1968, Siluro-Devonian rocks of Lake Memphremagog and their correlatives in the Eastern Townships: Ministère des Richesses naturelles du Québec Special Paper ES-1, 46 p.

Bradley, D. C., 1983, Tectonics of the Acadian Orogeny in New England and adjacent Canada: Journal of Geology, v. 91, p. 381–400.

Cobbold, P., and Quinquis, H., 1980, Development of sheath folds in shear regimes: Journal of Structural Geology, v. 2, p. 119–126.

Cousineau, P. A., 1990, Le Groupe de Caldwell et le domaine océanique entre Saint-Joseph-de-Beauce et Sainte-Sabine: Ministère de l'Energie et des Ressources du Québec Mémoire MM 87-02, 178 p.

——, 1991, The Rivière des Plante Ophiolitic Melange: Tectonic setting and melange formation in the Québec Appalachians: Journal of Geology, v. 99, p. 81–96.

——, 1992, The Saint-Daniel Melange: Evolution of an accretionary complex in the Dunnage terrane of the Québec Appalachians: Tectonics, v. 11, p. 898–909.

Cousineau, P. A., and St-Julien, P., 1985, Paleotectonic environment of the oceanic domain from southern Québec Appalachians: Geological Association of Canada Program with Abstract, v. 10, p. A12.

Cowan, D. S., 1985, Structural styles in Mesozoic and Cenozoic mélanges in western Cordillera of North America: Geological Society of America Bulletin, v. 96, p. 451–462.

Cowan, D. S., and Schilling, R. M., 1978, A dynamic, scaled model of accretion at trenches and its implications for the tectonic evolution of subduction complexes: Journal of Geophysical Research, v. 83, p. 5389–5396.

David, J., and Gariépy, C., 1990, Early Silurian andesites from the central Québec Appalachians: Canadian Journal of Earth Sciences, v. 27, p. 632–643.

David, J., Chabot, N., Marcotte, C., and Lajoie, J., 1985, Stratigraphy and sedimentology of the Cabano, Point aux Tremble and Lac Raymond Formations, Témiscouata and Rimouski counties, Québec, *in* Current research, part B: Geological Survey of Canada Paper 85-1B, p. 491–497.

De Broucker, G., 1986, Evolution tectonostratigraphique de la boutonnière Maquereau-Mictaw (Cambro-ordovicien), Gaspésie, Québec [Ph.D. thesis]: Québec, Université Laval, 322 p.

Doolan, B. L., Gale, M. H., Gale, P. N., and Hoar, R. S., 1982, Geology of the Québec re-entrant: Possible constraints from the early rifts and the Vermont-Québec serpentine belt, *in* St-Julien, P., and Béland, J., eds., Major structural

zones and faults of the northern Appalachians: Geological Association of Canada Special Paper 24, p. 87–115.

Dunning, G. R., Krogh, T. E., and Pederson, R. B., 1986, U/Pb zircon ages of Appalachian-Caledonian ophiolites: Sixth International Conference on Geochronology, Cosmochronology and Isotope Geology, IUGS Subcommittee on Geochronology, International Association of Volcanology and Geochemistry Society, Cambridge, U.K., June 30–July 4, p. 155.

Hamilton, W. B., 1979, Tectonics of the Indonesian region: U.S. Geological Survey Professional Paper 1078, 345 p., reprinted with corrections, 1981 and 1985.

——, 1988, Plate tectonics and island-arcs: Geological Society of America Bulletin, v. 100, p. 1503–1527.

Karig, D. E., Sarewitz, D. R., and Haeck, G. D., 1986, Role of strike-slip faulting in the evolution of allochthonous terranes in the Philippines: Geology, v. 14, p. 852–855.

Kelly, R., 1975, Région des monts Ste-Cécile et Saint-Sébastien: Ministère de l'Energie et des Ressources du Québec Rapport Géologique RG-176, 30 p.

Labbé, J.-Y., 1992, Géologie de la région de Weedon-Lingwick: Ministère de l'Energie et des Ressources du Québec Etude terminale (in press).

Labbé, J.-Y., and St-Julien, P., 1989, Failles de chevauchement acadiennes dans la région de Weedon, Estrie, Québec: Canadian Journal of Earth Sciences, v. 26, p. 2268–2277.

Laurent, R., 1975, Occurrence and origin of the ophiolites of southern Quebec, Northern Appalachians: Canadian Journal of Earth Sciences, v. 12, p. 443–445.

——, 1977, Ophiolites from the Northern Appalachians of Quebec, *in* Coleman, R. G., and Irwin, W. P., eds., North American ophiolites: Oregon Department of Geology and Mineral Industries Bulletin 95, p. 25–40.

Lucas, S. E., and Moore, J. C., 1986, Cataclastic deformation in accretionary wedges: Deep Sea Drilling Project Leg 66, southern Mexico, and on-land examples from Barbados and Kodiak Islands, *in* Moore, J. C., ed., Structural fabrics in Deep Sea Drilling Project cores from forearcs: Geological Society of America Memoir 166, p. 89–103.

Malo, M., and Béland, J., 1989, Acadian strike-slip tectonics in the Gaspé region, Québec Appalachians: Canadian Journal of Earth Sciences, v. 26, p. 1764–1777.

Moore, J. C., and Lundberg, N., 1986, Tectonic overview of Deep Sea Drilling Project transects of fore-arcs, *in* Moore, J. C., ed., Structural fabrics in Deep Sea Drilling Project cores from forearcs: Geological Society of America Memoir 166, p. 1–12.

Murphy, F. C., 1985, Non-axial planar cleavage and Caledonian sinistral transpression in eastern Ireland: Geological Journal, v. 20, p. 257–279.

Naylor, R. S., 1991, Role of the Connecticut Valley-Gaspé through in the Acadian Orogeny: Geological Society of America Northeastern Section Abstracts with Programs, v. 21, 54 p.

Nicolas, A., 1984, Principes de tectonique: Paris, Masson, 196 p.

Oliver, W. A., Jr., 1971, The coral fauna and age of the Famine Limestone in Québec, *in* Dutro, J. T., Jr., ed., Paleozoic perspectives: A paleantological tribute to G. Arthur Cooper: Smithsonian Contribution Paleo-biology, 3, p. 193–201.

Osberg, P. H., 1978, Synthesis of the geology of the Northern Appalachians, USA: Geological Survey of Canada Paper 78-13, p. 137–147.

Quinquis, H., Audren, C., Brun, J. P., and Cobbold, C. R., 1978, Intense progressive shear in Ile de Groix blueschist and compatibility with subduction or obduction: Nature, v. 273, p. 43–45.

Relevés géophysiques Inc., Les, 1978, Interprétation de données aéromagnétiques dans la région de Beauce-Charlevoix: Ministère des Richesses naturelles du Québec DPV-557, 82 p.

Robinson, P., and Hall, L., 1979, Tectonic synthesis of southern New England, *in* Wones, D., ed., The Caledonides in the USA: Blackburg, Virginia, Virginia Polytechnic Institute, p. 73–82.

Rowley, D. B., and Kidd, W.S.F., 1981, Stratigraphic relationships and detrital composition of the medial Ordovician flysch of western New England— Implication for the tectonic evolution of the Taconic orogeny: Journal of

Geology, v. 89, p. 199–218.

Roy, D. C., 1989, The Depot Mountain Formation: Transition from syn- to post-Taconian basin along the Baie Verte–Brompton line in Northwestern Maine: Maine Geological Survey Studies in Maine Geology, v. 2, p. 85–99.

Sanderson, D. J., Andrews, J. R., Philips, W.E.A., and Hutton, D.H.W., 1980, Deformation studies in the Irish Caledonides: Journal of the Geological Society of London, v. 137, p. 289–302.

Simonetti, A., and Doig, R., 1990, U-Pb and Rb-Sr geochronology of Acadian plutonism in the Dunnage zone of the southern Québec Appalachians: Canadian Journal of Earth Sciences, v. 27, p. 881–892.

Spencer, C., Green, A., Morel-à-l'Huissier, P., and Milkereit, B., 1989, The extension of Grenville basement beneath the Northern Appalachians: Results from the Québec-Maine seismic reflection and refraction surveys: Tectonics, v. 8, p. 677–696.

Stanley, R. S., and Ratcliffe, N. M., 1985, Tectonic synthesis of the Taconian orogeny in western New England: Geological Society of America Bulletin, v. 96, p. 1227–1250.

St-Julien, P., 1987, Région de Saint-Victor et de Thetford (est): Ministère de l'Energie et des Ressources du Québec Mémoire MM 86-01, 66 p.

St.-Julien, P., and Hubert, C., 1975, Evolution of the Taconic orogen in the Québec Appalachians: American Journal of Science, v. 275-A, p. 337–362.

St-Julien, P., Slivitzky, A., and Feininger, T., 1983, A deep structural profile across the Appalachians of southern Québec, *in* Hatcher, R. D., Williams, H., and Zietz, I., eds., Contributions to the tectonics and geophysics of mountain chains: Geological Society of America Memoir 158, p. 103–111.

Stockmall, G. S., Colman-Sadd, S. P., Keen, C. E., O'Brien, S. J., and Quinlan, G., 1987, Collision along an irregular margin: A regional plate tectonic interpretation of the Canadian Appalachians: Canadian Journal of Earth Sciences,

v. 24, p. 1098–1107.

Torrini, R., Speed, R. C., and Matiolli, G. S., 1985, Tectonic relationships between forearc-basin strata and accretionary complex at Bath, Barbados: Geological Society of America Bulletin, v. 96, p. 861–874.

Tremblay, A., and St-Julien, P., 1990, Structural style and evolution of a segment of the Dunnage Zone from the Quebec Appalachians and its tectonic implications: Geological Society of America Bulletin, v. 102, p. 1218–1229.

Tremblay, A., Hébert, Y., and Bergeron, M., 1989a, Le Complexe d'Ascot des Appalaches du sud du Québec: Pétrologie et géochimie: Canadian Journal of Earth Sciences, v. 26, p. 2407–2420.

Tremblay, A., St-Julien, P., and Labbé, J.-Y., 1989b, Mise à l'évidence et cinématique de la faille de la Guadeloupe, Appalaches du sud du Québec: Canadian Journal of Earth Sciences, v. 26, p. 1932–1943.

Vallières, A., 1984, Stratigraphie et structure de l'orogène taconique de la région de Rivière-du-Loup [Ph.D. thesis]: Québec, Université Laval, 302 p.

Westbrook, G. K., 1982, The Barbados Ridge Complex: Tectonics of a mature forearc system, *in* Leggett, J. K., ed., Trench forearc geology: Geological Society of London Special Publication 10, p. 275–291.

Williams, H., 1979, Appalachian Orogen in Canada: Canadian Journal of Earth Sciences, v. 16, p. 792–1003.

Williams, H., and St-Julien, P., 1982, The Baie Verte–Brompton line: Early Paleozoic continent-ocean interface in the Canadian Appalachians, *in* St-Julien, P., and Béland, J., eds., Major structural zones and faults of the northern Appalachians: Geological Association of Canada Special Paper 24, p. 177–207.

MANUSCRIPT ACCEPTED BY THE SOCIETY JUNE 8, 1992

Geological Society of America
Special Paper 275
1993

Timing of the deformation events from Late Ordovician to Mid-Devonian in the Gaspé Peninsula

Michel Malo*
INRS-Géoressources, Centre Géoscientifique de Québec, 2700 Einstein, C.P. 7500 Sainte-Foy, Québec G1V 4C7, Canada
Pierre-André Bourque*
Département de Géologie, Université Laval, Ste.-Foy, Québec, G1K 7P4, Canada

ABSTRACT

The Middle Paleozoic rocks deformed by the Acadian orogeny in the Gaspé Penin-sula are divided into three major structural zones, from north to south: (1) the Connecti-cut Valley–Gaspé synclinorium, (2) the Aroostook-Percé anticlinorium, and (3) the Chaleurs Bay synclinorium. These three structural zones were part of a single deposi-tional belt, the Gaspé Belt, located mainly to the south of the Baie Verte–Brompton Line over the Dunnage Zone.

The Gaspé Belt comprises four broad temporal and lithological packages: (1) Upper Ordovician–lowermost Silurian deep water fine-grained siliciclastic and carbonate fa-cies, (2) Silurian–lowermost Devonian shallow to deep shelf facies, (3) Lower Devonian mixed siliciclastic and carbonate fine-grained deep shelf and basin facies, and (4) upper Lower to Upper Devonian nearshore to terrestrial coarse-grained facies. Upper Ordovi-cian to Middle Devonian rocks of the Gaspé Belt are bracketed between the Taconian and the Acadian unconformities, whereas the Salinic unconformity is well recorded in the Gaspé Belt sequence. Three shallowing-upward phases, separated by two transgres-sive episodes, are recorded by the sequence, the shallowing phases occurring more or less in response to three tectonic pulses: the Middle to Late Ordovician Taconian orogeny, the Late Silurian Salinic disturbance, and the Middle Devonian Acadian orogeny.

The major structural trend of the Late Ordovician to Middle Devonian rocks of the Gaspé Belt is oriented roughly northeast. Major easterly striking dextral strike-slip faults (in the southeastern Aroostook-Percé anticlinorium) and northwesterly striking faults (in the northeastern Connecticut Valley–Gaspé synclinorium) transect this trend, where-as northeasterly striking high-angle reverse faults are present in the western and central regions. Two phases of folding are recorded in the Gaspé Belt. Northwest-southeast-striking F_1 folds recognized mainly in the Aroostook-Percé anticlinorium and synsedimentary faulting along the northwesterly trending faults of the northeastern Connecticut Valley–Gaspé synclinorium that affect the Silurian part of the sequence may be related to the Salinic disturbance. The northeast-southwest-striking F_2 folds corre-

*GIRGAB, Groupe Interuniversitaire de Recherches Géologiques en Analyse de Bassin.

Malo, M., and Bourque, P.-A., 1993, Timing of the deformation events from Late Ordovician to Mid-Devonian in the Gaspé Peninsula, *in* Roy, D. C., and Skehan, J. W., eds., The Acadian Orogeny: Recent Studies in New England, Maritime Canada, and the Autochthonous Foreland: Boulder, Colorado, Geological Society of America Special Paper 275.

spond to the major Acadian structural trend of the Gaspé Peninsula. Structural features related to the D$_2$ deformation can be integrated in a model of strike-slip tectonics. Shortening of the cover rocks of the Gaspé Belt and transcurrent motion along the major easterly striking strike-slip faults are related to local transpression in the northeastern part of the Québec Reentrant during the continued continental convergence and accretion of outboard terranes to the North American craton in post–Middle Devonian and pre-Carboniferous time.

INTRODUCTION

The Gaspé segment (Fig. 1) of the Canadian Appalachians has recorded much of the history of a mountain belt, from the opening of the Iapetus Ocean in Late Precambrian time to its closure in mid-Devonian. Two major orogenies have shaped the Gaspé Peninsula: the Middle to Late Ordovician Taconian orogeny and the Middle Devonian Acadian orogeny.

Paleozoic rocks of the Gaspé Peninsula can be divided into three temporal assemblages: (1) Latest Precambrian to Late Ordovician rocks (commonly referred to as the Cambro-Ordovician), mainly deformed by the Taconian orogeny, occupying the northern part of the region and some inliers in the southern part of the peninsula; (2) Late Ordovician to Middle Devonian rocks, deformed by the Acadian orogeny, making up the center and the southern part of the region; and (3) the flat-lying Late Devonian and Carboniferous rocks, disturbed by faulting only during the Alleghanian orogeny, occurring along the shore of Chaleurs Bay in the southern and eastern Gaspé Peninsula (Fig. 2).

Cambro-Ordovician rocks of the northern Appalachians were divided into a number of tectonostratigraphic zones by Williams (1979); from the northwest to the southeast they are the Humber, Dunnage, Gander, Avalon, and Meguma zones. In the Gaspé Peninsula, the Taconian deformed rocks belong to the Humber and Dunnage zones only (Fig. 2). The Baie Verte–Brompton Line (Williams and St-Julien, 1982) is the boundary between the two zones.

The Late Ordovician to Middle Devonian rocks deformed by the Acadian orogeny in the Gaspé Peninsula are divided into three major structural zones, from north to south: (1) the Connecticut Valley–Gaspé synclinorium, (2) the Aroostook-Percé anticlinorium, and (3) the Chaleurs Bay synclinorium (Fig. 2). Palinspastic reconstruction to pre-Acadian faulting (Bourque and others, 1990) demonstrates that the rocks of these three structural zones were part of a single depositional belt, referred to as the Gaspé Belt (Bourque and others, 1993), located mainly to the south of the Baie Verte–Brompton Line over the Dunnage Zone (Fig. 2).

Rocks of the Gaspé Belt in the Gaspé Peninsula can be viewed as four broad temporal and lithological packages (Figs. 3 and 4): (1) the *Honorat and Matapédia groups,* composed of Upper Ordovician–lowermost Silurian (Caradocian to Llandoverian) deep water fine-grained siliciclastic and carbonate facies and occurring chiefly but not exclusively in the Aroostook-Percé anticlinorium; (2) the *Chaleurs Group,* made up of Silurian-lowermost Devonian (Llandoverian to Lochkovian) shallow to deep shelf facies and occurring in both the Connecticut Valley–Gaspé and the Chaleurs Bay synclinoria; (3) the *Upper Gaspé Limestones and Fortin groups,* composed of Lower Devonian (Praguian-Emsian) mixed siliciclastic and carbonate fine-grained deep shelf and basin facies and occurring chiefly in the Connecticut Valley–Gaspé synclinorium; and (4) the *Gaspé Sandstones Group,* made up of upper Lower to Upper Devonian (Emsian to Frasnian) nearshore to terrestrial coarse-grained facies and occurring in the Connecticut Valley–Gaspé synclinorium, and its equivalent *La Garde and Pirate cove formations* of the Chaleurs Bay synclinorium; the younger (Frasnian) *Fleurant and Escuminac formations* of the Chaleurs Bay synclinorium are also included in this package.

Rocks of the Gaspé Belt and those of the Cambro-Ordovician are bounded by unconformities and faults. Undeformed Upper Devonian and Carboniferous strata in southern and eastern Gaspé unconformably overlie rocks of the Gaspé Belt as well as Cambro-Ordovician inliers (Fig. 2). Upper Ordovician to Middle Devonian rocks of the Gaspé Belt can therefore be bracketed between two major unconformities: the Taconian and the Acadian. The main deformation of the Gaspé Belt can be ascribed to the Acadian orogeny. However, the Salinic unconformity, well recorded in the Gaspé Belt sequence (see below), shows evidence that tectonic movements took place in the Gaspé Peninsula before Middle Devonian time and after the emplacement of the north-verging Taconian thrust sheets.

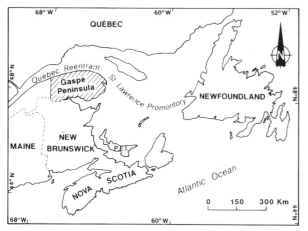

Figure 1. Map of southeastern Canada showing location of the Gaspé Peninsula.

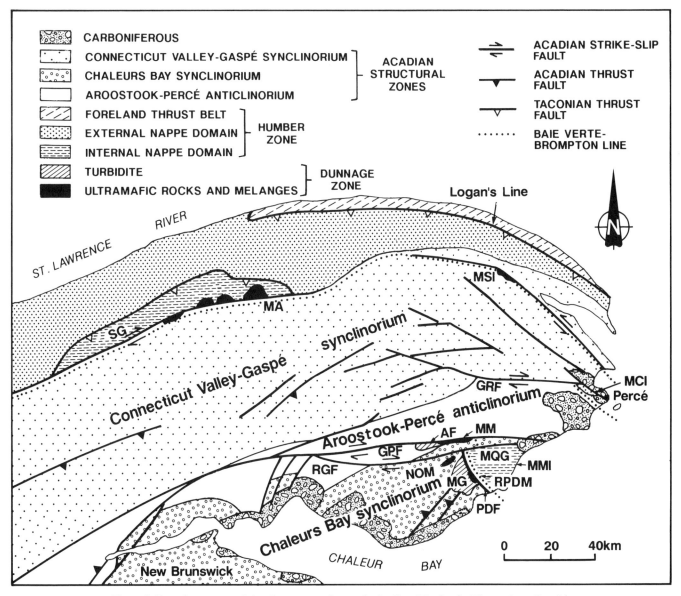

Figure 2. Taconian zones and Acadian structural zones in the Gaspé Peninsula. The northern Gaspé is from Slivitsky, St-Julien, and Lachambre (1991). AF: Arsenault Formation; GPF: Grand Pabos fault; GRF: Grande Rivière fault; MA: Mont Albert Complex; MCI: Murphy Creek inlier; MG: Mictaw Group; MM: McCrea mélange; MMI: Maquereau-Mictaw inlier; MQG: Maquereau Group; MSI: Mont Serpentine inlier; NOM: Nadeau ophiolitic Mélange; PDF: Port-Daniel fault; RGF: Rivière Garin fault; RPDM: Rivière Port-Daniel Mélange; SSF Shickshock-Sud fault; SG: Shickshock Group.

The Taconian and the Acadian unconformities have long been recognized in the Gaspé Peninsula (Skidmore, 1967; Béland, 1969), and the Taconian unconformity here was used as evidence for the Taconian orogeny throughout the northern Appalachians (Pavlides and others, 1968). The Late Ordovician to Middle Devonian evolution of the Gaspé Belt shows two distinct deformation events: the Late Silurian Salinic disturbance and the Middle Devonian Acadian orogeny. The Gaspé Peninsula is a key region for unraveling the evolution of the deformation events

that occurred during mid-Paleozoic times in the northern Appalachians.

The following analysis is based mainly on recent work (Bourque and others, 1990) on stratigraphy and structural geology of the Upper Ordovician to Middle Devonian rocks of the Gaspé Peninsula. Only a few recent studies dealing with volcanic rocks (Doyon and others, 1990; Bédard, 1986; Laurent and Bélanger, 1984) and plutonic rocks (Amireault and Valiquette, 1990; Whalen and Gariépy, 1986) are used to determine the

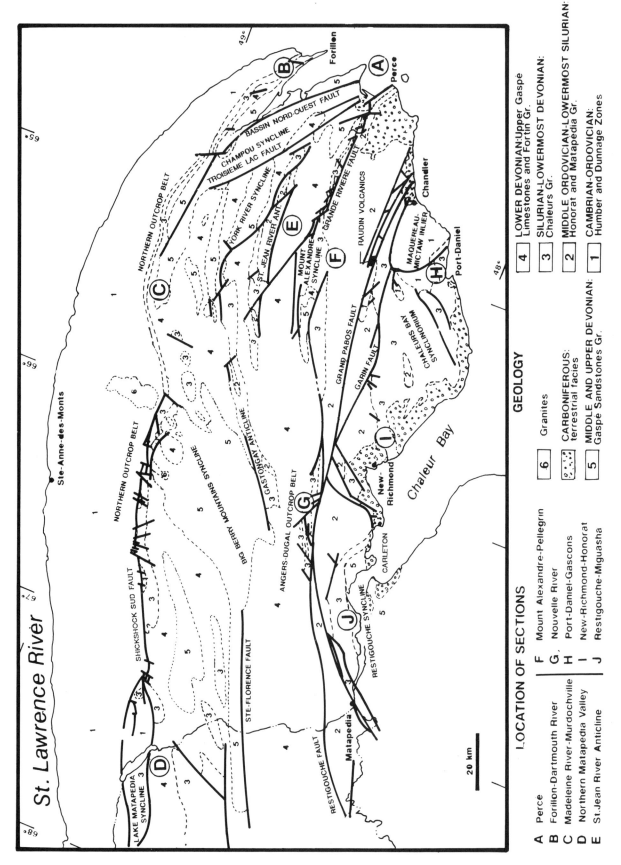

Figure 3. Geological map of Gaspé Belt in Gaspé Peninsula. Letters in circles refer to location of stratigraphic sections of Figure 4. From Bourque and others (1991).

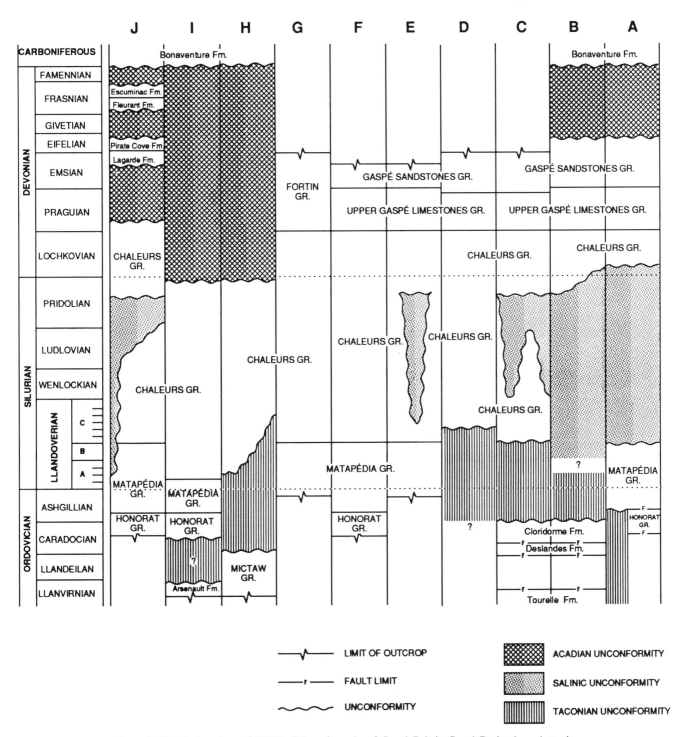

Figure 4. Correlation chart of Middle Paleozoic rocks of Gaspé Belt in Gaspé Peninsula and stratigraphic range of unconformities. Summarized from Bourque and others (1991). Letters at tops of columns refer to Figure 3.

tectonic setting of this part of the Appalachians during the Late Ordovician–Middle Devonian time interval. We should also mention that the regional metamorphic grade in the Gaspé Belt basin is very low (Hesse and Dalton, 1990). Illite crystallinity studies indicate anchimetamorphism grade for the Honorat and Matapédia groups in the Aroostook-Percé anticlinorium (Hesse and Dalton, 1990; Chagnon, 1988), whereas prehnite-pumpell-yite grade has been reported for Silurian volcanic rocks of the Chaleurs Group (Bédard, 1986).

CAMBRO-ORDOVICIAN BASEMENT

We will first briefly review the Cambro-Ordovician rocks of the Gaspé Peninsula, the Humber and Dunnage zones, because they constitute the immediate basement on which the sedimentation of the Gaspé Belt evolved. In the Gaspé Peninsula, the Humber and Dunnage zone are both allochthonous over the Grenville lower crustal block (Marillier and others, 1989).

The Humber zone

In the northern Gaspé, from north to south, the Humber Zone comprises three tectonic domains (Slivitsky and others, 1991; St-Julien and Hubert, 1975): (1) the parautochthon domain or the foreland thrust belt, (2) the external nappe domain, and (3) the internal nappe domain. The last two domains constitute the allochthon, thrusted to the northwest during the Taconian orogeny (Fig. 2). Logan's Line separates the nappe domains from the thrust-imbricated structures of the parautochthon.

In the southern Gaspé Peninsula, some other Taconian terranes are present in the Maquereau-Mictaw and the Murphy Creek inliers and along the Grand Pabos fault (Fig. 2). Rocks of the Murphy Creek inlier correlate with some of the external nappe domain of the Humber Zone (Kirkwood, 1989), whereas the Maquereau Group in the Maquereau-Mictaw inlier correlates with the Shickshock Group of the internal nappe domain of the Humber Zone (De Broucker, 1987).

The Baie Verte–Brompton Line

In Québec, the Baie Verte–Brompton Line is well defined in the Eastern Townships and in the Maquereau-Mictaw inlier of the southern Gaspé Peninsula. Between these two regions, the trace of the line is interpretative for most of its length under the cover rocks of the Gaspé Belt (Williams, 1979). Based on the presence of ultramafic bodies and mélanges, as well as on the reinterpretation of gravimetric and aeromagnetic data of the peninsula, the trace of the Baie Verte–Brompton Line is thought to follow the Shickshock-Sud and the Bassin Nord-Ouest faults and then is dextrally offset by at least three faults—the Grande Rivière, the Grand Pabos, and the Rivière Garin faults—and reappears in the Maquereau-Mictaw inlier (Fig. 2; De Broucker, 1987; Bourque and others, 1993; Malo and others, 1992). The Port-Daniel fault between the Maquereau Group to the northeast and the Mictaw Group to the southwest represents in part the Baie Verte–Brompton Line (De Broucker, 1987; Williams and St-Julien, 1982).

The Dunnage Zone

In the northern Gaspé Peninsula, most of the Dunnage Zone is concealed by the overlying Upper Ordovician to Middle Devonian rocks, except along the Shickshock-Sud fault and in the Mont Serpentine inlier (Fig. 2) where rocks that can be ascribed to the Dunnage Zone crop out. Klippes of ophiolitic rocks within the internal nappe domain of the Humber Zone constitute additional Dunnage Zone segments. The Mont Albert nappe, the upper thrust slice, is composed of alpine peridotite of the Mont Albert Complex (Fig. 2) and an unnamed amphibolite unit at the sole of the thrust (Beaudin, 1980), which was to the northwest of the Baie Verte–Brompton Line during the accretion of the Dunnage Zone to the Humber Zone.

In the southern Gaspé Peninsula (Fig. 2), the Mictaw Group of the Maquereau-Mictaw inlier and the Rivière Port-Daniel Mélange (De Broucker, 1987), just southwest of the Port-Daniel fault; the Nadeau ophiolitic Mélange (De Broucker, 1987); the Arsenault Formation; and the McCrea mélange along the Grand Pabos fault (Malo and Moritz, 1991) are considered to belong to the Dunnage Zone (Tremblay and others, 1993). All are remnants of the Iapetus ocean floor and of the fore-arc basin. The Dunnage Zone of the Gaspé Peninsula is equated with the Notre-Dame subzone of the Newfoundland Dunnage Zone (Tremblay and others, 1993).

STRATIGRAPHY OF THE GASPÉ BELT

The Aroostook-Percé anticlinorium

The Aroostook-Percé anticlinorium extends from Percé to Matapédia and across northwestern New Brunswick to Aroostook County, Maine. The northern boundary between rocks of the Aroostook-Percé anticlinorium and the Siluro-Devonian strata of the Connecticut Valley–Gaspé synclinorium is in most places a conformable contact (Fig. 3). Locally, as in the western part of the anticlinorium, the Restigouche fault juxtaposes rocks of the Matapédia Group with Devonian units of the Connecticut Valley–Gaspé synclinorium. Similarly, in the northeastern part of the anticlinorium, the Grande Rivière fault system separates the Connecticut Valley–Gaspé synclinorium and the Aroostook-Percé anticlinorium (Fig. 2). The southern limit of the Aroostook-Percé anticlinorium is also a conformable contact, locally faulted. In Québec, a major strike-slip fault, the Grand Pabos fault, crosses the Aroostook-Percé anticlinorium (Skidmore and McGerrigle, 1967) and merges with the Restigouche fault (Fig. 2).

The Murphy Creek inlier occurs at the eastern end of the Aroostook-Percé anticlinorium, in the Percé area (Fig. 2). On the southwestern side of the northwesterly trending inlier, the contact between the rocks of the anticlinorium and the Cambrian strata of the inlier is the Taconian unconformity; whereas on the northeastern side, it is a fault (Fig. 4, column A).

The Aroostook-Percé anticlinorium in the Gaspé Belt is composed of a lower siliciclastic assemblage (Honorat Group and part of the Matapédia Group) and an upper carbonate assemblage (upper part of the Matapédia Group) (Fig. 4). The lower siliciclastic assemblage is Late Ordovician in age, and the carbonate assemblage is mainly Early Silurian.

The 1,200 m of the Honorat Group (Garin Formation; Malo, 1988a) is made up of various chiefly terrigenous rocks: black claystone, noncalcareous gray mudstone, greenish-gray siltstone, calcareous siltstone, calcareous quartz-wacke, lithic wacke, conglomerate, and silty dolomitic limestone. The sandstones exhibit numerous sedimentary structures (graded bedding, parallel- and cross-laminations, load and flute casts).

The lower 1,400 m of the Matapédia Group (Pabos Formation of Kindle, 1936, redefined by Malo, 1988a) consists of laminated calcareous mudstone at its base, followed by a sequence of calcareous rocks: calcareous mudstone, argillaceous limestone, calcareous siltstone, silty limestone, calcareous sandstone, calcareous conglomerate, and sandy calcarenite and calcilutite. Siltstone and sandstone are thin to medium bedded and exhibit parallel- and cross-lamination as well as flute casts. Calcareous conglomerates are thick bedded and contain clasts composed of Pabos and Honorat lithologies together with exotic clasts of sericite and chlorite schists, milky quartz, and foliated sandstone.

The upper 880 m of the Matapédia Group (White Head Formation) is composed of thin-bedded gray or brown calcilutites with thiner interbeds of calcareous mudshales and a few thin beds of calcarenites, dark green calcareous mudstones, and thinly bedded silty and argillaceous limestone, calcareous shale and lenticular calcarenite with local limestone conglomerate, calcarenite, calcilutite, and sandy limestone in thin to very thick beds. Total thickness of both the Honorat and Matapédia groups is more than 4 km.

Dating of the Aroostook-Percé anticlinorium rocks units is based mainly on graptolites (Riva and Malo, 1988), brachiopods (Lespérance, 1968, 1974; Lespérance and Sheehan, 1976; Sheehan and Lespérance, 1981; Lespérance and Tripp, 1985; Lespérance and others, 1987), and conodonts (Nowlan, 1981, 1983, and in Malo, 1988b). Results are summarized in Figure 4.

There are two key areas to understanding the relationship between the rocks of the anticlinorium and the underlying Cambro-Ordovician basement: (1) the Percé area, and (2) the area south of the Grand Pabos fault, northwest of the Maquereau-Mictaw inlier. In the Percé area, the oldest unit of the anticlinorium, the Honorat Group, is absent on the southwestern side of the Murphy Creek inlier, and the Matapédia Group overlies unconformably the Cambrian Murphy Creek Formation, which correlates with rocks of the external nappe domain of the Humber Zone (Kirkwood, 1989). At the second area, the Llanvirnian Arsenault Formation, formerly the lower formation of the Honorat Group (Malo, 1988a), is exposed south of the Grand

Pabos fault and northwest of the Maquereau-Mictaw inlier (Fig. 2). The Arsenault Formation is composed mainly of thick-bedded, greenish-gray lithic wacke with black to olive-green claystone, minor gray to black fine-grained tuff interbeds, and some gabbroic sills. The Arsenault Formation can be geographically divided into an eastern Arsenault and a western Arsenault (Figs. 5 and 6).

In the eastern Arsenault area (Figs. 5 and 6), the Arsenault sits on the McCrea mélange and is in turn overlain unconformably by the Chaleurs Group. This situation is identical to that occurring 15 km to the southeast, where the Mictaw Group rests unconformably on the Rivière Port-Daniel Mélange and is in turn overlain unconformably by the Chaleurs Group. The Rivière Port-Daniel Mélange–Mictaw Group assemblage belongs to the Taconian Dunnage Zone. Owing to the compositional similarity between the McCrea and the Rivière Port-Daniel mélanges and the lithological equivalence between the flysch sequences of the Arsenault Formation and the Neckwick Formation of the Mictaw Group (De Broucker, 1987; Malo, 1988a), there are no reasons not to conclude that the McCrea mélange–Arsenault assemblage are part of the Dunnage Zone.

In the western Arsenault area (Figs. 5 and 6), the situation is

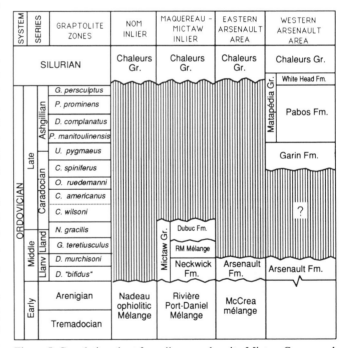

Figure 5. Correlation chart for mélange rock units, Mictaw Group, and the Arsenault Formation in the southeastern Gaspé Peninsula. NOM inlier: Nadeau ophiolitic Mélange inlier; RM Mélange: Rivière du Milieu Mélange. The graptolites zones are, from the top of the Ashgillian to the base of the Llanvirnian series: Glyptograptus persculptus, Paraclimacograptus prominens, Dicellograptus complanatus, Paraclimacograptus manitoulinensis, Uticagraptus pygmaeus, Climacograptus spiniferus, Ortograptus ruedemanni, Corynoides americanus, Climacograptus wilsoni, Nemagraptus gracilis, Glyptograptus teretiusculus, Didymograptus murchisoni, Didymograptus bifidus.

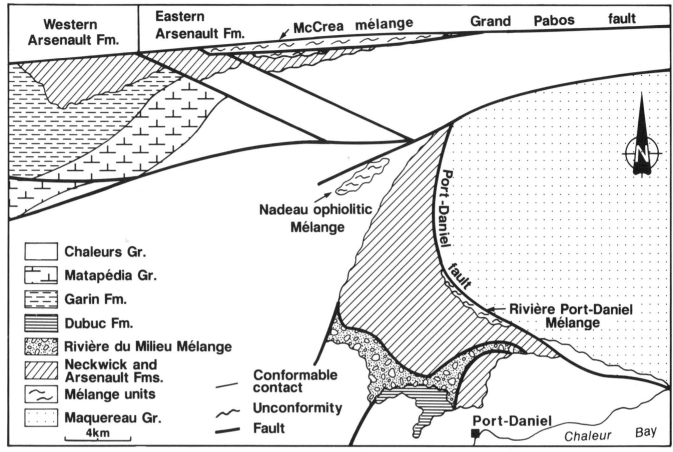

Figure 6. Localization of mélange rock units, the Mictaw Group, and the Arsenault Formation in the southeastern Gaspé Peninsula. From De Broucker (1987), Malo and Moritz (1991).

different. The McCrea mélange is not exposed, and the Arsenault Formation is there overlain by the Caradocian to Ashgillian Garin Formation of the Honorat Group, in turn conformably overlain by the complete Ashgillian to Llandoverian Matapédia Group and Silurian Chaleurs Group sequence. Field relationships show that, on the one hand, the contact between the Arsenault and the Garin is structurally concordant, but that, on the other hand, there is a hiatus between the two formations, since the age of the Arsenault is Llanvirnian and the oldest graptolite fauna recovered from the Garin is late Caradocian (*Climacograptus spiniferus* Zone; Riva and Malo, 1988). The Llandeilian and much of the Caradocian is therefore missing (Fig. 5). From the above it is concluded that: (1) the McCrea mélange and overlying Arsenault Formation are part of the Dunnage Zone of the Taconian basement; (2) a hiatus between the Arsenault and Garin formations in the western Arsenault area, as well as the unconformity between the Arsenault Formation and the Chaleurs Group in the eastern Arsenault area and the Mictaw and Chaleurs Group 15 km to the southeast, corresponds to the Taconian unconformity; and (3) therefore the Arsenault Formation would be better assigned to the Mictaw Group correlating to the Neck-

wick Formation. These two Middle Ordovician units represent together synorogenic turbidites in the fore-arc basin of the Taconian arc (De Broucker, 1987).

The Chaleurs Bay synclinorium

The Chaleurs Bay synclinorium (*sensu lato*) is located south of the Aroostook-Percé anticlinorium (Fig. 2). It is made of two main structures, the Chaleurs Bay synclinorium (*sensu stricto*) and the subsidiary smaller Restigouche syncline (Fig. 3). The boundary between the Aroostook-Percé anticlinorium and the Chaleurs Bay synclinorium (*s.l.*) is conformable, corresponding chiefly to the Matapédia–Chaleurs Group contact.

Of the four stratigraphic packages constituting the Gaspé Belt, three occur in the Chaleurs Bay synclinorium (*s.l.*) (Figs. 3 and 4): the Honorat-Matapédia Group package, the Chaleurs Group package, and the package composed of the Gaspé Sandstone Group equivalents and the younger Fleurant and Escuminac formations. However, the Silurian–lowermost Devonian Chaleurs Group package is predominant.

The Maquereau-Mictaw inlier occurs in the eastern end of

the Chaleurs Bay synclinorium (Fig. 2). The Chaleurs Group is either unconformable or faulted against the older rocks of the inlier. Much of the Taconian unconformity is between the Chaleurs Group and rocks of the Dunnage Zone (Mictaw Group) and also between the Chaleurs Group and rocks of the internal nappe domain of the Humber Zone, the Maquereau Group (Fig. 2). Carboniferous rocks rest unconformably on rocks of the Chaleurs Bay synclinorium (*s.l.*) in Chaleurs Bay.

The Middle Paleozoic rock sequence of the Chaleurs Bay synclinorium (*s.l.*) contains various sedimentary rocks, mainly siliciclastics with a few intercalated limestone units. Locally, felsic and mafic volcanics are abundant. Most sedimentary rocks of the synclinorium are fossiliferous, permitting reasonable biostratigraphic correlation.

In the eastern part of the Chaleurs Bay synclinorium (*s.s.*) (column H, Fig. 4), the sequence above the Mictaw Group (Middle Ordovician) and below the Bonaventure Formation (Carboniferous) belongs to the Chaleurs Group (Ami, 1990; Schuchert and Dart, 1926; Alcock, 1935; Northrop, 1939; Burk, 1964; Bourque, 1975; Bourque and Lachambre, 1980). In the western part of the synclinorium (New Richmond area), the Chaleurs Group has a conformable contact with the Matapédia Group.

The stratigraphy of the Chaleurs Bay synclinorium (*s.s.*) consists of three broad assemblages (Bourque and Lachambre, 1980; Bourque and others, 1986; Bourque and others, 1990): (1) a lower siliciclastic assemblage, ranging from fine to coarse grained (Clemville, Weir, and Anse Cascon formations); (2) a distinctive middle limestone assemblage (Anse à Pierre-Loiselle and La Vieille formations); and (3) an upper fine-grained siliciclastic *Zoophycos*-rich assemblage (Gascons and Indian Point formations) with intervening reef limestones and red beds (West Point Formation) and mafic lava flows (Black-Cape Volcanics). The overall thickness of the sequence ranges from 2,590 m in the eastern part of the synclinorium to 5,120 m in the western part.

All these lithostratigraphic units contain sufficient brachiopods (Boucot and Bourque, 1981) and some graptolites and conodonts to bracket their ages (Fig. 4). No Devonian rocks are known in the Chaleurs Bay synclinorium (*s.s.*).

Toward the west, however, in the Restigouche syncline (Fig. 3 and column J, Fig. 4), Devonian rocks occur above the Silurian sequence equivalent to that of the Chaleurs Bay synclinorium (*s.s.*). There, the Chaleurs Group is separated into two entities by a mappable unconformity (Bourque and Lachambre, 1980, map 1958). The succession between the Matapédia Group and the unconformity is composed of three rock assemblages. The lower assemblage (Mann Formation) consists of gray, structureless, fine-grained siliciclastites with intercalated thinly bedded parallel and convolute-laminated bioturbated fine-grained sandstones and of minor nodular calcilutites and calcarenites. It is locally rich in shallow shelf brachiopods. It correlates with the lower siliciclastic assemblage of the Chaleurs Group in the Chaleurs Bay synclinorium (*s.s.*). The overlying assemblage (Anse à Pierre-Loiselle and La Vieille formations) is the same limestone assemblage as in the Chaleurs Bay synclinorium (*s.s.*). It is composed of nodular lime-

stones, with locally abundant large smooth-shelled pentamerid brachiopods, and biohermal limestones. Mafic volcanic rocks (Restigouche Volcanics) occur within and above the limestone assemblage. They consist of basic lava flows, tuffs, and volcanic breccias. The part of the Chaleurs Group beneath the unconformity was affected by folding before Late Silurian erosion that locally has stripped off all units of the Chaleurs Group and part of the Matapédia Group. Beds under the unconformity are not younger than early Ludlovian.

The Chaleurs Group above the unconformity consists of two units. It starts with a 30- to 300-m-thick basal conglomerate (New Mills Formation; type-section in New Brunswick; correlation by Bourque and Gosselin, 1986) whose lithology is laterally variable. In the western part of the syncline, the group consists mainly of pebbles and boulders of dark mafic volcanic rocks in a volcanic sandy matrix; in the central part, pebbles and boulders (up to 1 m) are dark gray calcilutite, contained in red fine-grained sandstone. The limestone conglomerate forms very thick beds interlayered with red sandstones. Minor pink felsic volcanic rocks occur in the western part of the syncline. The conglomerate unit disappears toward the east. The second unit is a very thick sequence (Dalhousie Formation; type-section in the southern limb of the Restigouche syncline in New Brunswick) of sedimentary and volcanic rocks. In the western part of the syncline, 1,500 m of thin- to medium-bedded dark gray parallel-laminated argillaceous fine-grained sandstones interbedded with mudstones underlie approximately 5,000 m of mafic and minor felsic volcanic rocks. In the eastern part of the syncline, sedimentary units are more massive, muddier, and more calcareous and alternate regularly with the volcanic units. The volcanic rocks are vesicular and massive or brecciated lava flows and pyroclastic rocks varying in composition from basaltic to rhyolitic (Bélanger, 1982; Laurent and Bélanger, 1984). Andesites predominate. The pyroclastic rocks occur as thin layers of well-bedded pumice or scoria and ash deposits and as thick units of unsorted block and ash deposits. Consanguineous intrusive bodies are few and always of small size. Fossils of that unit in both Québec and New Brunswick indicate a Lochkovian-Praguian age interval (Boucot and Johnson, 1967; A. J. Boucot *in* Bourque and Lachambre, 1980).

At the western extremity of the Restigouche syncline (Fig. 3), a structural slice (Sellarsville slice; Bourque and Lachambre, 1980) contains sedimentary rocks of Ludlovian to Lochkovian age, consisting of approximately 4,000 m of mud-cracked red and green fine-grained siliciclastics, minor limestone conglomerates with marine fossils, and small stromatoporoid bioherms and biostromes on top of the sequence. The red beds, as well as the bioherms and biostromes, correspond in age and lithology to the reefs and red beds of the upper assemblage of the Chaleurs Group in the Chaleurs Bay synclinorium (*s.s.*). Above that 4,000-m-thick sequence are 500 m of massively bedded greenish-gray calcareous fine-grained siliciclastic rocks, locally rich in brachiopods. That 4,500-m-thick sequence of the Sellarsville slice is equivalent in time to the unconformity and the New Mills conglomerate Formation of the nearby Restigouche syncline, indicating the occurrence of local basins in which accelerated sedimentation took place in response to local uplift.

Above the Chaleurs Group, in the core of the Restigouche syncline, coarse-grained Emsian to Frasnian terrestrial deposits occur. They outcrop chiefly on the Québec side of Chaleurs Bay. Inland exposures are scarce and patchy. The sequence is divided into four units (Alcock, 1935; Williams and Dineley, 1966): the La Garde, Pirate Cove, Fleurant, and Escuminac formations (column J, Fig. 4).

The La Garde Formation (Alcock, 1935; Béland, 1958; Dineley and Williams, 1968) is believed to lie disconformably on the Chaleurs Group. It is composed of thick beds of well-rounded pebble and cobble conglomerates interlayered with gray or greenish-gray coarse- to fine-grained cross-bedded sandstone and a few mudstone beds. Clasts of Matapédia Group limestone were found in conglomerates in the eastern part of the outcrop area. The sequence coarsens upward and contains very few fossils. The thickness of the unit is estimated to be less than 1,500 m (Dineley and Williams, 1968; Rust, 1989). The conformably overlying Pirate Cove Formation (Alcock, 1935; Dineley and Williams, 1968; Zaitlin and Rust, 1983) consists predominantly of red sandy siltstone and mudstone. The most striking lithology is limestone conglomerate whose clasts were derived predominantly from the Matapédia Group outcrop area to the north (Fig. 3). Clasts of the underlying Devonian units also occur. Except for the occurrence of palynomorphs, the unit is unfossiliferous. The thickness of the formation was evaluated at about 500 m by Dineley and Williams (1968). The Frasnian Fleurant Formation (Alcock, 1935; Dineley and Williams, 1968; Zaitlin and Rust, 1983) lies unconformably upon the Pirate Cove Formation. It is a gray, well-rounded pebble and cobble conglomerate whose clasts are dominantly limestone, with less amounts of volcanics and sandstones and minor plutonic rocks (Rust, 1982). The unit is 18 m thick and unfossiliferous. The overlying Escuminac Formation rests conformably upon the Fleurant. It consists of greenish-gray thin- to thick-bedded sandstone, siltstone, and laminated shale. Sandstones and siltstones exhibit abundant sole marks as well as parallel lamination and current ripples. Fossil fishes and plants are common and have been the subject of several paleontological studies.

The Connecticut Valley–Gaspé synclinorium

In the Québec Appalachians, the Connecticut Valley–Gaspé synclinorium (Williams, 1978) extends from the eastern coast of the Gaspé Peninsula to the Québec–United States border in southern Québec. It lies between the Cambro-Ordovician allochthonous rocks of the Taconian orogen to the north and northwest and a number of Cambro-Ordovician anticlinoria to the south and southeast. In the Gaspé Peninsula, the northern contact of the synclinorium with older rocks is a fault in the west, the Shickshock-Sud fault (Fig. 2), and the Taconian unconformity in the east (Fig. 2). At the northeastern end of the Connecticut Valley–Gaspé synclinorium, the Chaleurs Group, which is Late

Silurian–Early Devonian in this area, rests unconformably on rocks of the external nappe domain of the Humber Zone. The southern limit of the synclinorium is either conformable or faulted against the Aroostook-Percé anticlinorium.

In western Gaspé, the Connecticut Valley–Gaspé synclinorium is bounded by the Shickshock-Sud fault, to the north, and by the Restigouche fault, to the south (Fig. 3). In this area, the Sainte-Florence fault divides two tectonostratigraphic domains: (1) a northern domain consisting of platformal limestones (Upper Gaspé Limestones) gently folded and lacking penetrative cleavage, and (2) a southern domain constituted by Fortin Group flysch folded in accordion style with an associated slaty cleavage. These two domains of the Connecticut Valley–Gaspé synclinorium are also recognized in the central Gaspé Peninsula but become unrecognizable in the eastern part of the peninsula. The southern domain is part of the Early Devonian "slate belt" of the Connecticut Valley–Gaspé synclinorium (Boucot, 1970).

The Connecticut Valley–Gaspé synclinorium includes fine to very coarse grained siliciclastic rocks, various types of limestones, felsic to mafic volcanic rocks, and intrusive rocks. The four stratigraphic packages previously described occur (Fig. 3), but the Devonian Upper Gaspé Limestones and Fortin groups predominate.

The more complete succession of the Chaleurs Group in the Connecticut Valley–Gaspé synclinorium occurs in the Mount Alexander–Pellegrin and Nouvelle River areas (Fig. 3 and columns F and G, Fig. 4), where the group rests conformably on the Matapédia Group of the Aroostook-Percé anticlinorium and is overlain—also conformably—by the Upper Gaspé Limestones or Fortin groups. Along the northern edge of the Connecticut Valley–Gaspé synclinorium (Northern Outcrop Belt and Lake Matapédia syncline; columns B, C, and D, Fig. 4), in the Percé area (column A, Fig. 4), and in the St. Jean River anticline (column E, Fig. 4), the Chaleurs Group succession is more or less complete, since the group lies unconformably on older rocks and since parts of it have been stripped off by local erosion during Ludlovian time. The Chaleurs Group is therefore best described in terms of two sequences: (1) a northern shallow water sequence, including the Northern Outcrop Belt and Matapédia syncline, the Percé area, and the eastern tip of the St. Jean River anticline; and (2) a southern deeper water sequence, comprising the western part of the St. Jean River anticline, the Gastonguay anticline, the Mount Alexander syncline, and the Angers-Dugal Outcrop Belt (Fig. 3).

The Chaleurs Group of the northern sequence (Bourque, 1975, 1977; Lachambre, 1987; Bourque and others, 1990) is composed predominantly of siliciclastic facies, from bottom to top: (1) a claystone (Awantjish and Burnt Jam Brook formations) or calcilutite (Sources Formation) horizon, (2) a thin quartzarenite unit (Val-Brillant Formation), (3) a nodular and biohermal limestone horizon (Sayabec Formation) correlating with the middle limestone assemblage of the Chaleurs Group in the Chaleurs Bay synclinorium, and (4) a thick fine-grained siliciclastic unit (Gascons, Indian Point, and Roncelles formations) correlating with the upper fine-grained siliciclastic assemblage of the Chaleurs Group in the Chaleurs Bay synclinorium. The last unit contains local reef limestone bodies (West Point Formation) and conglomerate lenses (Griffon Cove River and Owl Capes members). As in the Restigouche syncline, a mappable unconformity occurs within the Chaleurs Group. In the eastern part of the St. Jean River anticline (column E, Fig. 4), the unconformably is local and overlain by a distinctive thick-bedded coarse-grained petromictic conglomerate (Owl Capes Member). In the Madeleine River area (column C, Fig. 4), the erosion has stripped off variable portions of the Chaleurs Group. Reef limestone bodies (West Point Formation) reaching a thickness of up to 325 m occur above the unconformity (Bourque and others, 1986; Lachambre, 1987). In the easternmost part of the Northern Outcrop Belt (column B, Fig. 4), lenses of pebble and cobble petromictic conglomerate (Griffon Cove River Member) occur associated with mud-cracked red beds above the unconformity. This conglomerate–red bed association correlates in age and lithology with that of the Restigouche syncline above the unconformity (New Mills Conglomerate).

The Chaleurs Group of the southern sequence is more uniform in composition than that of the northern sequence. It is composed mostly of fine-grained siliciclastic rocks (Burnt Jam Brook and Saint-Léon formations), with the exceptions of a local thin lithoclastic and quartz sandstone and conglomerate unit (Laforce Formation) of Wenlockian–early Ludlovian age and mafic volcanic rock bodies (Lac McKay Volcanics) occurring in the upper portion of the group.

Thicknesses of the Chaleurs Group are variable. They range from a little more than 300 m to nearly 1,000 m from east to west in the northern sequence and up to 5,700 m in the southern sequence, if 3,000 m of thick volcanic rock bodies are included.

The Chaleurs Group of the Connecticut Valley–Gaspé synclinorium in Gaspé Peninsula is conformably overlain everywhere by the Upper Gaspé Limestones. These are in turn overlain by the Gaspé Sandstones, except for areas between the Sainte-Florence fault and the Gastonguay anticline to the north, between the Restigouche fault and the Angers-Dugal Belt to the south, and west of approximately longitude 66°W, where the Fortin Group is laterally equivalent to the Upper Gaspé Limestones and part of the Gaspé Sandstones (Bourque and others, 1993; Bourque, 1989a).

The Upper Gaspé Limestones Group is composed of three lithologic assemblages (Lespérance, 1980a, b; Rouillard, 1986): (1) a lower monotonous sequence of more or less shaly, dolomitic, and siliceous calcilutite or limy mudstone (Forillon Formation), which is more calcareous and more siliceous in the eastern part of the area; (2) a middle sequence of thinly to very thickly bedded siliceous and dolomitic limestone and mudstone, with minor calcarenite, sandstone, and bentonite beds (Shiphead Formation), which in the southern half of the area contains units of coarse-grained sandstone and mudstone with numerous slump structures, a lithology very similar to that of the Fortin Group of the Assémetquagan River area (Fig. 3); and (3) an upper homogeneous sequence of thin- to medium-bedded cherty to siliceous or silty calcilutite (Indian Cove Formation). Total thickness of the Upper Gaspé Limestones ranges from 710 m (maximum) in the easternmost Northern Outcrop Belt to 1,930 m in the vicinity of the St. Jean River anticline.

The Gaspé Sandstones of the Connecticut Valley–Gaspé synclinorium comprise four conformably superposed sandstone and conglomerate assemblages (McGerrigle, 1950). The first is a lower unit (York Lake Formation) of alternating siliceous calcilutites with minor quartz arenites and wackes and greenish-gray medium-grained feldspathic wackes, constituting a transition between the Upper Gaspé Limestones and Gaspé Sandstones groups. Next comes an overlying unit (York River Formation) composed of a lower mudstone-siltstone-sandstone assemblage with a few calcarenites and an upper sandstone assemblage with minor mudstones, the sandstones being medium- to thick-bedded, greenish-gray, medium- to coarse-grained, feldspathic, and lithic wackes with large-scale cross-bedding. The third unit (Battery Point Formation) consists at the base of a number of superposed fining-upward sequences of conglomeratic sandstone, medium- to coarse-grained sandstone, and minor siltstone and mudstone (Cant and Walker, 1976). These sequences are overlain by another assemblage of fining-upward sequences in which red mudstone and siltstone are abundant (Rust, 1981; Walker and Cant, 1979), which is overlain in turn by coarser sandstone with less clearly defined fining-upward sequences. The third unit ends with a red unit of sandstone, siltstone, and mudstone. The fourth—upper—unit (Malbaie Formation) is a very thick bedded conglomerate composed of pebbles and cobbles of limestone, siliciclastic, and volcanic rock fragments derived from the older Matapédia and Chaleurs groups, interbedded with medium- to coarse-grained red sandstone (Rust, 1976, 1981). The Gaspé Sandstones are unconformably overlain by Carboniferous rocks. Minimum total thickness of the Gaspé Sandstones in the eastern part of the Connecticut Valley–Gaspé synclinorium is 4,200 m.

The Fortin Group in the southwestern portion of the Connecticut Valley–Gaspé synclinorium in Gaspé Peninsula is a thick distinctive dark mudstone and fine- to medium-grained sandstone sequence (McGerrigle, 1946; Dalton, 1987; Hesse and Dalton, 1989). It conformably overlies the Chaleurs Group in the Nouvelle River area (column G, Fig. 4). The Fortin has a minimum thickness of 2,650 m in the Matapédia River Valley (Kirkwood and St-Julien, 1987).

SEDIMENTARY HISTORY OF GASPÉ BELT

The Upper Ordovician to Mid-Devonian rock succession of the Connecticut Valley–Gaspé synclinorium, the Aroostook-Percé anticlinorium, and the Chaleurs Bay synclinorium were part of a single depositional belt, the Gaspé Belt (Bourque and others, 1985; Kirkwood and others, 1988; Bourque and others, 1990; Bourque and others, 1993), that developed at the margin of the Québec Reentrant and the St. Lawrence Promontory. During Late Ordovician to Early Silurian time, the belt corresponded to an ocean basin that subsequently evolved as an unstable shelf (Fig. 7). We have shown above that (1) the lowest units of the succession (Honorat and Matapédia) occur mainly in the Aroostook-Percé anticlinorium, (2) units of the Chaleurs Group of both the Connecticut Valley–Gaspé and Chaleurs Bay synclinoria are easy to correlate, (3) the Devonian Upper Gaspé Limestones and Gaspé Sandstones groups chiefly occur in the Connecticut Valley–Gaspé synclinorium, and (4) unconformities occur at three distinct stratigraphic levels—at the end of Ordovician, the end of Silurian, and in the Middle Devonian times.

Basin filling

Sandstones of the Honorat Group and the lower siliciclastic part of the Matapédia Group commonly possess internal sedimentary structures typical of turbidites (Ducharme, 1979; Malo, 1988a). These sedimentary structures also characterize the silty limestones and calcarenites of the upper part of the Matapédia Group in the Gaspé Peninsula (Malo, 1988a). The equivalent units in adjacent New Brunswick, Grog Brook and Matapédia groups, also have sedimentary structures typical of turbidites

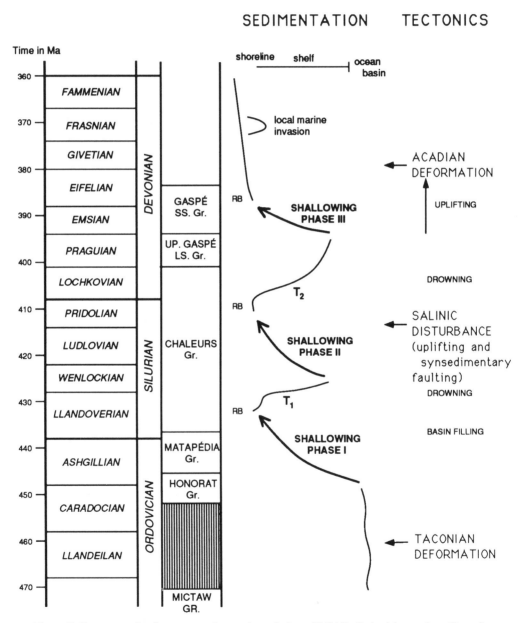

Figure 7. Summary of sedimentary and tectonic evolution of Middle Ordovician to Late Devonian sequence of Gaspé Belt in Gaspé Peninsula, Québec Appalachians. T_1 and T_2 are transgression phases. RB indicates occurrence of red beds. Minor local marine invasion occurred during Frasnian time. From Bourque and others (1991) and Bourque (1989b).

(Hamilton-Smith, 1972; St. Peter, 1977). The Late Ordovician–lowermost Silurian turbidites of the Aroostook-Percé anticlinorium occupied a relatively deep water marine basin (Fig. 7). This interpretation is supported by a general lack of shelly faunas (except for the Percé area), the presence of graptolites in the Honorat Group, and deep water trace fossil assemblages (Pickerill, 1980) in the Grog Brook and Matapédia groups of northern New Brunswick. A progressive change from deeper terrigenous (Honorat) to shallower limestone (Matapédia) facies reflects basin filling.

Following the basin infilling by fine-grained sediments and turbidites of the Honorat and Matapédia groups, the entire succession of the Chaleurs, Upper Gaspé Limestones, and Gaspé Sandstones groups is composed of three successive shallowing-upward phases (shallowing phases I, II, and III in Fig. 7), all three ending with nearshore and terrestrial facies separated by transgressive episodes (T_1 and T_2 in Fig. 7). The following is a summary of a more detailed sedimentary and vertical sequence analysis by Bourque and others (1990, 1993, 1989a).

Cyclic sedimentation

Following deposition of the lime turbidites of the Matapédia Group, shallowing phase I (Llandoverian–Early Wenlockian) was dominantly siliciclastic but ended with limestones. This phase is best exemplified by the superposition of the lower siliciclastic assemblage of the Chaleurs Bay synclinorium (*s.l.*) and the middle limestone assemblage of both the Chaleurs Bay and Connecticut Valley–Gaspé synclinoria. The facies evolved from terrigenous offshore muds with sandy storm layers, to lower shoreface sands (Bourque, 1981), and finally to peritidal limestones and subaerial facies (Bourque and others, 1986; Lavoie, 1988).

Following shallowing phase I, a transgressive episode (T_1) deposited deep shelf siliciclastic muds, represented by the upper fine-grained siliciclastic assemblage of the Chaleurs Group. In the middle part of that assemblage, reef limestone bodies and terrestrial red beds and conglomerate occur, testifying to the occurrence of a second shallowing phase (shallowing phase II). Several facies of the reefs recorded tectonic instability and in some cases developed in response to it (Bourque, 1979; Bourque and others, 1986). Terrestrial red beds and conglomerate are associated with uplifted areas (unconformity).

After that period of overall shallowing and local emergence, a second transgressive episode (T_2) resulted in a deepening-upward sequence represented by the fine-grained sediments and turbidites of the upper portion of the upper fine-grained siliciclastic assemblage of the Chaleurs Group and the fine-grained deep water limestones of the Upper Gaspé Limestones Group (Bourque, 1989b). In the Restigouche syncline, the second transgressive episode (T_2) was dominated by volcanism. In the eastern portion of the Connecticut Valley–Gaspé synclinorium, the Upper Gaspé Limestones has recorded a period of uniform sedimentation on a deep stable platform (Lespérance, 1980a). To the

southwest, the rather uniform and monotonous fine-grained, turbiditic, siliciclastic-dominated sediments of the Fortin Group suggest a deeper depositional setting (Dalton, 1987; Hesse and Dalton, 1989).

Shallowing-upward phase III is much better exposed in the Connecticut Valley–Gaspé synclinorium than in the Chaleurs Bay synclinorium. It is represented by the Gaspé Sandstones Group. It corresponds to a rapid shoaling trend that led from shallow marine facies to terrestrial facies upsection. Main facies are deltaic or estuarine environments (Mason, 1971; Sikander, 1975), distal braided stream deposits (Cant and Walker, 1976) probably periodically subjected to marine influx, meandering river deposits (Rust, 1981), and proximal braid-plain deposit (Rust, 1981).

In the Chaleurs Bay synclinorium, the only record left by shallowing phase III is the sands and gravels of the La Garde, Pirate Cove, Fleurant, and Escuminac formations that represent various aspects of terrestrial sedimentation: alluvial deposits (La Garde Formation; Rust, 1982); alluvial fan to flood plain deposits (Pirate Cove Formation; Dineley and Williams, 1968), gravelly flood plain deposits (Fleurant Formation; Zaitlin and Rust, 1983), and lake turbidites (Escuminac Formation; Dinely and Williams, 1968; Hesse and Sawh, 1982). In the case of the Escuminac Formation, its fossil fishes were commonly considered freshwater and the unit consequently interpreted as lacustrine deposits, but elements of the fauna have been described recently as brackish or marine (Schultze and Arsenault, 1985), implying possible local marine invasions (Fig. 7).

The unconformities

Three distinctive unconformities (Fig. 4) occur in the Middle Ordovician to Late Devonian sequence of the Gaspé Peninsula. The oldest unconformity is the Taconian unconformity discussed above (Figs. 4 and 5). The second unconformity is located within the Chaleurs Group. It is dated as Late Ludlovian–Early Pridolian and corresponds to the Salinic disturbance (Boucot, 1962). It is either an angular unconformity (Restigouche syncline in Chaleurs Bay synclinorium) or an erosional surface that in places cuts deeply into the underlying older Silurian and even Ordovician rocks (Northern Outcrop Belt in Connecticut Valley–Gaspé synclinorium). The third unconformity is angular and occurs between Middle or Late Devonian and Carboniferous rocks. It corresponds to the Acadian orogeny.

The shallowing phases occurred more or less in response to the three tectonic pulses recorded by the unconformities: the Taconian orogeny, the Salinic disturbance, and the Acadian orogeny. Figure 7 shows the main tectonic events with respect to sedimentary evolution of the Gaspé Belt. Shallowing phase I corresponds to infilling of the successor basin left after the Taconian orogeny, whereas shallowing phase II culminated with the Salinic disturbance. Shallowing phase III is a response to general uplifting of the basin that preceded the main Acadian deformation.

LATE ORDOVICIAN TO MID-DEVONIAN DEFORMATION EVENTS

The major structural trend of the Late Ordovician to Middle Devonian rocks of the Gaspé Belt is oriented roughly northeast-southwest. In the southeastern Gaspé, major easterly striking dextral strike-slip faults transect this trend, whereas northeasterly striking high-angle reverse faults are present in the western Gaspé (Fig. 8). The Connecticut Valley–Gaspé synclinorium has been interpreted as a fault-bounded, subsided trough that originated after the emplacement of the Taconian allochthon in northern Gaspé (Beaudin, 1980; Rodgers, 1970). This regime of normal faulting, which was probably active during the sedimentation of the whole Gaspé Belt, was followed by strike-slip movements along the same faults during the Acadian orogeny. Some Acadian faults may also have used older Taconian faults, and their history can be very complex.

Faults

Major faults of the Gaspé Peninsula are straight longitudinal lineaments easily visible on aerial photos and Landsat images. The faults follow corridors of high strain zone where the fabrics developed are indicative of a ductile-brittle regime of deformation (Kirkwood and Malo, 1991; Malo, 1987). In general, regional metamorphism and volcanism do not vary on approaching major faults. Locally, in the southeastern Gaspé, the Raudin volcanics of probably Devonian age are bounded by the major Grand Pabos fault and its subsidiary faults (Malo and Moritz, 1991). Mineral occurrences are also spatially associated with the Grand Pabos fault system (Malo and others, 1990; Malo and Moritz, 1991). Hydrothermal alteration, felsic dykes, and magmatic breccia are locally associated with mineralizations. The Gaspé Peninsula can be divided into three regions according to the dominant trend of Acadian faults. The western region, which comprises the western and central parts of the Connecticut Valley–Gaspé synclinorium, the western part of the Aroostook-Percé anticlinorium, and the Restigouche syncline, is characterized by northeasterly trending faults (Fig. 8, South-West, North-West, and Central rose diagrams). The southeastern region, which comprises the eastern part of the Aroostook-Percé anticlinorium and the Chaleurs Bay synclinorium (*s.s.*), is characterized by easterly trending faults (Fig. 8, South-East rose diagram). The northeastern region, which comprises the northeastern part of the Connecticut Valley–Gaspé

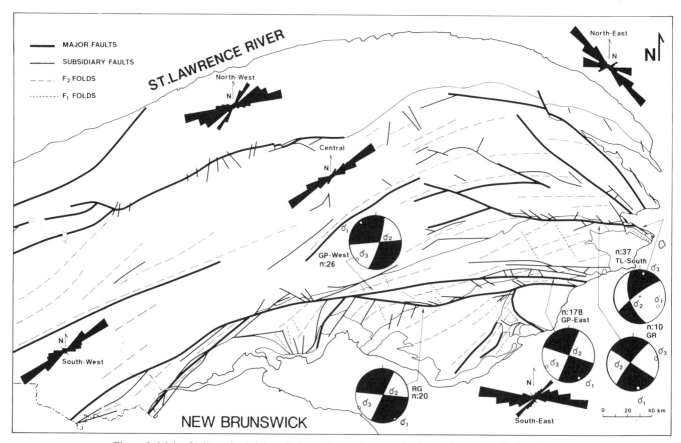

Figure 8. Major faults and subsidiary faults with rose diagrams for five regions, compilation of mesoscopic faults at five sites (GP: Grand Pabos, east and west; GR: Grande Rivière fault; RG: Rivière Garin fault; TL: Troisième Lac fault), and the traces of F_1 and F_2 folds. n: number of data of conjugate mesoscopic faults.

synclinorium, is characterized by northwesterly trending faults (Fig. 8, North-East rose diagram).

In the western region, the major fault trend is northeasterly (Fig. 8, North-West, South-West, and Central rose diagrams). These faults, subparallel to the regional folding, are thrust or high-angle reverse faults. Acadian thrusting is northwestward directed to the north of the Aroostook-Percé anticlinorium, whereas it is southeastward directed to the south (Fig. 2). In the middle part of the Connecticut Valley–Gaspé synclinorium (Fig. 8), the Sainte-Florence fault, a thrust to the northwest (Kirkwood and St-Julien, 1987), could represent the northeastern extension of the La Guadeloupe fault of the Québec Eastern Townships, which is a northwestward-directed Acadian thrust fault (Cousineau and Tremblay, this volume; Tremblay and others, 1989). In the central part of the Connecticut Valley–Gaspé synclinorium, the Marcil Nord fault, just south of the Big Berry syncline (Fig. 3), was also interpreted as a northwestward high-angle reverse fault (Carbonneau, 1959). Farther south, the Restigouche fault is a longitudinal fault that separates the Connecticut Valley–Gaspé synclinorium and the Aroostook-Percé anticlinorium. The Restigouche fault continues to the south in New Brunswick (St. Peter, 1977), where it trends north northeast compared with an east-northeast trend in the Matapédia area and an easterly one for its prolongation into the eastern Gaspé, the Grand Pabos fault. This different direction of the Grand Pabos–Restigouche fault suggests that the movement in its western end could be vertical as opposed to the horizontal strike-slip movement in its eastern end (see below). The Restigouche fault was described as a northwestward high-angle reverse fault in the Matapédia area (Lachance, 1979) and northern New Brunswick (St. Peter, 1977). In the Matapédia area, subvertical C-S fabrics in the fault zone indicate a dextral strike-slip component of movement, and the fault is there interpreted as a dextral oblique-slip fault (Kirkwood and St-Julien, 1987). In the southwestern Gaspé Peninsula (Fig. 8), the southern limit of the Aroostook-Percé anticlinorium is the Sellarsville fault, which separates the anticlinorium from the Restigouche syncline. This fault was considered as a southeastward thrust fault (Bourque and Lachambre, 1980). The study of mesoscopic brittle faults and shear-sense indicators along the Sellarsville fault zone confirms that this fault is a high-angle reverse fault with a movement to the southeast (Trudel, 1990). We should also note that, farther east, other southeastward-directed high-angle reverse faults are present in the Chaleurs Bay synclinorium (*s.s.*) (Fig. 2; De Broucker, 1987). The change in polarity of Acadian thrusting, from northwestward directed in the north to southeastward directed in the south, remains to be explained.

The Shickshock-Sud, which is the northern limit of the Gaspé Belt, is an example of a complex fault that was active during both the Taconian and Acadian orogenies. The movement along the Shickshock-Sud fault is interpreted as a succession of distinct normal and inverse movement during the Taconian and Acadian orogenies (Beaudin, 1980). Berger (1985) and Lebel (1985) have shown that the fault had a dextral strike-slip component of movement in Acadian time. Field observations along the Shickshock-Sud fault, where mélange-type rock units of the Dunnage Zone crop out (Fig. 2), help to understand the history of the fault. A vertical stretching lineation is observed in the Cambro-Ordovician rocks on the north side of the fault, but there is no such lineation in the Siluro-Devonian rocks on the south side of the fault. Vertical C-S fabrics indicating a horizontal dextral strike-slip movement are present on both sides of the fault (St-Julien and others, 1990). The stretching lineation in Cambro-Ordovician rocks is related to a Taconian thrust movement, whereas the C-S fabrics are associated with an Acadian strike-slip movement.

In the southeastern region, the major faults are the easterly trending Grand Pabos, Grande Rivière, and Rivière Garin faults (Fig. 8, South-East rose diagram). These long straight faults represent partially the boundaries of the Aroostook-Percé anticlinorium in its eastern end. They follow bands of intense deformation corresponding to zones of high internal strain that contrast with the mildly to moderately deformed intervals of the anticlinorium that separate them. The Grand Pabos fault is the most important of the Acadian fault system and can be traced over a distance of 140 km, from Chaleurs Bay in the east to a point northwest of Carleton (Fig. 3), where it joins the Restigouche fault and the southeast-verging Sellarsville thrust fault. The Grand Pabos–Restigouche fault is 225 km long in Québec and can be followed south in New Brunswick (St. Peter, 1977). A kinematic study of the Grand Pabos fault zone shows that the fault is a horizontal dextral strike-slip fault of Acadian age (Kirkwood and Malo, 1991; Malo, 1987). Offset of stratigraphic markers along the three major faults of the southern Gaspé indicates an apparent dextral displacement of 85 km for the Grand Pabos fault, 22 km for the Grande Rivière fault, and 10 km for the Rivière Garin fault (Malo and Béland, 1989). Since the markers are subvertical, apparent horizontal offsets are probably the true displacements. The combined right-lateral brittle displacement along these three faults is 117 km, to which can be added 38 km of ductile movement along the fault zones themselves and in the intervals between the faults for a total of 155 km (Malo and Béland, 1989). Subsidiary faults are associated with these three faults. the sense of movement observed on these subsidiary faults, as well as their angular relationships with the master east-west faults, shows that they are riedel shears compatible with major dextral easterly trending strike-slip faults (see, for example, the low- and high-angle faults in the area along the Grand Pabos north of the Maquereau-Mictaw inlier, Fig. 8). Conjugate mesoscopic faults observed in the fault zones of the southeastern region faults (Fig. 8, GP-West, GP-East, RG, and GR stereograms) are vertical as is the intermediate stress σ_2, which is typical of vertical strike-slip faults.

The northeastern region, the eastern Connecticut Valley–Gaspé synclinorium, is characterized by northwesterly striking faults (Fig. 8, North-East rose diagram). The most important ones are the Bassin Nord-Ouest, the Troisième Lac, and the Gastonguay faults. These are Acadian dextral strike-slip faults (Béland, 1980), probably corresponding to the last Acadian movement.

For instance, in the Percé area, it has been shown that the southern extension of the Troisième Lac fault crosscuts easterly striking faults associated with the Grande Rivière fault (Kirkwood, 1989), which is one of the major Acadian fault (Malo and Béland, 1989). These northwesterly striking faults are probably older than late Acadian and are thought to have been active during sedimentation of the Silurian and Lower Devonian facies of the Gaspé Belt. The lines of evidence pointing to that conclusion are: (1) marked change in thickness of the formations of the Chaleurs Group observed on either side of a line running from the Gastonguay fault straightforward to the northwest to the Northern Outcrop Belt (Bourque and others, 1990); (2) platformal affinities of the facies northeast of that line, as compared to basinal affinities of those southwest of that line (Bourque and others, 1990); and (3) significant wedging of the Silurian–Lower Devonian facies as observed along the seismic profiles that crosscut the Bassin Nord-Ouest and Troisième Lac faults, indicating synsedimentary block rotations along these faults (Roksandic and Granger, 1981; Société québécoise d'initiative pétrolière (SOQUIP), unpublished data; P. A. Bourque, work in progress).

Folds

Two phases of folding are recorded in the Aroostook-Percé anticlinorium (Malo and Béland, 1989). A mild northwesterly trending F_1 folding, with no penetrative cleavage, precedes the northeasterly trending F_2 folding accompanied by a well-developed cleavage (S_2). The result of these two successive phases of folding produced a dome and basin interference pattern that is well illustrated in the area north of Chandler where the core of the anticlinorium is occupied by a large dome of the oldest rocks of the Gaspé Belt, the Honorat Group (Malo, 1988a, Fig. 5; Malo and Béland, 1989, Fig. 2). In the eastern part of the Aroostook-Percé anticlinorium, F_1 folds are large open structures with wavelengths of 15 to 20 km (Malo, 1986), whereas in the Matapédia they are upright to slightly inclined, open to tight, with wavelengths ranging from 2 to 6 km (Théberge, 1979). The fact that the S_2 cleavage remains unaffected by this folding clearly indicates the succession of events. The transverse F_1 folding is observed throughout the Aroostook-Percé anticlinorium (Kirkwood, 1989; Gauthier, 1986; Malo, 1988b; Simard, 1986; Théberge, 1979; Vennat, 1979). In the Restigouche syncline, earlier or pre-F_2 folds of indeterminate trend occur in the Chaleurs Group under the Salinic unconformity. Since northwesterly trending F_1 folds affected the Chaleurs Group (Simard, 1986) in the Carleton area, which is the eastern part of the Restigouche syncline, it is believed that the earlier folds of the Restigouche syncline belong to the F_1 folding episode.

Still in the Aroostook-Percé anticlinorium, the northeast-southwest–striking F_2 folds are doubly plunging as a result of superposition on F_1 folds. F_2 folds are open and upright in the core of the anticlinorium but tight, upright, and inclined to the northeast or the southwest near the major faults. The northeasterly trending F_2 folds are oblique to the major east-west

strike-slip faults. The coeval regional northeast-southwest–striking S_2 cleavage, possibly a transecting cleavage (Powell, 1974), is well developed throughout the Aroostook-Percé anticlinorium and is usually rotated toward an easterly direction in structural domains that border the Grande Rivière and Grand Pabos faults (Malo and Béland, 1989). Within zones bordering the major faults, the S_2 cleavage is deformed by late S_3 riedel-shear fabrics in the semibrittle to brittle deformation zone, whereas some C-S fabrics and shear bands are well developed in the ductile deformation zone. In the Aroostook-Percé anticlinorium, Acadian structural elements pertaining to the D_2 deformation, rotated oblique F_2 folding, and S_2 cleavage as well as to the major east-west striking faults, their subsidiary riedel-type faults, and the late shear fabrics can be integrated into a model of strike-slip tectonics (Malo and Béland, 1989).

Outside the Aroostook-Percé, the most important folds are F_2; F_1 folds are not recognized in the Connecticut Valley–Gaspé synclinorium (Fig. 8). In most of this latter Acadian structural zone, except in the area south of the Sainte-Florence fault, F_2 folds are large, low-amplitude, open, and upright, trending northeasterly. South of the largest syncline, the Big Berry Mountains syncline (Fig. 3), the Gastonguay anticline, which is bounded by faults, is tighter and inclined to the northwest. The Gaspé part of the Devonian slate belt, south of the Sainte-Florence fault and the Gastonguay anticline, is structurally characterized by folds having a well-developed slaty cleavage. These folds are northeasterly trending, gently plunging, open to tight, and upright.

F_2 folds of the Chaleurs Bay synclinorium are open to moderate northeast-southwest striking with subhorizontal hinge lines. Axial-plane cleavage is not a penetrative fabric in the rocks of the Chaleurs Group within the synclinorium, except for the northeasternmost tip of the Chaleurs Bay synclinorium, south of the Grand Pabos fault and north of the Maquereau-Mictaw inlier (Fig. 3), where a strong penetrative cleavage is well developed.

Faulting and folding relationship

Northwest-southeast–striking F_1 folds are not developed throughout the Gaspé Belt, but they are recognized mainly in the Aroostook-Percé anticlinorium. The northeast-southwest–striking F_2 folding corresponds to the major northeasterly structural trend of the Gaspé Belt. The D_2 deformation is the major Middle Devonian Acadian event in the Gaspé Belt.

F_2 folds are generally open and upright, northeasterly trending, doubly plunging in the Aroostook-Percé anticlinorium and have subhorizontal hinge lines in the Connecticut Valley–Gaspé and Chaleurs Bay synclinoria. In the northeastern part of the Connecticut Valley–Gaspé synclinorium, the general northeasterly trend varies from east-northeast to east (the Saint-Jean River anticline for example; Fig. 3). Near the major faults, F_2 folds are tighter and inclined, and they have a rotation from northeast to east-northeast like the regional S_2 cleavage (Malo and Béland, 1989). This rotation is compatible with a dextral sense of shear along these faults. Major F_2 folds are usually oblique to major east-west–striking faults or parallel to some northeasterly trending faults (Fig. 5). The obliquity to major east-west–striking faults suggests that the folds possibly originate from the dextral strike-slip movement along the faults. Since the D_2 deformation implies horizontal shortening (F_2 folds, S_2 cleavage, and high-angle reverse faults) across the Gaspé Belt accompanied by transcurrent motion along major strike-slip faults, it is believed that the D_2 deformation is the result of a transpressional regime (Sanderson and Marchini, 1984) during Middle Devonian time. Transpressive deformation is also recognized elsewhere in the northern Appalachians, particularly in New Brunswick. For example, in northern New Brunswick, Late Silurian and/or Early Devonian structural evolution of the Dunnage Zone and Chaleurs Bay synclinorium is related in part to a dextral transpressive deformation associated with the Rocky Brook–Millstream and Catamaran faults (van Staal and Fyffe, 1993). Along the Bay of Fundy in southern New Brunswick, Carboniferous sedimentation and structure are explained in term of dextral transpression along the western part of the Cobequid-Chedabucto fault system (Nance, 1987).

DISCUSSION

Basement of the Gaspé Belt

The depositional Gaspé Belt developed as a successor basin over a Cambro-Ordovician Taconian basement after the Taconian orogeny, at the margin of the Québec Reentrant and St. Lawrence Promontory (Bourque and others, 1990; Bourque and others, 1993; Bourque, 1989a). The contact between the strata of the Gaspé Belt and the older Cambro-Ordovician rocks is either faults or the Taconian unconformity.

Pre-Acadian palinspastic reconstruction (Bourque and others, 1990; Kirkwood and others, 1988) that takes into account lateral displacements along the Acadian strike-slip faults of the Gaspé Peninsula shows that the Baie Verte–Brompton Line was a continuous suture line that parallels the Québec Reentrant and St. Lawrence promontory margin (Fig. 9). This reconstruction shows also that the Cambrian strata of the Murphy Creek inlier and rocks of the Maquereau Group of the Maquereau-Mictaw inlier are part of the same belt, the Humber Zone, north of the Baie

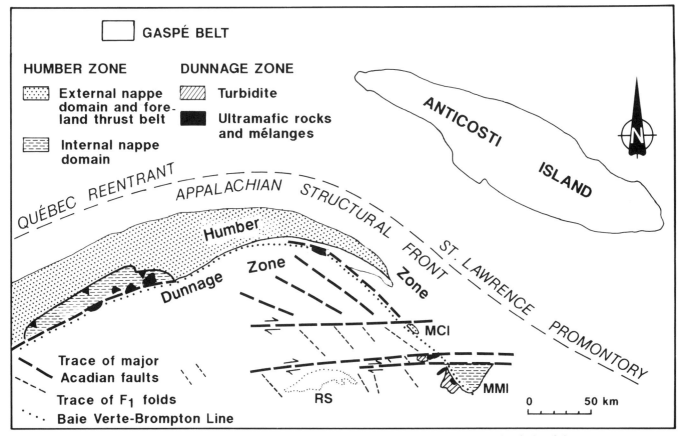

Figure 9. Pre-D_2 palinspastic position of F_1 folds and traces of the northwest-trending faults of the northeastern part of the Connecticut Valley–Gaspé synclinorium and of the major dextral strike-slip faults of the southeastern Gaspé region. MCI: Murphy Creek inlier, MMI: Maquereau-Mictaw inlier; RS: Restigouche syncline. Palinspastic base map from Bourque and others (1991) and Kirkwood and others (1988).

Verte–Brompton Line (Fig. 9). In the Gaspé Peninsula, the Taconian unconformity is observed over both the Humber and Dunnage zones (Fig. 2), but the Gaspé Belt evolved mostly over the Dunnage Zone south of the Baie Verte–Brompton Line (Fig. 9).

In the Gaspé Peninsula, the Dunnage Zone probably forms a continuous basement under the Gaspé Belt cover rocks and is in turn underlain by the Grenville lower crustal block (Marillier and others, 1989). In adjacent northern New Brunswick where the Gaspé Belt extends (Bourque and others, 1990), the situation is very much like that of the Gaspé Peninsula: The Middle Ordovician volcanic and sedimentary rocks of the Dunnage Zone constitute the basement of the Aroostook-Percé anticlinorium and the Chaleurs Bay synclinorium (van Staal and Fyffe, 1993). In the Popelogan anticline (Ruitenberg and others, 1977), a hiatus similar to the one between the Arsenault and the Garin formations exists between the Middle Ordovician Balmoral Group and the Late Ordovician Grog Brook Group (van Staal and Fyffe, 1993). The Grog Brook Group, equivalent to the Garin Formation (Malo, 1988a), is the basal unit of the Aroostook-Percé anticlinorium in northern New Brunswick, whereas the Balmoral Group represents the Popelogan subzone of the New Brunswick Dunnage Zone (van Staal and Fyffe, 1993).

Timing and tectonic setting of the deformation events

A hiatus between Middle Ordovician and Upper Ordovician rocks (Arsenault-Garin in the western Arsenault area and Balmoral–Grog Brook in the Popelogan anticline), an unconformity between the Llanvirnian Arsenault and the lowermost Silurian Chaleurs Group, and an unconformity between the Middle Ordovician Mictaw Group and lowermost Silurian Chaleurs Group (Fig. 6) all show evidence of uplift, erosion, and tectonic movement during late Middle Ordovician to early Late Ordovician times before the onset of the sedimentation of the Gaspé Belt in Late Ordovician. The late Caradocian to Ashgillian Garin Formation (Honorat Group) should be considered as the basal unit of the Gaspé Belt over rocks of the Dunnage Zone. The oldest rocks of the Garin contain graptolites of the *Climacograptus spiniferus* Zone, which is the age of the younger rocks of the Cloridorme Formation of the Humber Zone (Riva and Malo, 1988). Although the Acadian deformation had some influence on the Cambro-Ordovician Humber Zone (Beutner, 1989) (for example, some Acadian faults crosscut the Humber Zone in northern Gaspé (Bourque and others, 1990), the main deformation in the Humber Zone is related to the Taconian orogeny (Slivitsky and others, 1991). The last Taconian nappe emplacement occurred in post-Caradocian time, since the parautochthon represented by the Cloridorme Formation is overlain by thrust sheets of the external and internal domains. To the south of the Baie Verte–Brompton Line, the main deformation of the Cambro-Ordovician Dunnage Zone in the Maquereau-Mictaw inlier—the northeasterly trending structures, folds, and associated penetrative cleavage—is related to the same Acadian D_2 deformation that affects the Gaspé Belt. However, northwesterly trending pre-F_2

folds are recognized in the Mictaw Group (De Broucker, 1987). The angular unconformity between Silurian strata of the Chaleurs Group and the northwesterly folded Middle Ordovician Mictaw Group implies folding during Late Ordovician time. Deformation related to these northwesterly folds is a late Taconian event probably contemporaneous with the last nappe emplacement in the Humber Zone (De Broucker, 1987).

The timing and the cause of the northwesterly F_1 folds of the Gaspé Belt are not well understood. They are recognized mainly in the Late Ordovician–lowermost Silurian rocks of the Honorat and Matapédia groups in the Aroostook-Percé anticlinorium. The earlier folds recognized in Silurian rocks of the Chaleurs Group under the Salinic unconformity in the Restigouche syncline are probably related to the same F_1 folding event. The Honorat-Matapédia-Chaleurs groups represent a continuous sequence, and the first significant break in the sedimentation is the Ludlovian-Pridolian unconformity; the F_1 folds in the lower Gaspé Belt and the earlier folds in the Restigouche syncline can be ascribed to a folding event associated with the Salinic disturbance. The pre-Acadian palinspastic reconstruction of the Gaspé Belt is in fact a pre-D_2 palinspastic reconstruction (Kirkwood and others, 1988). F_1 folds have preceded F_2 folds as well as the motion along the major strike-slip faults. The observed F_1 folds on a pre-D_2 map are located in the eastern part of the Québec Reentrant, along the eastern limit of the Gaspé Belt (Fig. 9), which is also the location of the northwesterly trending faults of the eastern part of the Connecticut Valley–Gaspé synclinorium that were active during the sedimentation of the Gaspé Belt. Block-faulting tectonics probably affected that part of the basin in Late Silurian to Early Devonian times (Bourque, 1990). The large F_1 folds could be extensional drape folds in the response of this block faulting, the lower part of the sequence (Honorat-Matapédia-Chaleurs) being passively folded over large blocks. It is also not impossible that a local northeast-southwest–oriented compressional event affected this part of the Gaspé Belt during Late Silurian to Early Devonian times. The Silurian tectonism during the sedimentation of the Gaspé Belt and the Salinic disturbance in the Gaspé Peninsula could also be related to the closure of Iapetus II, to the south, in northern New Brunswick (van der Pluijm and van Staal, 1988). According to van Staal (1987) a marginal basin, Iapetus II, opened in Lower-Middle Ordovician times in the Dunnage Zone of northern New Brunswick and closed in Late Ordovician–Silurian times (van der Pluijm and van Staal, 1988).

The main compressional event that occurred later in post–Middle Devonian and pre-Carboniferous time is referred to as the Acadian orogeny. In the northern Appalachians, the Acadian orogeny is related to continued continental convergence after the collision between the Taconian-arc and the North American craton in the Middle to Late Ordovician (Williams and Hatcher, 1983; Keppie, 1985). It should be pointed out that the collision was along an irregular continental margin (Stockmal and others, 1987, 1990), of which the Québec Reentrant–St. Lawrence Promontory represents a major irregularity. The Acadian collision induced deformation characterized by west-directed thrusting and faulting on the northeastern side of the promontory in Newfoundland (Cawood and Williams, 1988) as well as in the southwestern part of the Québec Appalachians (Tremblay and others, 1989; Cousineau and Tremblay, this volume). In the Gaspé Peninsula, between these two latter regions on the southwestern side of the promontory, the shortening caused by the Acadian collision was first achieved by F_2, the development of the S_2 regional cleavage, and some northeast-southwest–striking high-angle reverse faults. This deformation in the post-Taconian cover rocks was synchronous with the initiation of the strike-slip faulting in the basement. A second stage of deformation is expressed by the deformation structures typical of a brittle-ductile regime of deformation (C-S fabric, shear band, riedel-shear) along the fault zones and the rotation of F_2 folds and its associated S_2 cleavage in the high strain zone. Displacement along the master east-west–striking fault was the subsequent event, followed by the late movement along the northwest-southeast–striking faults in the northeastern part of the belt. The succession of events for the D_2 deformation is similar to the one for wrench tectonics models (see Wilcox and others, 1973), where the weaker cover rocks are detached from the mechanically stronger basement. Strike-slip faults in the northernmost part of the Quebec Reentrant have accommodated late regional shortening during a period of lateral tectonic escape that was favored by the detachment of the cover rocks of the Gaspé Belt.

Acadian strike-slip faults of the Gaspé Peninsula are transcurrent faults in an intraplate tectonic environment, according to Sylvester's (1988) classification. These transcurrent faults were formed in the foreland of the Acadian mountain belt, the Gaspé Peninsula region, in response to collision of the outboard terranes to the south. These faults could also be related to a major transform fault along the St. Lawrence Promontory (Stockmal and others, 1987). The volcanic rocks within the Gaspé Belt are interbedded with shallow marine deposits on top of the continental crust (Laurent and Dostal, 1990). They were generated in an intraplate tectonic environment, unrelated to subduction, and no trench environments appear to have existed in the Gaspé region during Siluro-Devonian time (Laurent and Dostal, 1990). A transtensional episode during the sedimentation of the Gaspé Belt may have favored this volcanism (Laurent and Dostal, 1990; Bédard, 1986). Block faulting, the Salinic unconformity, and other deformation events pertaining to the D_1 deformation could be related to this transtensional regime in Late Silurian time, which was followed by a transpressional regime in Middle Devonian time. In the Gaspé Peninsula, the deformation was roughly continuous from Middle Ordovician to Middle Devonian times or from the Taconian orogeny to the end of the Acadian orogeny. The deformation events were diachronous, and it is sometimes difficult to separate the late Taconian features from the earliest Acadian ones.

The Acadian fault system of the Gaspé Peninsula was probably reactivated during post-Carboniferous time, since the Grand Pabos fault affects Carboniferous strata at its east end (Ayrton, 1967). This post-Carboniferous movement is not important com-

pared to the 85-km dextral displacement of pre-Carboniferous rocks to the west. In general, the Carboniferous rocks of the Chaleur Bay are horizontal and undisturbed. However, in the Carleton and Percé areas, some minor normal faults affect the Carboniferous rocks (Bernard and St-Julien, 1986; Kirkwood, 1989).

ACKNOWLEDGMENTS

We wish to acknowledge the Ministère de l'Energie et des Ressources du Québec, which defrayed most of the field expenses and permitted the use of the data. We benefited from discussions with Jacques Béland, Daniel Brisebois, Donna Kirkwood, and Pierre St-Julien. The chapter benefited from reviews by Brian H. O'Brien and Walter E. Trzcienski, Jr. The Natural Sciences and Engineering Research Council of Canada is acknowledged for operating grants to M. Malo (GP1908) and P. A. Bourque (A9142). Thanks are also due to Yvon Houde and Luce Dubé for drafting the figures.

REFERENCES CITED

Alcock, F. J., 1935, Geology of Chaleurs Bay region: Geological Survey of Canada Memoir 183, 146 p.

Ami, H. M., 1900, Synopsis of the geology of Canada (being a summary of the principal terms employed in the Canadian geological nomenclature): Transactions Royal Society of Canada, Ser. 2, v. 4, p. 187–225.

Amireault, S., and Valiquette, G., 1990, Comparaisons géochimiques des suites felsiques en Gaspésie, *in* Malo, M., Lavoie, D.,and Kirkwood, D., eds., Program with abstracts of the Québec–Maine–New Brunswick Appalachian workshop: Geological Survey of Canada Open-file 2235, p. 72–76.

Ayrton, W. G., 1967, Chandler–Port-Daniel area: Québec, Department of Natural Resources Geological Report 120, 97 p.

Beaudin, J., 1980, Région du Mont-Albert et du Lac Matapédia: Ministère de l'Energie et des Ressources du Québec Rapport intérimaire DPV-705, 83 p.

Bédard, J. H., 1986, Pre-Acadian magmatic suites of the southeastern Gaspé Peninsula: Geological Society of America Bulletin, v. 97, p. 1177–1191.

Béland, J., 1958, Oak Bay map-area: Québec Department of Mines Preliminary Report 375, 12 p.

—— , 1969, The geology of Gaspé: Canadian Mining and Metallurgical Bulletin, v. 62, p. 811–818.

—— , 1980, Faille du Bassin Nord-Ouest et faille du Troisième Lac dans la partie est de la Gaspésie: Ministère de l'Energie et des Ressources du Québec Rapport intérimaire DP-740, 20 p.

Bélanger, J., 1982, Roches volcaniques dévoniennes de la bande de Restigouche: Ministère de l'Energie et des Ressources du Québec Rapport intérimaire DP-939, 13 p.

Berger, J., 1985, Analyse structurale de la faille Shickshock-Sud en Gaspésie occidentale, Québec [M.Sc. thesis]: Montréal, Université de Montréal, 29 p.

Bernard, D., and St-Julien, P., 1986, Analyse structurale du Siluro-Dévonien du centre de la Gaspésie et du Carbonifère du sud et de l'est de la Gaspésie: Ministère de l'Energie et des Ressources du Québec Rapport intérimaire MB 86-36, 33 p.

Beutner, E. C., 1989, Tectonics of the frontal Québec Reentrant on the Gaspé Peninsula: Geological Association of Canada Program with Abstracts, v. 14, p. A26.

Boucot, A. J., 1962, Appalachian Siluro-Devonian, *in* Coe, K., ed., Some aspects of the Variscan fold belt: Manchester, England, Manchester University Press, p. 155–163.

—— , 1970, Devonian slate problems in the Northern Appalachians: Maine

Geological Survey Bulletin, v. 23, p. 42–48.

Boucot, A. J., and Bourque, P. A., 1981, Brachiopod biostratigraphy of the Llandoverian rocks of the Gaspé Peninsula, *in* Lespérance, P. J., ed., Field Meeting Anticosti-Gaspé, Québec, Subcommission on Silurian Stratigraphy, Ordovician-Silurian Boundary Working Group, v. 2: Stratigraphy and paleontology: Montréal, Département de geologie, Université de Montréal, p. 315–321.

Boucot, A. J., and Johnson, J. G., 1967, Paleogeography and correlation of Appalachian Province Lower Devonian sedimentary rocks: Tulsa Geological Society, Digest 35, p. 35–87.

Bourque, P. A., 1975, Lithostratigraphic framework and unified nomenclature for Silurian and basal Devonian rocks in eastern Gaspé Peninsula, Québec: Canadian Journal of Earth Sciences, v. 12, p. 858–872.

——, 1977, Silurian and basal Devonian of northeastern Gaspé Peninsula: Ministère de l'Energie et des Ressources du Québec Etude spéciale ES-29, 232 p.

——, 1979, Facies of the Silurian West Point reef complex, Baie des Chaleurs, Gaspésie, Québec: Geological Association of Canada, Annual Meeting, Québec City, Guidebook to field trip B-2, 29 p.

——, 1981, Introduction to Gaspé Peninsula and Baie des Chaleurs area, *in* Lespérance, P. J., ed., Field Meeting Anticosti-Gaspé, Québec, Subcommission on Silurian Stratigraphy, Ordovician-Silurian Boundary Working Group, v. 1: Guidebook: Montréal, Département de Géologie, Université de Montréal, p. 27–29, 42–53.

——, 1989a, Upper Ordovician to Middle Devonian rocks of Gaspé Peninsula, *in* Bourque, P. A., Hesse, R., and Rust, B., eds., Sedimentology, paleoenvironments and paleogeography of the Taconian to Acadian rock sequence of Gaspé Peninsula, Montréal, Geological Association of Canada–Mineralogical Association of Canada, Guidebook to field trip B8, p. 1–55.

——, 1989b, The Late Silurian–Early Devonian transgression of eastern Gaspé Peninsula, *in* Bourque, P. A., Hesse, R., and Rust, B., eds., Sedimentology, paleoenvironments and paleogeography of the Taconian to Acadian rock sequence of Gaspé Peninsula: Montréal, Geological Association of Canada–Mineralogical Association of Canada, Guidebook to field trip B8, p. 159–184.

——, 1990, La pulsation salinienne en Gaspésie-Témiscouata: Nature de la déformation et contrôle de la distribution des récifs de la fin Silurien–début Dévonien, *in* Malo, M., Lavoie, D., and Kirkwood, D., eds., Program with abstracts of the Québec–Maine–New Brunswick Appalachian workshop: Geological Survey of Canada Open-File 2235, p. 25–26.

Bourque, P. A., and Gosselin, C., 1986, Stratigraphie du Silurien et du Dévonien basal du bassin de Gaspésie: Ministère de l'Energie et des Ressources du Québec Rapport intérimaire MB 86-34, 49 p.

Bourque, P. A., and Lachambre, G., 1980, Stratigraphie du Silurien et du Dévonien basal du Sud de la Gaspésie: Ministère de l'Energie et des Ressources du Québec Etude spéciale ES-30, 123 p.

Bourque, P. A., Laurent, R., and St-Julien, P., 1985, Acadian wrench faulting in Southern Gaspé Peninsula, Québec Appalachians: Geological Association of Canada Program with Abstracts, v. 10, p. A6.

Bourque, P. A., and six others, 1986, Silurian and Lower Devonian reef and carbonate complexes of the Gaspé Basin, Québec—A summary: Bulletin of Canadian Petroleum Geology, v. 34, p. 452–489.

Bourque, P. A., Gosselin, C., Kirkwood, D., Malo, M., and St-Julien, P., 1990, Le Silurien du segment appalachien Gaspésie-Matapédia-Témiscouata, Québec: Stratigraphie, géologie structurale et paléogéographie: Ministère de l'Energie et des Ressources du Québec Mémoire MM 90-01.

Bourque, P. A., Brisebois, D., and Malo, M., 1993, Middle Paleozoic rocks of Québec and adjacent New Brunswick, *in* Williams, H., ed., Geology of the Appalachian-Caledonian Orogen in Canada and Greenland: Geological Society of America, Decade of North American Geology, v. F-1 (also, Geological Survey of Canada, Geology of Canada, no. 6) (in press).

Burk, C. F., Jr., 1964, Silurian stratigraphy of Gaspé Peninsula, Québec: Bulletin of American Association of Petroleum Geologists, v. 48, p. 437–464.

Cant, D. J., and Walker, R. G., 1976, Development of a braided-fluvial facies model for the Devonian Battery Point Sandstone, Quebec: Canadian Journal of Eearth Sciences, v. 13, p. 102–119.

Carbonneau, C., 1959, Région de Richard-Gravier, Péninsule de Gaspé: Ministère des Richesses Naturelles du Québec Rapport Géologique 90, 75 p.

Cawood, P. A., and Williams, H., 1988, Acadian basement thrusting, central delamination and structural styles in and around the Humber Arm Allochthon, western Newfoundland: Geology, v. 16, p. 370–374.

Chagnon, A., 1988, Géologie des argiles, diagenèse et altération hydrothermal, dans l'anticlinorium d'Aroostook-Percé, Québec (Canada) [D.Sc. thesis]: Neuchâtel, Switzerland, Université de Neufchâtel, 367 p.

Dalton, E., 1987, Sedimentary facies and diagenesis of the Lower Devonian Temiscouata and Fortin Formations, Northern Appalachians, Québec and New Brunswick [M.Sc. thesis]: Montréal, McGill University, 228 p.

De Broucker, G., 1987, Stratigraphie, pétrographie et structure de la boutonnière de Maquereau-Mictaw (région de Port-Daniel, Gaspésie): Ministère de l'Energie et des Ressources du Québec Mémoire MM 86-03, 160 p.

Dineley, D. L., and Williams, B.P.J., 1968, The Devonian continental rocks of the lower Restigouche River, Québec: Canadian Journal of Earth Sciences, v. 5, p. 945–953.

Doyon, M., Dalpé, C., and Valiquette, G., 1990, Silurian and Devonian volcanic rocks of the Gaspé Peninsula, *in* Malo, M., Lavoie, D., and Kirkwood, D., eds., Program with abstracts of the Québec–Maine–New Brunswick Appalachian workshop: Geological Survey of Canada Open-file 2235, p. 67–71.

Ducharme, D., 1979, Pétrographie du flysch de l'Ordovicien supérieur et du Silurien inférieur—Anticlinorium d'Aroostook-Percé, Gaspésie, Québec [M.Sc. thesis]: Montréal, Université de Montréal, 92 p.

Gauthier, L., 1986, Analyse stratigraphique et structurale de l'anticlinorium d'Aroostook-Percé, au nord de Port-Daniel [M.Sc. thesis]: Montréal, Université de Montréal, 89 p.

Hamilton-Smith, T., 1972, Stratigraphy and structure of Silurian rocks of the McKenzie Corner area, New Brunswick: New Brunswick Department of Natural Resources Report of Investigation 15, 26 p.

Hesse, R., and Dalton, E., 1989, The Devonian Fortin Formation, *in* Bourque, P. A., Hesse, R., and Rust, B., eds., Sedimentology, paleoenvironments and paleogeography of the Taconian to Acadian rock sequence of Gaspé Peninsula: Montréal, Geological Association of Canada–Mineralogical Association of Canada, Guidebook to field trip B8, p. 57–69.

——, 1990, Diagenetic–low grade metamorphic trends in the Acadian belt of Gaspé Peninsula, *in* Malo, M., Lavoie, D., and Kirkwood, D., eds., Program with abstracts of the Québec–Maine–New Brunswock Appalachian workshop: Geological Survey of Canada Open-file 2235, p. 17–20.

Hesse, R., and Sawh, H., 1982, Escuminac Formation, *in* Hesse, R., Middleton, G. V., and Rust, B. R., eds., Paleozoic continental margin sedimentation in the Québec Appalachians: International Association of Sedimentologists, 11th International Congress, Excursion Guidebook 7B, p. 72–80.

Keppie, J. D., 1985, The Appalachian collage, *in* Gee, D.G., and Sturt, B. A., eds., The Caledonide Orogen—Scandinavia and related areas: New York, John Wiley, p. 1217–1226.

Kindle, C. H., 1936, A geologic map of southeastern Gaspé: The Eastern Geologist, v. 1, 8 p.

Kirkwood, D., 1989, Géologie structurale de la région de Percé: Ministère de l'Energie et des Ressources du Québec Etude ET 87-17, 42 p.

Kirkwood, D., and Malo, M., 1991, Geometry and strain patterns of the Grand Pabos fault, Gaspé Peninsula, Québec Appalachians: Geological Society of America, Abstracts with Programs, v. 23, p. 53.

Kirkwood, D., and St-Julien, P., 1987, Analyse structurale du Siluro-Dévonien dans la vallée de la Matapédia: Ministère de l'Energie et des Ressources du Québec Rapport intérimaire MB 87-33, 17 p.

Kirkwood, D., Malo, M., and St-Julien, P., 1988, Palinspastic reconstruction of the Gaspé Basin during the Silurian time: Geological Association of Canada Program with Abstracts, v. 13, p. A66.

Lachambre, G., 1987, Le Silurien et le Dévonien basal du Nord de la Gaspésie: Ministère de l'Energie et des Ressources du Québec Etude ET 84-06, 83 p.

Lachance, S., 1979, Géologie de la région de Saint-André-de-Ristigouche: Ministère des Richesses Naturelles du Québec Rapport intérimaire DPV-667,

19 p.

Laurent, R., and Bélanger, J., 1984, Geochemistry of Silurian-Devonian alkaline basalt suites from the Gaspé Peninsula, Québec Appalachians: Maritime Sediments and Atlantic Geology, v. 20, p. 67–78.

Laurent, R., and Dostal, J., 1990, The Siluro-Devonian basalt-tholeiitic andesite association from the Gaspé Peninsula, *in* Malo, M., Lavoie, D., and Kirkwood, D., eds., Program with abstracts of the Québec–Maine–New Brunswick Appalachian workshop: Geological Survey of Canada Open-file 2235, p. 77.

Lavoie, D., 1988, Stratigraphie, sédimentologie et diagenèse du Wenlockien (Silurien) du Bassin de Gaspésie-Matapédie [Ph.D. thesis]: Québec, Université Laval, 330 p.

Lebel, D., 1985, Analyse structurale de la déformation acadienne, principalement la faille de Shickshock-Sud dans la région de Rimouski-Matapédia [M.Sc. thesis]: Montréal, Université de Montréal, 130 p.

Lespérance, P. J., 1968, Ordovician and Silurian trilobite faunas of the White Head Formation, Percé region, Québec: Journal of Paleontology, v. 143, p. 811–826.

—— , 1974, The Hirnantian fauna of the Percé area (Québec) and the Ordovician-Silurian boundary: American Journal of Science, v. 274, p. 10–30.

—— , 1980a, Calcaires supérieurs de Gaspé: Les aires-types et le prolongement vers l'ouest: Ministère de l'Energie et des Ressources du Québec Rapport intérimaire DPV-595, 92 p.

—— , 1980b, Les calcaires supérieurs de Gaspé (Dévonien Inférieur) dans le nord-est de la Gaspésie: Ministère de l'Energie et des Ressources du Québec Rapport intérimaire DPV-751, 35 p.

Lespérance, P. J., and Sheehan, P. M., 1976, Brachiopods from the Hirnantian stage (Ordovician-Silurian) at Percé, Québec: Paleontology, v. 19, p. 719–731.

Lespérance, P. J., and Tripp, R. P., 1985, Encrinurids (Trilobita) from the Matapédia Group (Ordovician), Percé, Québec: Canadian Journal of Earth Sciences, v. 22, p. 205–213.

Lespérance, P. J., Malo, M., Sheehan, P. M., and Skidmore, W. B., 1987, A stratigraphical and faunal revision of the Ordovician-Silurian strata of the Percé area, Québec: Canadian Journal of Earth Sciences, v. 24, p. 117–134.

Malo, M., 1986, Stratigraphie et structure de l'anticlinorium d'Aroostook-Percé en Gaspésie, Québec [Ph.D. thesis]: Montréal, Université de Montréal, 280 p.

—— , 1987, Structural evidences for Acadian wrench faulting in the southeastern Gaspé Peninsula, Québec: Geological Association of Canada Program with Abstracts, v. 12, p. 70.

—— , 1988a, Stratigraphy of the Aroostook-Percé anticlinorium in the Gaspé Peninsula, Québec: Canadian Journal of Earth Sciences, v. 25, p. 893–908.

—— , 1988b, L'anticlinorium d'Aroostook-Percé de l'est de la Gaspésie: Ministère de l'Energie et des Ressources du Québec Etude ET 87-06, 42 p.

Malo, M., and Béland, J., 1989, Acadian strike-slip tectonics in the Gaspé region, Québec Appalachians: Canadian Journal of Earth Sciences, v. 26, p. 1764–1777.

Malo, M., and Moritz, R., 1991, Géologie et métallogénie de la faille du Grand Pabos, région de Raudin-Weir, Gaspésie: Ministère de l'Energie et des Ressources du Québec Rapport intérimaire MB 91-03, 47 p.

Malo, M., Moritz, R., Roy, F., Chagnon, A., and Bertrand, R., 1990, Géologie et métallogénie de la faille du Grand Pabos, région de Robidoux-Reboul, Gaspésie: Ministère de l'Energie et des Ressources du Québec Rapport intérimaire MB 90-09, 76 p.

Malo, M., Kirkwood, D., De Broucker, G., and St-Julien, P., 1992, A reevaluation of the position of the Baie Verte-Brompton Line in the Quebec Appalachians: The influence of Middle-Devonian strike-slip faulting in Gaspé Peninsula: Canadian Journal of Earth Sciences, v. 29, p. 1265–1273.

Marillier, F., and six others, 1989, Crustal structure and surface zonation of the Canadian Appalachians: Implications of deep seismic reflection data: Canadian Journal of Earth Sciences, v. 26, p. 305–321.

Mason, G. D., 1971, A stratigraphical and paleoenvironmental study of the Upper Gaspé Limestone and Lower Gaspé Sandstone Groups (Lower Devonian) of Eastern Gaspé peninsula, Québec [Ph.D. thesis]: Ottawa, Carleton University, 191 p.

McGerrigle, H. W., 1946, A revision of the Gaspé Devonian: Transactions Royal Society of Canada, Ser. 3, sec. 4, v. 11, p. 41–54.

—— , 1950, The geology of eastern Gaspé: Québec Department of Mines Geological Report 35, 168 p.

Nance, R. D., 1987, Dextral transpression and Late Carboniferous sedimentation in the Fundy coastal zone of southern New Brunswick, *in* Beaumont, C., and Tankard, A. F., eds., Sedimentary basins and basin-forming mechanisms: Canadian Society of Petroleum Geologists Memoir 12, p. 363–377.

Northrop, S. A., 1939, Paleontology and stratigraphy of the Silurian rocks of the Port-Daniel–Black Cape region, Gaspé: Geological Society of America Special Paper 21, 302 p.

Nowlan, G. S., 1981, Late Ordovician-Early Silurian conodont biostratigraphy of the Gaspé Peninsula—A preliminary report, *in* Lespérance, P. J., ed., Field Meeting Anticosti-Gaspé, Québec, Subcommission on Silurian Stratigraphy, Ordovician-Silurian Boundary Working Group, v. 2: Stratigraphy and paleontology: Montréal, Département de geologie, Université de Montréal, p. 251–291.

—— , 1983, Early Silurian conodonts of eastern Canada: Fossils and Strata, v. 15, p. 95–110.

Pavlides, L., Boucot, A. J., and Skidmore, W. B., 1968, Stratigraphic evidence for the Taconic Orogeny in the Northern Appalachians, *in* Zen, E-an, White, W. S., Hadley, J. B., and Thompson, J. B., Jr., eds., Studies of Appalachian geology: Northern and maritime: Billing volume, New York, Interscience, p. 61–82.

Pickerill, R. K., 1980, Phanerozoic flysch trace fossil diversity—Observations based on an Ordovician flysch ichnofauna from the Aroostook-Matapedia Carbonate Belt of northern New Brunswick: Canadian Journal of Earth Sciences, v. 17, p. 1259–1270.

Powell, Mc. A., 1974, Timing of the slaty cleavage during folding of Precambrian rocks, north-west Tasmania: Geological Society of America Bulletin, v. 85, p. 1043–1060.

Riva, J., and Malo, M., 1988, Age and correlations of the Honorat Group, southern Gaspé Peninsula: Canadian Journal of Earth Sciences, v. 25, p. 1618–1628.

Rodgers, J., 1970, The tectonics of the Appalachians: New York, John Wiley and Sons, 271 p.

Roksandic, M. M., and Granger, B., 1981, Structural styles of Anticosti Island, Gaspé Passage, and eastern Gaspé Peninsula inferred from reflection seismic data, *in* Lespérance, P. J., ed., Field Meeting Anticosti-Gaspé, Québec, Subcommission on Silurian Stratigraphy, Ordovician-Silurian Boundary Working Group, v. 2: Stratigraphy and paleontology, Montréal, Département de geologie, Université de Montréal, p. 211–221.

Rouillard, M., 1986, Les calcaires supérieurs de Gaspé (Dévonien inférieur), Gaspésie: Ministère de l'Energie et des Ressources du Québec Rapport intérimaire MB 86-15, 94 p.

Ruitenberg, A. A., Fyffe, L. R., McCutcheon, R. S., St. Peter, C. J., Irrinki, R. R., and Venugopal, D. V., 1977, Evolution of pre-Carboniferous tectonostratigraphic zones in the New Brunswick Appalachians: Geoscience, v. 14, p. 171–181.

Rust, B. R., 1976, Stratigraphic relationships of the Malbaie Formation (Devonian), Gaspé, Québec: Canadian Journal of Earth Sciences, v. 13, p. 1556–1559.

—— , 1981, Alluvial deposits and tectonic style: Devonian and Carboniferous successions in Eastern Gaspé, *in* Miall, A. D., ed., Sedimentology and tectonics in alluvial basins: Geological Association of Canada Special Paper 23, p. 49–76.

—— , 1982, Continental Devonian and Carboniferous sedimentation of eastern Gaspé Peninsula, *in* Hesse, R., Middleton, G. V., and Rust, B. R., eds., Paleozoic continental margin sedimentation in the Québec Appalachians: International Association of Sedimentologists, 11th International Congress, Guidebook to field excursion 7B, p. 107–125.

——, 1989, Proximal gravelly braidplain deposit of the La Garde Formation, Highway 132, Ristigouche, Stop 1.3, *in* Bourque, P. A., Hesse, R., and Rust, B., eds., Sedimentology, paleoenvironments and paleogeography of the Taconian to Acadian rock sequence of Gaspé Peninsula: Geological Association of Canada–Mineralogical Association of Canada, Guidebook to field trip B8, p. 75–77.

Sanderson, D. J., and Marchini, W.R.D., 1984, Transpression: Journal of Structural Geolgoy, v. 8, p. 449–458.

Schuchert, C., and Dart, J. D., 1926, Stratigraphy of the Port-Daniel–Gascons area of southeastern Québec: Geological Survey of Canada Bulletin 44, Geological Series 46, p. 35–38, 116–121.

Schultze, H.-P, and Arsenault, M., 1985, The Panderichthyid fish Elpistostege: A close relative of tetrapods? Palaeontology, v. 28, p. 293–309.

Sheehan, P. M., and Lespérance, P. J., 1981, Brachiopods from the White Head Formation (Late Ordovician–Early Silurian) of the Percé region, Québec, Canada, *in* Lespérance, P. J., ed., Field Meeting Anticosti-Gaspé, Québec, Subcommission on Silurian Stratigraphy, Ordovician-Silurian Boundary Working Group, v. 2: Stratigraphy and paleontology: Montréal, Département de geologie, Université de Montréal, p. 247–256.

Sikander, A. H., 1975, Geology and hydrocarbon potential of the Berry Mountain Syncline, Central Gaspé (Matane, Matapédia, Gaspé W and Bonaventure counties): Ministère des Richesses Naturelles du Québec Rapport intérimaire DP-376, 119 p.

Simard, M., 1986, Géologie et évaluation du potential minéral de la région de Carleton: Ministère de l'Energie et des Ressources du Québec Etude ET 84-11, 27 p.

Skidmore, W. B., 1967, The Taconic Unconformity in the Gaspé Peninsula and neighboring regions, *in* Clark, T. H., ed., Appalachian tectonics: Royal Society of Canada Special Publication 10, p. 26–32.

Skidmore, W. B., and McGerrigle, H. W., 1967, Geological map, Gaspé Peninsula: Québec Department of Natural Resources Map 1641, scale 1:253,440.

Slivitsky, A., St-Julien, P., and Lachambre, G., 1991, Synthèse géologique du Cambro-Ordovicien du Nord de la Gaspésie: Ministère de l'Energie et des Ressources du Québec ET-88-14, 61 p.

St-Julien, P., and Hubert, C., 1975, Evolution of the Taconian Orogen in the Quebec Appalachians: American Journal of Science, v. 275-A, p. 337–362.

St-Julien, P., Trzcienski, W. E., and Wilson, C., 1990, A structural, petrological and geochemical traverse of the Shickshock Terrane, Gaspésie, *in* Trzcienski, W. E., ed., Guidebook for field trips in La Gaspésie, Québec: New England Intercollegiate Geological Conference, 82nd Annual Meeting, Gîte du Mont Albert, Gaspésie, Québec, p. 248–285.

Stockmal, G. S., Colman-Sadd, S. P., Keen, C. E., O'Brien, S. J., and Quinlan, G., 1987, Collision along an irregular margin: A regional plate tectonic interpretation of the Canadian Appalachians: Canadian Journal of Earth Sciences, v. 24, p. 1098–1107.

Stockmal, G. S., Colman-Sadd, S. P., Keen, C. E., Marillier, F., O'Brien, S. J., and Quinlan, G., 1990, Deep seismic structure and plate tectonic evolution of the Canadian Appalachians: Tectonics, v. 9, p. 45–62.

St. Peter, C. J., 1977, Geology of parts of Restigouche, Victoria and Madawaska Counties, Northwestern New Brunswick: Mineral Resources Branch, New Brunswick Department of Natural Resources Report of Investigation 17, 69 p.

Sylvester, A. G., 1988, Strike-slip faults: Geological Society of America Bulletin, v. 100, p. 1666–1703.

Théberge, R., 1979, Etude tectono-stratigraphique des roches ordoviciennes et siluriennes de l'anticlinorium d'Aroostook-Percé à Matapédia [M.Sc. thesis]:

Montréal, Université de Montréal, 63 p.

Tremblay, A., St-Julien, P., and Labbé, J. Y., 1989, Mise à l'évidence et cinématique de la faille de La Guadeloupe, Appalaches du sud du Québec: Canadian Journal of Earth Sciences, v. 26, p. 1932–1943.

Tremblay, A., Malo, M., and St-Julien, P., 1993, Dunnage Zone of Québec, *in* Williams, H., ed., Geology of the Appalachian-Caledonian Orogen in Canada and Greenland: Geological Society of America, Decade of North American Geology, v. F-1 (also, Geological Survey of Canada, Geology of Canada, no. 6) (in press).

Trudel, C., 1990, Analyse structurale des failles acadiennes, région de Matapédia, Gaspésie: [M.Sc. thesis]: Québec, Université Laval, 36 p.

van der Pluijm, B. A., and van Staal, C. R., 1988, Characteristics and evolution of the Central Mobile Belt, Canadian Appalachians: Journal of Geology, v. 96, p. 535–548.

van Staal, C. R., 1987, Tectonic setting of the Tetagouche Group in northern New Brunswick: Implications for plate tectonic models of the northern Appalachians: Canadian Journal of Earth Sciences, v. 24, p. 1329–1351.

van Staal, C., and Fyffe, L. R., 1993, Dunnage Zone of New Brunswick, *in* Williams, H., ed., Geology of the Appalachian-Caledonian Orogen in Canada and Greenland: Geological Society of America, Decade of North American Geology, v. F-1 (also, Geological Survey of Canada, Geology of Canada, no. 6) (in press).

Vennat, G., 1979, Structure et stratigraphie de l'anticlinorium d'Aroostook-Percé dans la région de Saint-Omer-Carleton, Gaspésie, Appalaches du Québec [M.Sc. thesis]: Montréal, Université de Montréal, 67 p.

Walker, R. G., and Cant, D. J., 1979, Facies model 3: Sandy fluvial systems, *in* Walker, R. G., ed., Facies models: Geoscience Canada Reprint Series 1, p. 23–31.

Whalen, J. B., and Gariépy, C., 1986, Petrogenesis of the McGerrigle plutonic complex, Gaspé, Québec: A preliminary report, *in* Current research, part A: Geological Survey of Canada Paper 86-1A, p. 265–274.

Wilcox, R. E., Harding, T. P., and Seeley, D. R., 1973, Basic wrench tectonics: Bulletin of American Association of Petroleum Geologists, v. 57, p. 74–96.

Williams, B.P.J., and Dineley, D. L., 1966, Studies of the Devonian strata of Chaleurs Bay, Québec: Maritime Sediments, v. 2, p. 7–10.

Williams, H., comp., 1978, Tectonic lithofacies map of the Appalachian Orogen: Memorial University of Newfoundland Map 1, scale 1:2,000,000.

——, 1979, Appalachian Orogen in Canada: Canadian Journal of Earth Sciences, v. 16, p. 792–807.

Williams, H., and Hatcher, R. D., Jr., 1983, Appalachian suspect terranes, *in* Hatcher, R. D., Jr., Williams, H., and Zietz, I., eds., Contributions to tectonics and geophysics of mountains chains: Geological Society of America Memoir 158, p. 33–53.

Williams, H., and St-Julien, P., 1982, The Baie Verte–Brompton Line: An Early Paleozoic continent ocean interface in the Canadian Appalachians, *in* St-Julien, P., and Béland, J., eds., Major structural zones and faults of the Northern Appalachians: Geological Association of Canada Special Paper 24, p. 177–208.

Zaitlin, B. A., and Rust, B. R., 1983, A spectrum of alluvial deposits in the Lower Carboniferous Bonaventure Formation of western Chaleurs Bay area, Gaspé and New Brunswick, Canada: Canadian Journal of Earth Sciences, v. 20, p. 1098–1110.

MINISTÈRE DE L'ENERGIE ET DES RESSOURCES DU QUÉBEC CONTRIBUTION NO. 90-5110-05

MANUSCRIPT ACCEPTED BY THE SOCIETY JUNE 8, 1992

Geological Society of America
Special Paper 275
1993

Acadian orogeny in Newfoundland

Harold Williams*
Department of Geology, St. Francis Xavier University, Antigonish, Nova Scotia B2G 1C0, Canada

ABSTRACT

The climactic, middle Paleozoic event that affected the Newfoundland Appalachians has been referred to traditionally as Acadian orogeny. The latest field studies and isotopic ages indicate that it began in the Early Silurian and continued into the Devonian. It affected the Newfoundland Humber, Dunnage, and Gander zones and western parts of the Avalon Zone. Intensities of the effects of the orogeny decrease westward across the Humber Zone (Appalachian miogeocline) and eastward across the onland Avalon Zone. Offshore, the wide Avalon Zone is virtually unaffected by Paleozoic deformation.

The most intense regional metamorphism coincides mainly with the Gander Zone. Plutonism affected a wider area—from the eastern Humber Zone of White Bay to the western Avalon Zone of Placentia Bay. Deformation affected the widest area, from the Appalachian Structural front, which defines the western boundary of the Humber Zone, to the Avalon Peninsula of the western Avalon Zone.

A change from marine to terrestrial conditions preceded the Silurian-Devonian deformation. Uninterrupted shallow marine conditions prevailed in bordering regions outside the Acadian deformed zone.

The Acadian orogen spans the eastern portion of the Grenville lower crustal block that underlies the Humber Zone and western parts of the Dunnage Zone. It spans the Central lower crustal block that underlies eastern parts of the Dunnage Zone and the Gander Zone, and it spans the western part of the Avalon lower crustal block. Orogenic effects are most intense above the narrow Central lower crustal block and diminish outward across the margins of the opposing Grenville and Avalon lower crustal blocks. This spatial relationship between the surface orogen and lower crustal blocks implies that collisional interaction among lower crustal blocks controlled the tectonothermal effects of Acadian orogeny.

INTRODUCTION

Acadian orogeny takes its name from Acadia, the former French name for the Canadian Maritime Provinces. Where middle Paleozoic stratigraphic sections are most complete, as in the Canadian Appalachian miogeocline of Gaspe, Quebec, the event is dated as middle to late Devonian (Neale and others, 1961; Rodgers, 1967). There, Silurian and Devonian rocks lie upon the Taconian (Ordovician) deformed zone, and deformed Silurian

and Devonian rocks are overlain unconformably by Carboniferous strata. Elsewhere the stratigraphic definition of Acadian orogeny is less sharp.

The climatic event in the middle Paleozoic structural evolution of the Newfoundland Appalachians has been referred to traditionally as Acadian orogeny. Its age was interpreted as Devonian because sparse, widely separated, fossiliferous latest Silurian or Devonian strata are deformed and/or cut by granites, whereas nearby Carboniferous rocks are unaffected. For example, at Port au Port Peninsula in western Newfoundland (Fig. 1), the Clam Bank Group is dated as latest Silurian (Pridoli; Berry and Boucot, 1970) in the central part of a 500-m section, and thrusting and overturning that affected the Clam Bank Group are

*Present address: Department of Earth Sciences, Memorial University of Newfoundland, St. John's, Newfoundland A1B 3X5, Canada.

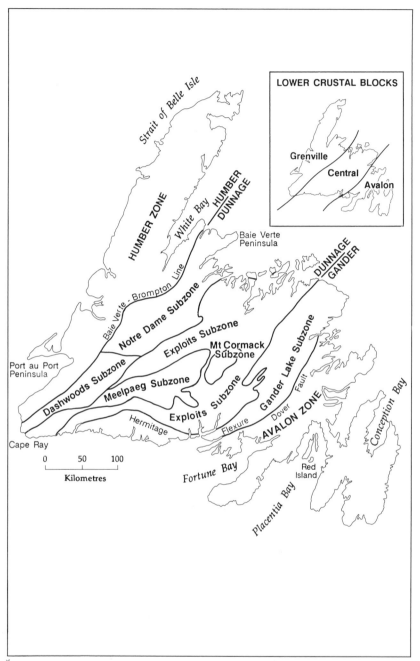

Figure 1. Zonal subdivision and lower crustal blocks of the Newfoundland Appalachians (modified after Williams and others, 1988, and Marillier and others, 1989).

not recorded in nearby subhorizontal Carboniferous strata (Rodgers, 1965; Cawood, this volume). Also, at Cape Ray in southwest Newfoundland, Middle Devonian rocks of the Windsor Point Group record intense deformation, mylonitization, and intrusion that do not affect nearby Carboniferous strata (Wilton, 1983). In Fortune Bay of eastern Newfoundland, the Silurian-Devonian Cinq Isles and Pools Cove formations are folded, faulted, and cut by Devonian granite, and the Late Devonian Great Bay de l'Eau Formation is cut by the Belleoram Granite, generally considered Devonian (Williams, 1971).

However, recent U-Pb zircon ages of 415 to 430 Ma on syntectonic intrusions indicate a major Silurian orogenic event in central Newfoundland that involved regional metamorphism, deformation, and plutonism. The event affected Ordovician and Silurian rocks, and it predated the emplacement of granites dated at about 400 Ma (Dunning and others, 1988; Currie and van

Berkel, 1989; Dunning and O'Brien, 1989; Dunning and others, 1990; Elliott and others, 1991; O'Brien and others, 1991).

All of the pre-Carboniferous deformation that affected Silurian and Devonian rocks or is dated isotopically as Silurian or Devonian is considered in this analysis under the broad heading "Acadian orogeny." Used in this way, Acadian orogeny is not a brief, singular Devonian event but rather a prolonged episode or sequence of events that began in the early to middle Silurian and continued through the Devonian. Other names may be required

to distinguish Devonian and Silurian events as the data base expands.

The objective of this chapter is to compare the distribution of structural, metamorphic, and plutonic effects of Acadian orogeny (Figs. 2, 3, and 4, respectively); the surface tectonostratigraphic zonation of the Newfoundland Appalachians; and the subsurface boundaries of lower crustal blocks—with the aim of establishing spatial relationships that may provide insight into orogenic controls. The zonal subdivision of Newfoundland

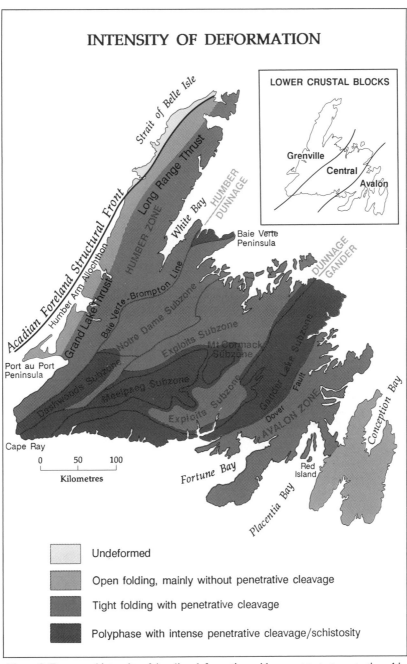

Figure 2. Extent and intensity of Acadian deformation with respect to tectonostratigraphic zones and lower crustal blocks.

(Fig. 1)—Humber Zone, Dunnage Zone (Notre Dame, Exploits, and Dashwoods subzones), Gander Zone (Meelpaeg, Mount Cormack, and Gander Lake subzones), and Avalon Zone from west to east—follows the divisions of Williams and others (1988, 1989a) and of Piasecki and others (1990). The definition of lower crustal blocks (Fig. 1, inset)—Grenville, Central, and Avalon, from west to east—follows Keen and others (1986) and Marillier and others (1989). The Grenville lower crustal block underlies

the Humber Zone and western parts of the Dunnage Zone. The Central lower crustal block underlies eastern parts of the Dunnage Zone and the Gander Zone. The Avalon lower crustal block underlies the Avalon Zone.

Brief descriptions are provided to demonstrate general structural, metamorphic, and plutonic styles. More complete referencing and descriptions are given in an imminent Decade of North American Geology volume (Williams, in press).

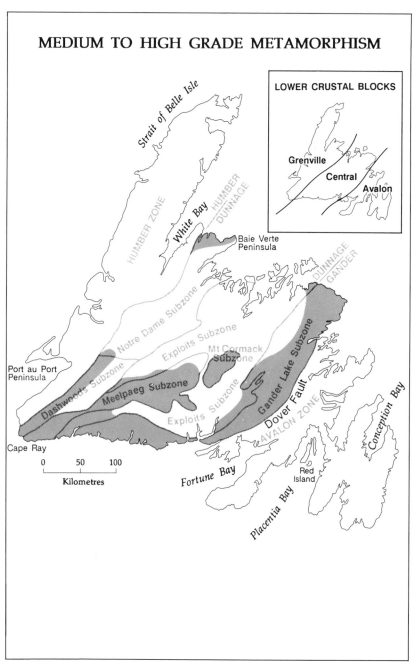

Figure 3. Extent of Acadian upper greenschist and amphibolite facies regional metamorphism (shaded area) with respect to tectonostratigraphic zones and lower crustal blocks.

GENERAL EFFECTS OF ACADIAN OROGENY IN NEWFOUNDLAND

Deformation

The effects of Acadian deformation extend all the way across Newfoundland from the Humber Zone in the west to the Avalon Zone in the east (Fig. 2). At the surface, the Acadian foreland structural front of the Humber Zone coincides roughly with Logans Line or the foreland limit of Taconic allochthons. Offshore to the west, seismic reflection profiles suggest the presence of a triangle zone similar to those developed at the foreland edges of other thrust belts (Stockmal and Waldron, 1990). The morphologic Acadian front is a west-directed thrust at Port au Port Peninsula (Round Head Thrust; Williams, 1985) and a thrust or steep reverse fault farther north (Long Range Front;

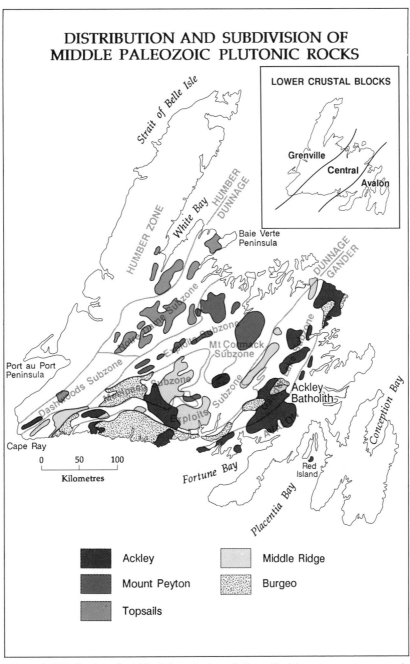

Figure 4. Distribution of middle Paleozoic plutonic "types" with respect to tectonostratigraphic zones and lower crustal blocks.

Grenier and Cawood, 1988; Grenier, 1990). The Devonian age of the front is defined only at Port au Port, where the Clam Bank Group is affected. Structures interpreted as coeval bring Grenville basement above the Cambrian-Ordovician carbonate sequence at the Grand Lake and Long Range thrusts (Williams and Cawood, 1989). Polarity of structures is westward toward the foreland, except in the vicinity of the Humber Arm Allochthon, where thrusts and folds are east directed (Lilly, 1967; Bosworth, 1985; Waldron, 1985; Cawood and Williams, 1988; Cawood, this volume). Intensity of deformation increases eastward toward the Baie Verte–Brompton Line, or Humber-Dunnage zone boundary. At Baie Verte Peninsula, structural styles of Silurian rocks vary from mild open folding to intense polyphase deformation, and the latest structures are east directed (Kidd, 1974; Hibbard, 1983). This variation in tectonic style appears to be controlled by the Baie Verte Flexure, a major bend in the Humber-Dunnage zone boundary (Hibbard, 1982, 1983). Rocks along the north-northeast–trending limb of the flexure, such as the Mic Mac Lake Group, have been subjected to mild Acadian deformation, whereas rocks along the east-northeast–trending limb, such as the Cape St. John Group, display intense Acadian deformation and amphibolite facies metamorphism.

Silurian rocks of the western Dunnage Zone (Notre Dame Subzone) are mildly deformed in most places, and they lie unconformably upon deformed Ordovician or older rocks (e.g., Neale and others, 1975). A broad continuous belt of volcanic and sedimentary rocks in the northeast (Springdale Group and southward correlatives) is affected by open folding (Neale and Nash, 1963). Deformation is generally more intense throughout the eastern Dunnage Zone (Exploits Subzone). There, Ordovician and Silurian rocks are in places conformable and deformed together (Karlstrom and others, 1982). Locally, a mild unconformity separates upward-shoaling Ordovician-Silurian marine rocks (Samson and Goldson formations) and overlying terrestrial volcanic rocks and red beds of the Botwood Group in the northeast (van der Pluijm and others, 1989). Intensity of deformation increases southwestward along the Exploits Subzone where it narrows toward Cape Ray. Devonian rocks at Cape Ray are polydeformed (Wilton, 1983). Mylonitic fabrics in the region indicate two phases of thrusting: southeast over northwest movement that predates the Devonian Windsor Point Group and east over west oblique thrusting that produced mylonites in the Windsor Point Group (Williams and others, 1989a; Currie and Piasecki, 1989). The polarity of transport corresponds to that along the Exploits-Meelpaeg subzone boundary to the northeast (Williams and others, 1989a) and also corresponds to northwest overthrusting in the Silurian La Poile Group about 60 km to the east (O'Brien, 1988).

Most Dunnage-Gander boundaries are tectonic, and some are marked by zones of orogen-parallel ductile shearing (Hanmer, 1981; Piasecki, 1988; Williams and others, 1989a; Currie and Piasecki, 1989; Piasecki and others, 1990; Williams and others, 1991). Intensity of deformation in rocks of the Dunnage Zone increases toward the Gander boundaries. The ductile shearing

followed local formation of ophiolitic melanges at some boundaries that are interpreted as features of Ordovician Dunnage-Gander juxtapositioning (Piasecki and others, 1990; Williams and Piasecki, 1990). Most evidence suggests that the ductile shearing is Silurian (Dunning and others, 1988; Currie and Piasecki, 1989; Dunning and others, 1990; O'Brien and others, 1991).

Gander Zone rocks are mainly metaclastics that are everywhere penetratively deformed; most show more than one phase of folding. Intensity of deformation increases eastward and southeastward across the Gander Lake Subzone toward the Avalon zone boundary (Blackwood, 1978, 1985; Colman-Sadd, 1982; Piasecki, 1988). U/Pb zircon isotopic dates of 428 and 415±2 Ma for pre- and syntectonic plutons and 423±3 Ma for migmatitic rocks suggest a Silurian age for this deformation where rocks and structures extend westward around the Hermitage Flexure (Dunning and others, 1988, 1990; O'Brien and others, 1991).

In the north-central Exploits Subzone, a 423±3 Ma age for silicic volcanic rocks at Stony Lake and the likelihood of an unconformity between the subhorizontal Stony Lake volcanics and underlying steeply dipping folded rocks of the Silurian Botwood Group is further evidence for Silurian deformation (Dunning and others, 1988, 1990). Also, a 408±2 Ma age for the Loon Bay intrusive suite places a Late Silurian upper limit on the timing of thrust faulting and three generations of folding in the New World Island area of Notre Dame Bay (Elliott and others, 1991).

The northern Gander-Avalon zone boundary (Dover Fault; Blackwood and Kennedy, 1975) is marked by a wide zone of mylonitization that affects granite and metamorphic rocks on the Gander side and less-deformed late Precambrian sedimentary and volcanic rocks on the Avalon side. $^{40}Ar/^{39}Ar$ geochronologic data indicate development of mylonites at about 365 to 395 Ma (Dallmeyer and others, 1981).

Deformation within the Avalon zone is mild compared to that of the adjacent Gander Zone (Williams and others, 1972). Tight upright folds affect parts of the western Avalon Zone, whereas eastern parts have open folds. Some areas of Cambrian-Ordovician rocks in the eastern Avalon Zone, such as those at Conception Bay, are subhorizontal and lack penetrative deformation. Offshore, Paleozoic deformation is uniformly mild or absent across the wide area of the Grand Banks. A mild erosional unconformity separates marine Silurian and terrestrial Devonian rocks (King and others, 1986; Durling and others, 1987).

Metamorphism

Upper greenschist to amphibolite facies regional metamorphism of Silurian-Devonian age affects parts of the eastern Humber Zone, parts of the Dunnage Zone, and most of the Gander Zone (Williams and others, 1972; Hibbard, 1983). Rocks of the Avalon Zone are lower grade, regardless of age (Fig. 3). The Dover Fault—the Gander-Avalon zone boundary—is the sharpest metamorphic break of all Newfoundland zone boundaries.

The Silurian Cape St. John Group is locally metamorphosed to amphibolite facies at the northern Baie Verte Peninsula, and amphibolites of the Ordovician Pacquet Harbour Group have Devonian cooling ages (Hibbard, 1983). Conodonts from the Silurian Sops Arm Group on the west side of White Bay have a color alteration index of 8, implying upper greenschist conditions (Nowlan and Barnes, 1987). The highest metamorphic grades recorded in schists and gneisses of the intervening Fleur de Lys Supergroup of the western Baie Verte Peninsula and southern correlatives are interpreted as Ordovician (Hibbard, 1983). Some of the high-grade regional metamorphism throughout the Dashwoods Subzone was interpreted as Acadian (Wilton, 1983), but Silurian isotopic ages for posttectonic mafic plutons suggest it is mainly Ordovician (Dunning and others, 1988; Currie and van Verkel, 1989). Dunnage Zone rocks reach upper greenschist and amphibolite facies near Gander zone boundaries, as at the periphery of the Meelpaeg Subzone (Williams and others, 1988). Dunnage Zone rocks (Bay du Nord Group) also reach amphibolite facies where the Dunnage Zone is narrow in southwestern Newfoundland (O'Brien, 1988).

Regional metamorphism increases in intensity away from the Gander Lake–Dunnage zone boundary, reaching upper amphibolite facies with sillimanite, andalusite, and kyanite all present locally (Jenness, 1963; Williams, 1964; O'Neill and Knight, 1988). Upper greenschist–amphibolite facies metamorphic rocks are also present throughout the Meelpaeg and Mount Cormack divisions of the Gander Zone (Colman-Sadd and Swinden, 1984). Within the Mount Cormack Subzone, a concentric arrangement of metamorphic isograds suggests metamorphism contemporaneous with doming and emplacement of the Mount Cormack Subzone through its Dunnage (Exploits Subzone) cover (Colman-Sadd and Swinden, 1984). Although interpreted as middle Paleozoic (Colman-Sadd and Swinden, 1984), recent isotopic ages of synchronous and posttectonic granites suggest Mount Cormack metamorphism is Ordovician (G. R. Dunning, personal communication, 1990).

Plutonism

Silurian-Devonian plutons make up about 30 percent of the internal portion of the Newfoundland Appalachians (Williams and others, 1989b; Fig. 4). Large granite batholiths cut eastern parts of the Humber Zone, all of the Dunnage and Gander zones, and western parts of the Avalon Zone. The most westerly Acadian pluton is the Devils Room granite (391±3 Ma) that cuts Grenvillian rocks of the Long Range inlier on the west side of White Bay (Baadsgaard, personal communication, 1990). A similar pluton (372±10 Ma) cuts the Silurian Sops Arm Group nearby to the south (Baadsgaard, personal communication, 1990). The most easterly Acadian pluton occurs at Red Island of Placentia Bay in the western Avalon Zone (Anderson, 1965). A granite with a K-Ar biotite age of 376 Ma recovered from the Jaeger A-49 drill hole on the Grand Banks (Jansa and Wade, 1975) appears to lie south of the Avalon Zone, where it may be a sampling of Devonian granite like that so widespread throughout the Meguma Zone.

Alkali plutons that span the eastern Humber Zone and Notre Dame Subzone of the western Dunnage Zone (Topsails type of Williams and others, 1989b; Fig. 4) are interpreted as coeval with Silurian volcanism and a series of nested calderas that cross the Humber-Dunnage boundary (Coyle and Strong, 1987). Acadian plutons of the Exploits Subzone of the eastern Dunnage Zone are composite with early mafic phases and later granitic phases (Mount Peyton type of Fig. 4). Those of the Gander Zone are large deformed feldspar-phyric biotite granites (Burgeo type of Fig. 4) and deformed to undeformed garnet muscovite leucogranites (Middle Ridge type of Fig. 4). Massive megacrystic biotite granites (Ackley type of Fig. 4) cut the Burgeo and Middle Ridge varieties. They are commonest in the Gander Zone but extend well eastward into nearby parts of the Avalon Zone. The large Ackley batholith is composite and comprises several intrusions with $^{40}Ar/^{39}Ar$ cooling ages between 410 and 355 Ma (Tuach and others, 1986; Kontak and others, 1988). In general, the oldest Ackley intrusions occur northwest of the projection of the Gander-Avalon zone boundary, with the youngest Ackley intrusions to the southeast. One intrusion, the Hungry Grove (367.7±4.2 Ma), straddles the projected boundary and provides a probable upper limit for the age of latest significant movement on the Gander-Avalon zone boundary. The Middle Devonian Windsor Point Group is cut by Devonian granite at Cape Ray (Wilton, 1983). Rocks as young as Late Devonian are cut by the Belleoram granite, dated at 400 Ma and 342 Ma (Wanless and others, 1965, 1967), in the western Avalon Zone of Fortune Bay (Williams, 1971).

Plutons of Burgeo type are deformed with their surrounding rocks. Those of Middle Ridge type are pre-, syn-, and postdeformation. Others of Mount Peyton and Ackley types everywhere truncate deformed Silurian and older rocks (Williams and others, 1989b).

The distribution of middle Paleozoic plutons has a spatial relationship to lower crustal blocks (Fig. 4). The Topsails plutons are confined to the Grenville lower crustal block. The Burgeo and Middle Ridge plutons are confined to the Central lower crustal block, and their structural and metamorphic relationships imply a major tectonothermal episode. The posttectonic Mount Peyton plutons are confined to the western portion of the Central lower crustal block. Only the Ackley suite of plutons clearly crosses block boundaries, concentrated in the Central-Avalon boundary area. Petrological and age differences among intrusions of the Ackley suite on opposite sides of the Gander-Avalon boundary may reflect underlying crustal contrasts and are thus in accord with the seismic data for a vertical boundary between the Central and Avalon lower crustal blocks (Marillier and others, 1989).

DISCUSSION

The contrasts among marine lower Paleozoic rocks that allow a zonal subdivision of the Newfoundland Appalachians are less apparent in Silurian and Devonian rocks that are mainly terrestrial. In some places, Silurian conglomerates and volcanic rocks overlie the Taconian deformed zone (e.g., eastern Humber and western Dunnage zones). In other places, there is a transition from marine Ordovician rocks through upward-shoaling sequences of Late Ordovician–Early Silurian rocks to Silurian terrestrial volcanic rocks and red beds (e.g., northeast Exploits Subzone of the Dunnage Zone). The change from Ordovician marine to Silurian-Devonian terrestrial conditions, recorded everywhere across the Acadian deformed zone, contrasts with uninterrupted shallow marine conditions that existed in places outside the Acadian deformed zone, west of the Appalachian structural front (Berry and Boucot, 1970) and across the wide offshore Avalon Zone (King and others, 1986; Durling and others, 1987). The environmental change reflects uplift, probably the result of crustal thickening by events that preceded Acadian deformation, metamorphism, and intrusion.

The lack of complete Silurian and Devonian sections in western Newfoundland precludes a stratigraphic analysis of the sedimentological effects of Acadian orogeny in the local foreland of the Humber Zone. Possibly, easterly derived red beds of the Clam Bank Group are a latest Silurian molasse related to uplift in interior parts of the orogen. Likewise, Silurian-Devonian red beds and conglomerates of the Avalon Zone at Fortune Bay and the Devonian red beds that overlie Silurian rocks with mild unconformity on the Grand Banks may represent molasse deposits on the opposing side of the orogen.

The shape of the ancient continental margin of eastern North America also had an important control on the local extent and intensity of Acadian deformation. At the St. Lawrence Promontory (Thomas, 1977; Williams, 1978; Stockmal and others, 1987) of southwest Newfoundland, deformation in the Devonian Windsor Point Group is more intense than that in Silurian-Devonian rocks toward the Newfoundland Reentrant to the north. Furthermore, Cambrian-Ordovician rocks of the Newfoundland Reentrant at the Strait of Belle Isle are locally outside the Appalachian deformed zone, although the rocks are part of the Appalachian system by other considerations. A smaller promontory at Baie Verte has been proposed to explain local contrasts in intensity of deformation and metamorphism (Hibbard, 1982).

Linkages between the Humber Zone and the Notre Dame Subzone of the western Dunnage Zone were established by the Early-Middle Ordovician. Linkages between the Exploits Subzone of the eastern Dunnage Zone and Gander Zone were also established in the Early-Middle Ordovician (Williams and Piasecki, 1990; Dec and Colman-Sadd, 1990; Williams and others, 1991). The presence of continuous Ordovician-Silurian sections in the northeast Exploits Subzone and the lack of Taconian deformation (Karlstrom and others, 1982) suggest that central parts of the Dunnage Zone remained undisturbed. Deformation, regional metamorphism, and plutonism at eastern Dunnage-Gander boundaries and throughout the Gander Zone roughly coincide with imbrication and development of Silurian melanges throughout the northeast Exploits Subzone of the Dunnage Zone

(Reusch, 1987; Elliott and others, 1991). The situation suggests that the Grenville and Central lower crustal blocks were not fully juxtaposed in a deep collisional zone until the middle Paleozoic or that compression and shortening continued into the middle Paleozoic (Colman-Sadd, 1982).

Juxtaposed lower crustal blocks beneath the Dunnage Zone rationalize the presence and abundance of large granitic plutons that cut its oceanic rocks. The localization of deformed middle Paleozoic plutons of Burgeo and Middle Ridge types to the Gander Zone may reflect diapiric emplacement during doming of a lighter Gander crust through a heavier Dunnage cover (Colman-Sadd and Swinden, 1984). The spatial correspondence between these plutons and the Central lower crustal block supports genetic links between surface rocks of the Gander Zone and the Central lower crustal block.

The earliest Gander-Avalon zone linkage is established by metamorphic and plutonic detritus derived from the Gander Zone in Silurian-Devonian conglomerates of the Cinq Isle and Pools Cove formations of the adjacent Avalon Zone (Williams, 1971). Juxtapositioning of the Central and Avalon lower crustal blocks therefore seems to have been later than juxtapositioning of the Grenville and Central lower crustal blocks.

CONCLUSIONS

The area affected by Acadian orogeny is symmetrically disposed between the edges of the wide opposing Grenville and Avalon lower crustal blocks (Fig. 5). Deformation spans the eastern portion of the Grenville lower crustal block, all of the Central lower crustal block, and western parts of the Avalon lower crustal block. Plutonism also spans the eastern portion of the Grenville lower crustal block, all of the Central lower crustal block, and western parts of the Avalon lower crustal block, but within more narrow limits. High-grade regional metamorphism is more restrictive, mainly above the Central lower crustal block.

These spatial relations imply that Acadian orogeny resulted from the middle Paleozoic collisional interaction between Grenville-Central lower crustal blocks and between Central-Avalon lower crustal blocks. Only eastern parts of the Grenville lower crustal block and western parts of the opposing Avalon lower crustal block were involved. The main zone of Acadian deformation, regional metamorphism, and plutonism expectably coincides with the narrow, intervening Central lower crustal block and its boundaries.

In Nova Scotia, Acadian orogeny affected the Mira Terrane (Avalon Zone and Avalon lower crustal block), but intensity of deformation, regional metamorphism, and middle Paleozoic plutonism are less important than in the Bras d'Or and Aspy terranes to the northwest (Central lower crustal block; Barr and Raeside, 1989; Loncarevic and others, 1989). This relationship resembles the situation in Newfoundland. However, the Meguma Zone of Nova Scotia, outboard of the Avalon, exhibits the full effects of Acadian deformation, metamorphism, and plutonism. This difference may mean that the Meguma Zone is part of a separate lower crustal block (Keen and others, 1990).

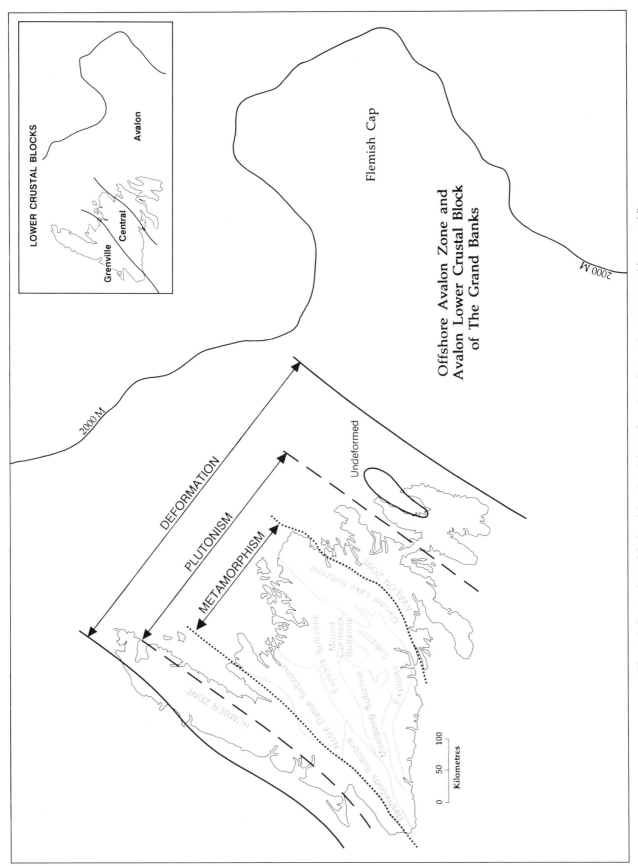

Figure 5. Comparison of limits of Acadian deformation, plutonism, and upper greenschist to amphibolite facies regional metamorphism across the Newfoundland Appalachians and Atlantic continental margin.

The overall situation suggests Acadian controls through the collisional interaction of lower crustal blocks (see also Rast and Skehan, this volume). Where the Avalon Zone (block) is especially wide in Newfoundland (more than twice the width of the rest of the orogen), it behaved rigidly, with interior parts inert from orogenic effects.

ACKNOWLEDGMENTS

Thanks are extended to the Natural Sciences and Engineering Research Council of Canada and the Department of Energy, Mines, and Resources of Canada for support of fieldwork through Operating and Lithoprobe Grants and Research Agreements, respectively. The manuscript has benefited from reviews by P. A. Cawood, G. R. Dunning, J. P. Hibbard, and Brendan Murphy.

REFERENCES CITED

Anderson, F. D., 1965, Belleoram, Newfoundland: Geological Survey of Canada Map 8-1965, scale 1:253,440.

Barr, S. M., and Raeside, R. P., 1989, Tectono-stratigraphic terranes in Cape Breton Island, Nova Scotia: Implications for the configuration of the northern Appalachian orogen: Geology, v. 17, p. 822–825.

Berry, W.B.N., and Boucot, A. J., 1970, Correlation of the North American Silurian rocks: Geological Society of America Special Paper 102, 289 p.

Blackwood, R. F., 1978, Northeastern Gander Zone, Newfoundland, *in* Report of activities for 1977: Newfoundland Department of Mines and Energy Mineral Development Division Report 78-1, p. 72–79.

—— , 1985, Geology of the Facheux Bay area (11P/9), Newfoundland: Newfoundland Department of Mines and Energy Mineral Development Division Report 85-4, 56 p.

Blackwood, R. F., and Kennedy, M. J., 1975, The Dover Fault: Western boundary of the Avalon Zone in northeastern Newfoundland: Canadian Journal of Earth Sciences, v. 12, p. 320–325.

Bosworth, W., 1985, East-directed imbrication and oblique-slip faulting in the Humber Arm Allochthon of western Newfoundland: Structural and tectonic significance: Canadian Journal of Earth Sciences, v. 22, p. 1351–1360.

Cawood, P. A., and Williams, H., 1988, Acadian basement thrusting, crustal delamination, and structural styles in and around the Humber Arm Allochthon, Western Newfoundland: Geology, v. 16, p. 370–373.

Colman-Sadd, S. P., 1982, Two stage continental collision and plate driving forces: Tectonophysics, v. 90, p. 263–282.

Colman-Sadd, S. P., and Swinden, H. S., 1984, A tectonic window in central Newfoundland? Geological evidence that the Appalachian Dunnage Zone is allochthonous: Canadian Journal of Earth Sciences, v. 21, p. 1349–1367.

Coyle, M., and Strong, D. F., 1987, Geology of the Springdale Group: A newly recognized Silurian epicontinental-type caldera in Newfoundland: Canadian Journal of Earth Sciences, v. 24, p. 1135–1148.

Currie, K. L., and Piasecki, M.A.J., 1989, A kinematic model for southwestern Newfoundland based upon Silurian sinistral shearing: Geology, v. 17, p. 938–941.

Currie, K. L., and van Berkel, J. T., 1989, Geochemistry of post-tectonic mafic intrusions in the Central Gneiss Terrane of southwestern Newfoundland: Atlantic Geology, v. 25, p. 181–190.

Dallmeyer, R. D., Blackwood, R. F., and Odom, A. L., 1981, Age and origin of the Dover Fault: Tectonic boundary between the Gander and Avalon zones of the northeastern Newfoundland Appalachians: Canadian Journal of Earth Sciences, v. 18, p. 1431–1442.

Dec, T., and Colman-Sadd, S. P., 1990, Timing of ophiolite emplacement onto the Gander Zone: Evidence from provenance studies in the Mount Cormack Subzone, *in* Current research (1990): Newfoundland Department of Mines and Energy Geological Survey Branch Report 90-1, p. 289–303.

Dunning, G. R., and O'Brien, S. J., 1989, Late Proterozoic–early Paleozoic crust in the Hermitage Flexure, Newfoundland Appalachians: U/Pb ages and tectonic significance: Geology, v. 17, p. 548–551.

Dunning, G. R., Krogh, T. E., O'Brien, S. J., Colman-Sadd, S. P., and O'Neill, P., 1988, Geochronologic framework for the Central Mobile Belt in southern Newfoundland and the importance of Silurian orogeny: Geological Association of Canada Program with Abstracts, v. 13, p. 34.

Dunning, G. R., and six others, 1990, Silurian Orogeny in the Newfoundland Appalachians: Journal of Geology, v. 98, p. 895–913.

Durling, P. W., Bell, J. S., and Fader, G.B.J., 1987, The geological structure and distribution of Paleozoic rocks on the Avalon Platform, offshore Newfoundland: Canadian Journal of Earth Sciences, v. 24, p. 1412–1420.

Elliott, C. G., Dunning, G. R., and Williams, P. F., 1991, New U/Pb zircon age constraints on the timing of deformation in north-central Newfoundland and implications for early Paleozoic Appalachian orogenesis: Geological Society of America Bulletin, v. 103, p. 125–135.

Grenier, R., 1990, The Appalachian fold and thrust belt, northwestern Newfoundland [M.Sc. thesis]: St. John's Memorial University of Newfoundland, 214 p.

Grenier, R., and Cawood, P. A., 1988, Variations in structural style along the Long Range Front, western Newfoundland, *in* Current research, part 1B: Geological Survey of Canada Paper 88-1B, p. 127–133.

Hanmer, S., 1981, Tectonic significance of the northeasterly Gander Zone, Newfoundland: An Acadian ductile shear zone: Canadian Journal of Earth Sciences, v. 18, p. 120–135.

Hibbard, J. P., 1982, Significance of the Baie Verte Flexure, Newfoundland: Geological Society of America Bulletin, v. 93, p. 790–797.

—— , 1983, Geology of the Baie Verte Peninsula, Newfoundland: Newfoundland Department of Mines and Energy Mineral Development Division Memoir 2, 279 p.

Jansa, L., and Wade, J. A., 1975, Geology of the continental margin off Nova Scotia and Newfoundland: Geological Survey of Canada Paper 74-30, p. 51–105.

Jenness, S. E., 1963, Terra Nova and Bonavista map areas, Newfoundland: Geological Survey of Canada Memoir 327, 184 p.

Karlstrom, K. E., van der Pluijm, B. A., and Williams, P. F., 1982, Structural interpretation of the eastern Notre Dame Bay area, Newfoundland: Regional post–Middle Silurian thrusting and asymmetrical folding: Canadian Journal of Earth Sciences, v. 19, p. 2325–2341.

Keen, C. E., and 10 others, 1986, Deep seismic reflection profile across the northern Appalachians: Geology, v. 14, p. 141–145.

Keen, C. E., Kay, W. A., Keppie, D., Marillier, F., Pe-Piper, G., and Waldron, J.W.F., 1990, Deep seismic reflection data from the Bay of Fundy and Gulf of Maine: Lithoprobe East Transect Report 13, p. 11–116.

Kidd, W.S.F., 1974, The evolution of the Baie Verte Lineament, Burlington Peninsula, Newfoundland [Ph.D. thesis]: Cambridge, Cambridge University, 294 p.

King, L. H., Fader, G.B.J., Jenkins, W.A.M., and King, E. L., 1986, Occurrence and regional geological setting of Paleozoic rocks on the Grand Banks of Newfoundland: Canadian Journal of Earth Sciences, v. 23, p. 504–526.

Kontak, D. J., Tuach, J., Strong, D. F., Archibald, D. A., and Farrar, E., 1988, Plutonic and hydrothermal events in the Ackley Granite, southeast Newfoundland, as indicated by total-fusion $^{40}Ar/^{39}Ar$ geochronology: Canadian Journal of Earth Sciences, v. 25, p. 1151–1160.

Lilly, H. D., 1967, Some notes on stratigraphy and structural styles in central-west Newfoundland, *in* Neale, E.R.W., and Williams, H., eds., Geology of the Atlantic Region: Geological Association of Canada Special Paper 4, p. 201–212

Marillier, F., and six others, 1989, Crustal structure and surface zonation of the Canadian Appalachians: Implications of deep seismic reflection data: Canadian Journal of Earth Sciences, v. 26, p. 305–321.

Neale, E.R.W., and Nash, W. A., 1963, Sandy Lake (east half) map-area, Newfoundland: Geological Survey of Canada Paper 62-28, 40 p.

Neale, E.R.W., Beland, J., Potter, R. R., and Poole, W. H., 1961, Preliminary tectonic map of the Canadian Appalachian Region based on age of folding: Transactions of the Canadian Institute of Mining and Metallurgy, v. 64, p. 405–412.

Neale, E.R.W., Kean, B. F., and Upadhyay, H. D., 1975, Post-ophiolite unconformity, Tilt Cove–Betts Cove area, Newfoundland: Canadian Journal of Earth Science, v. 12, p. 880–886.

Nowlan, G. S., and Barnes, C. R., 1987, Thermal maturation of Paleozoic strata in eastern Canada from conodont colour alteration index (CAI) data with implications for burial history, tectonic evolution, hotspot tracks and mineral and hydrocarbon exploration: Geological Survey of Canada Bulletin 367, 47 p.

O'Brien, B. H., 1988, Relationships of phyllite, schist and gneiss in the La Poile Bay–Roti Bay area (parts of 110/9 and 110/16), southwestern Newfoundland, *in* Current research: Newfoundland Department of Mines Mineral Development Division Report 88-1, p. 109–125.

O'Brien, B. H., O'Brien, S. J., and Dunning, G. R., 1991, Silurian cover, late Precambrian–Early Ordovician basement, and the chronology of Silurian orogenesis in the Hermitage Flexure (Newfoundland Appalachians): American Journal of Science, v. 291, p. 760–799.

O'Neill, P. P., and Knight, I., 1988, Geology of the east half of the Wier's Pond (2E/1) map area and its regional significance, *in* Current research (1988): Newfoundland Department of Mines Mineral Development Division Report 88-1, p. 165–176.

Piasecki, M.A.J., 1988, A major ductile shear zone in the Bay d'Espoir area, Gander Terrane, southeastern Newfoundland, *in* Current research: Newfoundland Department of Mines Mineral Development Division Report 88-1, p. 135–144.

Piasecki, M.A.J., Williams, H., and Colman-Sadd, S. P., 1990, Tectonic relationships along the Meelpaeg, Burgeo and Burlington Lithoprobe transects in Newfoundland, *in* Current research (1990): Newfoundland Department of Mines, and Energy Geological Survey Branch Report 90-1, p. 327–339.

Reusch, D., 1987, Silurian stratigraphy and melanges, New World Island, North central Newfoundland, *in* Roy, D. C., ed., Geological Society of America Northeastern Section centennial field guide, p. 463–466.

Rodgers, J., 1965, Long Point and Clam Bank Formations, Western Newfoundland: Geological Association of Canada Proceedings, v. 16, p. 83–94.

—— , 1967, Chronology of tectonic movements in the Appalachian region of eastern North America: American Journal of Science, v. 265, p. 408–427.

Stockmal, G. S., and Waldron, J.W.F., 1990, Structure of the Appalachian deformation front in western Newfoundland: Implications of industrial seismic reflection data: Geology, v. 18, p. 765–768.

Stockmal, G. S., Colman-Sadd, S. P., Keen, C. E., O'Brien, S. J., and Quinlan, G., 1987, Collision along an irregular margin: A regional plate tectonic interpretation of the Canadian Appalachians: Canadian Journal of Earth Sciences, v. 24, p. 1098–1107.

Thomas, W. A., 1977, Evolution of Appalachian-Ouachita salients and recesses from reentrants and promontories in the continental margin: American Journal of Science, v. 277, p. 1233–1278.

Tuach, J., Davenport, P. H., Dickson, W. L., and Strong, D. F., 1986, Geochemical trends in the Ackley Granite, southeast Newfoundland: Their relevance to magmatic-metallogenic processes in high-silica granitoid systems: Canadian Journal of Earth Sciences, v. 23, p. 747–765.

van der Pluijm, B. A., van der Voo, R., and Johnson, R.J.E., 1989, Middle Ordovician to Lower Devonian evolution of the Northern Appalachians:

The Acadian phase? Geological Society of America Abstracts with Programs, v. 21, p. 72–73.

Waldron, J.W.F., 1985, Structural history of continental margin sediments beneath the Bay of Islands Ophiolite, Newfoundland: Canadian Journal of Earth Sciences, v. 22, p. 1618–1632.

Wanless, R. K., Stevens, R. K., Lachance, G. R., and Rimsaite, J.Y.H., 1965, Age determinations and geological studies, K-Ar isotopic ages, Report 5: Geological Survey of Canada Paper 64-17, part 1.

Wanless, R. K., Stevens, R. D., Lahance, G. R., and Edmond, C. M., 1967, Age determinations and geological studies, K-Ar ages, Report 7: Geological Survey of Canada Paper 66-17.

Williams, H., 1964, The Appalachians in northeastern Newfoundland—A two-sided symmetrical system: American Journal of Science, v. 262, p. 1137–1158.

—— , 1971, Geology of Belleoram map-area, Newfoundland (1M/11): Geological Survey of Canada Paper 70-65, 39 p.

—— , 1978, Tectonic lithofacies map of the Appalachian Orogen: Memorial University of Newfoundland Map 1, scale 1:1,000,000.

—— , 1985, Geology, Stephenville map-area, Newfoundland: Geological Survey of Canada Map 1579A, scale 1:100,000, with descriptive notes and bibliography.

—— , 1993, Geology of the Appalachian/Caledonian Orogen in Canada and Greenland; Geological Survey of Canada, Geology of Canada, no. 6 (also Geological Society of America, The Geology of North America, v. F-1) (in press).

Williams, H., and Cawood, P. A., 1989, Geology, Humber Arm Allochthon, Newfoundland: Geological Survey of Canada Map 1678A, scale 1:250,000.

Williams, H., and Piasecki, M.A.J., 1990, The Cold Spring Melange and its significance in central Newfoundland: Canadian Journal of Earth Sciences, v. 27, p. 1126–1134.

Williams, H., Kennedy, M. J., and Neale, E.R.W., 1972, The Appalachian structural province, *in* Price, R. A., and Douglas, R.J.W., eds., Variations in tectonic styles in Canada, Geological Association of Canada Special Paper 11, p. 181–261.

Williams, H., Colman-Sadd, S. P., and Swinden, H. S., 1988, Tectonic-stratigraphic subdivisions of central Newfoundland, *in* Current research, part B: Geological Survey of Canada Paper 88-1B, p. 91–98.

Williams, H., Piasecki, M.A.J., and Colman-Sadd, S. P., 1989a, Tectonic relationships along the proposed central Newfoundland Lithoprobe Transect and regional correlations, *in* Current research, part B: Geological Survey of Canada Paper 89-1B, p. 55–66.

Williams, H., Dickson, W. L., Currie, K. L., Hayes, J. P., and Tuach, J., 1989b, Preliminary report on a classification of Newfoundland granitic rocks and their relations to tectonostratigraphic zones and lower crustal blocks, *in* Current research, part B: Geological Survey of Canada Paper 89-1B, p. 47–53.

Williams, H., Piasecki, M.A.J., and Johnston, D., 1991, The Carmanville Melange and Dunnage-Gander relationships in northeast Newfoundland, *in* Current research, part D: Geological Survey of Canada Paper 91-1D, p. 15–23.

Wilton, D.H.C., 1983, The geology and structural history of the Cape Ray fault zone in southwestern Newfoundland: Canadian Journal of Earth Sciences, v. 20, p. 1119–1133.

LITHOPROBE CONTRIBUTION NO. 131

MANUSCRIPT SUBMITTED NOV. 1989

MANUSCRIPT ACCEPTED BY THE SOCIETY JUNE 8, 1992

Geological Society of America
Special Paper 275
1993

Acadian orogeny in west Newfoundland:
Definition, character, and significance

Peter A. Cawood
Department of Earth Sciences, Memorial University of Newfoundland, St. John's, Newfoundland A1B 3X5, Canada

ABSTRACT

Western Newfoundland underwent widespread deformation and metamorphism following emplacement of the Taconian allochthons but prior to deposition of onlapping Carboniferous sediments. Stratigraphic and isotopic data indicate that this deformation and metamorphism took place during the latest Silurian to early Devonian Acadian orogeny.

Acadian deformation is characterized by an overall pattern of west-directed thrusting. Grenville basement is thrust above its carbonate cover sequence, and both are locally thrust over the Taconian allochthons. Between the Long Range and Indian Head inliers, cover and basement are probably delaminated with a series of east-verging folds and thrusts developed above an inferred west-directed passive roof duplex in Grenville basement. Metamorphism accompanying deformation locally reaches upper greenschist to lower amphibolite facies.

West Newfoundland lies at the Acadian orogenic front. The intensity of deformation and metamorphism decreases and dies out westward. Structural and topographic relief across the frontal zone resulted in gravitational collapse and the generation of a series of extensional structures. The basal detachment of west Newfoundland ophiolites truncates Acadian compressive structures and probably reflects late-Acadian extensional tectonics rather than initial Taconian emplacement.

INTRODUCTION

Orogenies are periods of regional deformation and thermal activity, concentrated at, or near, plate boundaries. They are a consequence of the impingement of buoyant lithosphere on a plate margin, resulting in crustal and lithospheric thickening and ultimately in the development of major linear mountain belts, or orogens. Structural and topographic relief produced during orogenesis results in major gravitational instabilities that can lead to late-stage structural collapse of segments of an orogen through extensional faulting (e.g., Dewey, 1988). The duration of an orogenic event is generally short, on the order of a few million years, as opposed to the history of the associated orogen that may span tens, or hundreds, of millions of years.

A number of major orogenic episodes are recognized in the late Precambrian to late Paleozoic Appalachian-Caledonian orogen. These include the latest Precambrian Avalonian orogeny, the

early Paleozoic Taconian orogeny, the mid-Paleozoic Salinic and Acadian orogenies, and the late Paleozoic Alleghanian orogeny. The Avalonian orogeny is restricted to the outboard Avalonian Zone. It largely predates, and may be unrelated to, the main Appalachian-Caledonian orogenic cycle. The other orogenic events correspond with the suturing of oceanic and continental fragments to the eastern North American margin of the Laurentian craton. This chapter outlines the effects of the Acadian orogeny at its orogenic front in the west Newfoundland segment of the orogen. Structural and topographic relief at the orogenic front has resulted in an orogenic event involving both early contractional and late extensional phases of deformation.

REGIONAL SETTING

West Newfoundland lies within the Humber Zone of the Appalachian-Caledonian orogen (Williams, 1978). The Humber Zone represents the ancient eastern continental margin, or mio-

Cawood, P. A., 1993, Acadian orogeny in west Newfoundland: Definition, character, and significance, *in* Roy, D. C., and Skehan, J. W., eds., The Acadian Orogeny: Recent Studies in New England, Maritime Canada, and the Autochthonous Foreland: Boulder, Colorado, Geological Society of America Special Paper 275.

geocline, of the Laurentian craton and extends from the western limit of Appalachian deformation to the Baie Verte–Brompton Line, a steep structural belt marked by discontinuous, dismembered ophiolite fragments (Williams, 1979, 1984; Williams and St. Julien, 1982). Rocks of the Humber Zone are divisible into three distinct pre–Late Ordovician elements (Fig. 1): Precambrian crystalline basement, a latest Precambrian to Middle Ordovician cover sequence that unconformably overlies basement, and a series of thrust sheets of sedimentary and igneous rocks that structurally overlie the cover sequence. The transported rocks comprise the Humber Arm, Hare Bay, and Southern White Bay allochthons (Williams and Cawood, 1989; Williams and Smyth, 1983; Smyth and Schillereff, 1982). The allochthons were emplaced during the Early to Middle Ordovician Taconian orogeny. Late Ordovician, Late Silurian, and Carboniferous successor basin sequences and a Silurian-Devonian igneous stitching assemblage occur both within and along the margin of the zone (Fig. 1).

The evolution of the Humber Zone records the birth and destruction of an ocean basin(s), represented by the Dunnage Zone which lies largely to the east of the Baie Verte–Brompton Line (Williams and Stevens, 1974; Cawood and others, 1988). Rocks immediately overlying Grenville basement represent a rift facies assemblage of latest Precambrian to early Cambrian clastic and igneous lithologies (Williams and Hiscott, 1987). This assemblage is overlain by a laterally continuous, Early Cambrian to Early Ordovician drift facies assemblage. Passive continental margin sedimentation was terminated in the Early-Middle Ordovician by emplacement of the imbricate thrust stack of Taconian allochthons. This caused drowning and collapse of the continental margin and establishment of a deep-water foreland basin that was infilled by sediment derived from the advancing thrust sheets (Stevens, 1970). The lower slices of the thrust stack consist of rocks from the continental margin. The highest slices of the allochthonous sequence consist of ophiolite suites, which represent oceanic crust and mantle (Church and Stevens, 1971). They are the farthest traveled of the slices within the thrust stack and represent fragments of the Dunnage Zone that have been thrust over the miogeocline (Fig. 1).

The structural fabric of the Humber Zone reflects the combined effects of the Taconian, Acadian, and Alleghanian orogenies. Although this chapter concentrates on the Acadian event, general features of the Taconian and Alleghanian events are briefly outlined below, so as to provide a background against which to assess the character of the Acadian orogeny.

Taconian deformation

Taconian deformation established the gross structural style of the Humber Zone. This deformation involved assembly and emplacement of thin slices of oceanic lithosphere and deep-water continental margin sequences onto the adjacent continental shelf (Williams and Stevens, 1974). Thrusting was dispersed through incompetent, shale-dominated sequences, resulting in widespread development of melanges (Stevens, 1970). In the western parts of the Humber Zone, Taconian thrusting and penetrative deformation are restricted to the uppermost crustal levels and only involve the transported slices of the allochthon. Underlying collapsed shelf and crystalline basement are largely undeformed. This relationship is well displayed in the Humber Arm and Hare Bay allochthons. The base of the Humber Arm allochthon north of the isthmus of Port au Port Peninsula is a sharp planar contact separating basal melange of the allochthon from the underlying, undeformed foreland basin sequence at the stratigraphic top of the shelf carbonates. Shearing in the melange and bedding in the underlying succession are subparallel to the basal contact, indicating that the allochthon was emplaced along a more or less flat surface (Cawood and others, 1988). A gradation between melange and underlying shale marks the same boundary at the northern end of the Humber Arm allochthon and at the base of the Hare Bay allochthon. Folding associated with Taconian thrusting is characteristically west-directed recumbent structures.

The extent of Taconian deformation and metamorphism in the eastern, interior parts of the Humber Zone is poorly constrained. Previous workers have considered the shelf carbonates below the Southern White Bay allochthon to be penetratively deformed prior to accumulation of the Silurian Sops Arm Group (e.g., Lock, 1972). This deformation has been related to Taconian allochthon emplacement (Williams, 1977; Smyth and Schillereff, 1982) and has been interpreted to reflect movement of the sole thrust for Taconian deformation to deeper crustal levels toward the more internal parts of the orogen. Recent work in this region indicates that the platform carbonates contain the same deformation sequence as the Sops Arm Group and thus did not undergo penetrative deformation until the Silurian or later (P. Cawood, unpublished data).

The Taconian structural front is the present exposed foreland (western) extent of the Humber Arm allochthon on the Port au Port Peninsula (Cawood and Williams, 1988; Stockmal and Waldron, 1990).

Alleghanian deformation

Alleghanian deformation is relatively mild and concentrated along the Cabot Fault System (Fig. 1), which runs from west of Cape Ray northeast to White Bay. The fault lies along the eastern side of the Humber Zone and has reactivated and largely overprinted the Baie Verte–Brompton Line. Dextral strike-slip movements occurred along the fault in the Early Carboniferous and controlled sedimentation in the Deer Lake and Bay St. George basins (Bradley, 1982; Hyde and others, 1988; Knight, 1983). Pre-Visean rock units within the basin are locally isoclinally folded but generally lack a penetrative cleavage, whereas unconformably overlying Visean and younger rock units are only deformed into broad open folds. Thermal maturation studies suggest temperatures of up to 200°C for the pre-Visean rock units and up to 100°C for the younger units (Hyde and others, 1988). Carboniferous strata outside the zone of the Cabot Fault system

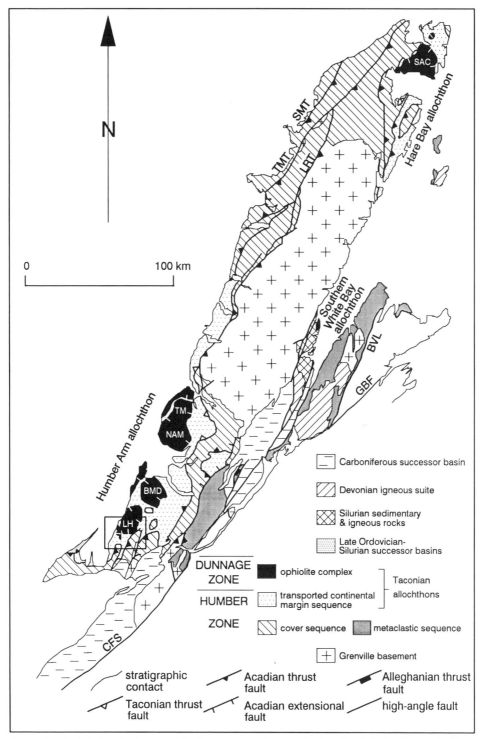

Figure 1. Simplified geologic map of the Humber Zone in west Newfoundland. Abbreviations: SMT—St. Margaret Bay thrust; TMT—Ten Mile Lake thrust; LRT—Long Range thrust; BVL—Baie Verte line; GBF—Green Bay fault; CFS—Cabot Fault system; SAC—St. Anthony complex; TM—Table Mountain; NAM—North Arm Mountain; BMD—Blow-Me-Down Mountain; LH—Lewis Hills. Box at southern end of Lewis Hills outlines area of Figure 5.

are essentially undeformed and unmetamorphosed (Williams and Cawood, 1989).

ACADIAN OROGENY

In Newfoundland, the term *Acadian* has generally been used to encompass any deformation, plutonism, and metamorphism of mid-Paleozoic age. The effects of Acadian orogeny are concentrated in central Newfoundland, decreasing in intensity to the east and west (Williams, this volume). Recent work by Dunning and others (1990), based on precise U-Pb age determinations on igneous and metamorphic rocks in southern Newfoundland, has questioned the grouping of all mid-Paleozoic orogenic activity under a single Acadian banner. Their data indicate a discrete and widespread orogenic event of Silurian age. They include this event within the Salinic orogeny (Boucot, 1962) and contest that it is temporally discrete from the Acadian (*sensu strictu*), which they argue should be restricted to events of Devonian age as originally defined (Williams, 1893; Boucot and others, 1964). Currently available data suggest that the majority of the mid-Paleozoic orogenesis in the Humber Zone is associated with the Acadian *sensu strictu*. Evidence for a Salinic orogeny in west Newfoundland and its effects are discussed at the end of this section.

Definition and timing

West Newfoundland has undergone widespread and in many places intense Acadian deformation and metamorphism (Cawood, 1989; Cawood and Williams, 1988; Smyth and Schillereff, 1982; Stockmal and Waldron, 1990; Williams, this volume; Williams and Bursnall, 1982; Williams and Cawood, 1989; Williams and Smyth, 1983). The intensity of orogenesis increases across the zone toward the orogenic hinterland in the east. Rocks in the west are essentially undeformed and unmetamorphosed, whereas those in the east show a penetrative fabric(s) and metamorphism(s) up to upper greenschist or lower amphibolite facies. Igneous activity is limited to a few posttectonic plutons (Devils Room Granite, and possibly parts of the Gull Lake Intrusive Suite, and the Wild Cove Pond Intrusive Suite) along the eastern side of the zone. No sedimentation associated with the Acadian orogeny is preserved in west Newfoundland, but part of the thick sedimentary sequences recognized on seismic lines from the Gulf of St. Lawrence may include detritus shed from the Acadian hinterland.

Stratigraphic age constraints on the Acadian orogeny in west Newfoundland are outlined in Figure 2. Acadian deformational and metamorphic features are bracketed by their overprinting of Ordovician Taconian structural elements associated with allochthon emplacement, the presence of sub-Carboniferous unconformities, posttectonic Devonian granites cutting deformed rocks of Silurian and older age, and Siluro-Devonian K-Ar and $^{40}Ar/^{39}Ar$ cooling ages.

The southern White Bay region (Fig. 1) provides the tightest stratigraphic controls on the age of Acadian orogeny in west Newfoundland. Here, deformed and metamorphosed rocks of the Silurian Sops Arm Group are intruded by the undeformed and posttectonic granite of the Gull Lake Intrusive Suite. Sedimentary rocks from the Natlins Cove Formation at the top of the Sops Arm Group (Lock, 1969) contain a conodont fauna of upper Ludlow age (F. O'Brien, personal communication, 1990). There are no age data for the base of the Sops Arm Group, but a microgranite intruding rocks of the adjacent southern White Bay allochthon, and probably representing part of the intrusive equivalent of the Sops Arm volcanics, has yielded a U-Pb zircon and titanite age of 432 ±2 Ma (Dunning, 1987; Fig. 3). This age corresponds with an early Silurian age on the time scales of Harland and others (1982) and Palmer (1983). Volcanic rocks within the Sops Arm Group are correlated with widespread epicontinental Silurian igneous activity along the Appalachian-Caledonian orogen (Coyle and Strong, 1987). Rhyolites extruded during this event in central Newfoundland have yielded U-Pb zircon ages of 429 +6/–5 Ma (Springdale Group; Chandler and others, 1987), 429 ±3 Ma to 427 ±3 Ma (Topsails; Whalen and others, 1987), and 429 +7/–3 Ma (Bear Pond Rhyolite; Dunning and others, 1990). An upper age limit on deformation of the Sops Arm Group is provided by the crosscutting posttectonic granites of the Gull Lake Intrusive Suite, which have yielded a U-Pb zircon age of around 390 Ma (H. Baadsgaard, P. Erdmer, and J. V. Owen, unpublished data). Carboniferous rocks of the Deer Lake Basin unconformably overlie both the Gull Lake Intrusive Suite and the Sops Arm Group (Fig. 2).

Age and structural relationships in the Humber Arm region are consistent with those from the White Bay region in constraining the age of the Acadian orogeny. On the Port au Port Peninsula, the Clam Bank Group of Pridoli age (Berry and Boucot, 1970; F. O'Brien, personal communication, 1990) lies in a footwall syncline, structurally overlain along the Round Head thrust by Ordovician shelf carbonates (Cawood and Williams, 1988; Stockmal and Waldron, 1990; Williams and Cawood, 1989). The Round Head thrust is interpreted to form part of a linked system lying at, or near, the Acadian thrust front. Nearby Carboniferous strata are undeformed and unconformably overlie the deformed Silurian and older sequence (Fig. 2).

In the Hare Bay region, the age of Acadian orogeny is poorly constrained because of the paucity of Siluro-Devonian strata. An isolated and unfossiliferous sequence of deformed and metamorphosed sedimentary rocks on St. Julian Island, just east of Hare Bay, is correlated with the Silurian Taylors Pond Conglomerate of the Sops Arm Group (Williams and Smyth, 1983). The rocks on the island contain the same penetrative fabric elements as those overprinting the adjoining rocks of the Taconian Hare Bay allochthon. Carboniferous strata in this region, although in fault contact with surrounding rock units, lack these deformational and metamorphic elements, which suggests that deformation is pre-Carboniferous in age and ascribable to the Acadian orogeny.

Additional constraints on the timing of Acadian orogenic

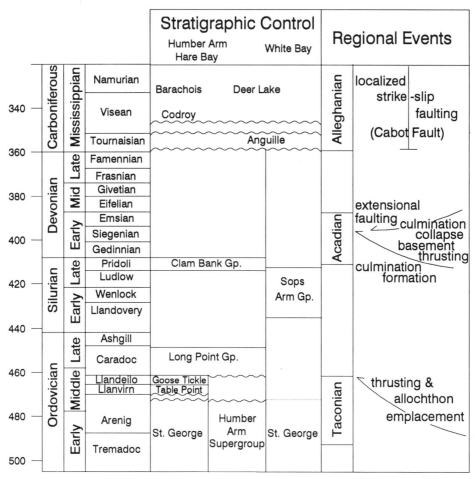

Figure 2. Stratigraphic constraints on timing of Paleozoic orogenies in west Newfoundland. Time scale from Harland and others (1982), except for Ordovician period, which is taken from Tucker and others (1990).

movements are provided by K-Ar and ^{40}Ar/^{39}Ar age data (Fig. 3; Appendix). Increasing metamorphic grade toward the eastern part of the Humber Zone (see below) has resulted in the partial to complete resetting of K-Ar and ^{40}Ar/^{39}Ar isotopic systems in preexisting mineral grains or the growth of new mineral phases along the eastern side of the zone. Grenvillian gneisses in the Indian Head Complex or western Long Range Complex give Precambrian K-Ar and ^{40}Ar/^{39}Ar cooling ages (Fig. 3; Dallmeyer, 1978; Bostock and others, 1983). Along the eastern side of the Long Range, the gneisses show a greenschist facies overprint. Hornblende from gneisses within this zone maintain Grenvillian-type K-Ar cooling ages, whereas biotites give Paleozoic cooling ages (Bostock and others, 1983). Two samples of Precambrian Long Range dykes from within the greenschist facies zone give ^{40}Ar/^{39}Ar whole rock apparent ages of 380 ± 14 Ma and 395 ± 5 Ma (Stukas and Reynolds, 1974, new decay constant), reflecting Acadian orogenic reset. On the Baie Verte Peninsula, the early Paleozoic or older Fleur de Lys Supergroup and East Pond metamorphic suite give hornblende, biotite, and

muscovite mineral ages that range from around 430 Ma to 340 Ma (Dallmeyer, 1977; Dallmeyer and Hibbard, 1984) but with discrete maxima of around 400 to 420 Ma and 340 to 360 Ma (Fig. 3; Appendix). The younger ages are restricted to the northern part of the peninsula and suggest the presence of a localized Carboniferous (Alleghanian) thermal overprint. There are insufficient data to resolve the question of whether the spread of older ages reflects incomplete resetting of old ages by the Acadian orogeny or the competing effects of protracted or multiple mid-Paleozoic orogenic movements (e.g., Acadian and Salinic). The K-Ar and ^{40}Ar/^{39}Ar isotopic data, particularly the Ar/Ar data from the Long Range dykes and Baie Verte Peninsula, indicate widespread cooling of minerals and whole rocks below their blocking temperature at around 400 ± 20 Ma. This is consistent with stratigraphic constraints on the timing of the Acadian orogeny and is interpreted to represent cooling due to rapid exhumation of the region.

Thus, the Acadian orogeny in west Newfoundland is constrained, largely on the basis of stratigraphic and structural rela-

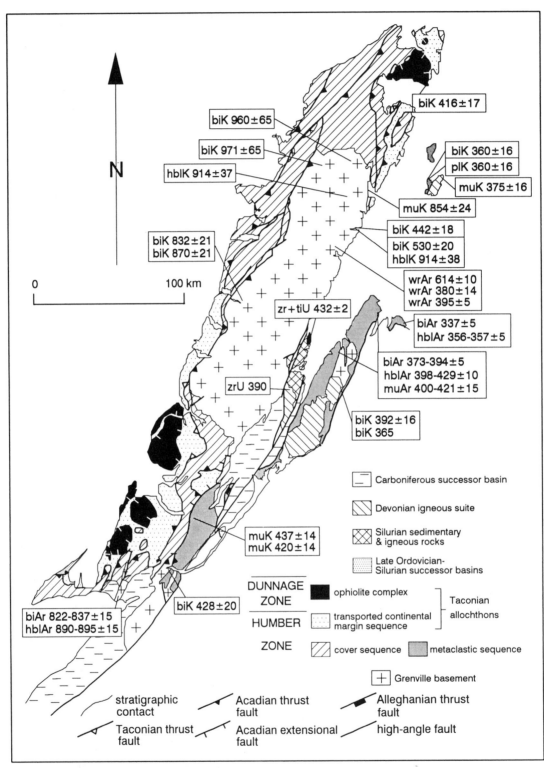

Figure 3. Map of west Newfoundland showing radiometric age data that place constraints on the timing of Acadian orogenesis in west Newfoundland. Source of data listed in the Appendix. Abbreviations: bi—biotite; hbl—hornblende; mu—muscovite; ti—titanite; wr—whole rock; zr—zircon; Ar—^{40}Ar/^{39}Ar; K—K/Ar; U—U/Pb.

tions in the White Bay region, to the Early Devonian (post–405 Ma to pre–390 Ma on the basis of the Harland and others [1982] and Palmer [1983] time scales). This is in agreement with stratigraphic and age data reviewed by McKerrow (1988) that showed that the climax of Caledonian deformation in the British Isles and Ireland and the Acadian orogeny in the Canadian Appalachians occurred during the Early Devonian Emsian stage (390 Ma to 400 Ma on the McKerrow and others [1985] time scale or around 390 Ma to 395 Ma on the Harland and others [1982] and Palmer [1983] time scales).

Character

Acadian deformation overprinted the Taconian regional fabric in west Newfoundland and is characterized by early contractional and late extensional structures (Cawood and Williams, 1988; Cawood, 1989, 1990). Whereas Taconian structural elements are restricted to the three allochthons in west Newfoundland, Acadian compressive deformation extended to deeper crustal levels, into the previously undeformed shelf carbonates and Grenville basement. Extensional structures are largely restricted to high-level crustal segments. Acadian metamorphism was synchronous with compressive deformation and increases in grade to the east. The following discussion will concentrate on the character of contractional structures within the western Humber Zone and will outline sructural elements from south to north along the zone. Along the eastern side of the zone, within the metaclastic sequence, it is difficult to establish the age of fabric elements and ascribe them to specific orogenic events.

Contractional structures

Acadian contractional structures are characterized by thrust faults and asymmetric folds, along with associated foliation development. Structures developed within a regime of overall west-directed compressive stress, but conjugate east-directed structures are locally developed.

In the vicinity of the Humber Arm allochthon, Acadian deformational style varies both along and across strike. At the northern and southern ends of the allochthon, west-directed thrusts and west-verging folds are the main structural elements. In the intervening central and eastern region of the allochthon, east-directed thrusts and folds are widespread (Bosworth, 1985; Cawood and Williams, 1988; Gillespie, 1983; Hibbard, 1983a; Lilly, 1967; Waldron, 1985; Fig. 4, sections C-C′, D-D′, and E-E′).

Around the southern end of the Humber Arm allochthon, shelf carbonates and locally Grenville basement lie in the core of west-verging anticlines. Rocks of the Humber Arm allochthon occupy the intervening synclines. The anticlines preserve a partial box-fold morphology with steep western limbs and flat-lying axial zones. The eastern margin of the anticlines is truncated by down-to-the-east normal faults. The anticlines lie in the hanging-wall to west-directed thrusts (Fig. 4, section E-E′). The presence

of Grenville basement in the core of the Phillips Brook and North Brook anticlines (Williams and Cawood, 1989) and in the hanging wall of the Grand Lake thrust indicates that at least in the central and eastern parts of the area the basal detachment of the thrust system extends to basement depths. Throw on the thrust faults decreases to the west. In the east, Grenville basement showing a well-developed Acadian shear fabric and amphibolite facies metamorphism is thrust over platformal carbonates of the cover sequence, requiring a minimum throw of around 5 km and perhaps considerably more. In the vicinity of the Table Mountain and Phillips Brook anticlines, the upper stratigraphic sections of the cover sequence are thrust over melanges from the lower thrust slices of the allochthon, suggesting a throw of around a few kilometers. Grenville basement within the core of these anticlines lacks a penetrative Paleozoic fabric and maintains late Proterozoic $^{40}Ar/^{39}Ar$ cooling ages (Dallmeyer, 1978). In the west, along the Round Head thrust, upper parts of the cover sequence are thrust over neoautochthonous strata of the Long Point Group, implying a throw of less than a few kilometers. Restoration of this 75-km-wide section suggests a minimum foreshortening on the order of 20 to 25 km. The basal detachment is inferred to cut upsection to the west (in the transport direction) via a series of ramps and flats. The anticlines developed in the hanging wall to the imbricate splays off the basal detachment are interpreted as fault bend folds (Suppe, 1983) formed through movement over ramps in the basal detachment. However, the presence of a well-developed footwall syncline indicates that folding is not all of fault bend origin but also involved some rock units prior to, or during, faulting. An east-dipping foliation, axial planar to the folds and parallel to the imbricate fault planes, is developed in the eastern and central parts of the region. Intensity of the cleavage fabric increases to the east along with increasing throw on the thrust faults.

In the central region of the allochthon, between Serpentine Lake and Bonne Bay, Acadian deformational style is dominated by a downward-converging cleavage fan (Cawood and Williams, 1988). Orientation of the cleavage fabric changes across this region from shallowly to moderately west dipping in the east, increasing to subvertical in the central part of the region, and then decreasing to east dipping in the west. Cleavage progressively decreases in intensity to the west, and rock units at, or east of, the Bay of Islands ophiolite complex lack an Acadian cleavage. Structural vergence is symmetrical across the cleavage fan. Cleavage is axial planar to folds that change progressively from west dipping, east verging on the eastern side of the cleavage fan to east dipping, west verging in the west. East-directed thrusting is widespread within the eastern part of the fan structure (Fig. 4, section D-D′; Bruckner, 1966; Bosworth, 1985; Williams and Cawood, 1986).

The northwestern and southwestern margins of the cleavage fan adjoin, and probably merge with, the east-dipping cleavage with associated west-verging folds and thrusts that are developed in and around the southern and northern ends of the allochthons. As pointed out by Cawood and Williams (1988), the absence of any overprinting relations at these transition zones in structural style indicates that along-strike variations in vergence reflect changes in the response of rock units to the same deformational event, rather than to separate, unrelated events. Thus, along- and across-strike changes in Acadian structural vergence developed in response to a regime of regional west-directed thrusting (Williams and Cawood, 1986; Cawood and Williams, 1988). To account

for the localized east-directed thrusting and east-verging folds developed in, and east of, the central parts of the allochthon, Cawood and Williams (1988) proposed that cover and basement were delaminated, with the latter stacked in a west-directed duplex wedge below the cover sequence (Fig. 4, section D–D'). The absence of Grenville basement from the central region and its presence to the north (Long Range Inlier) and south (Indian

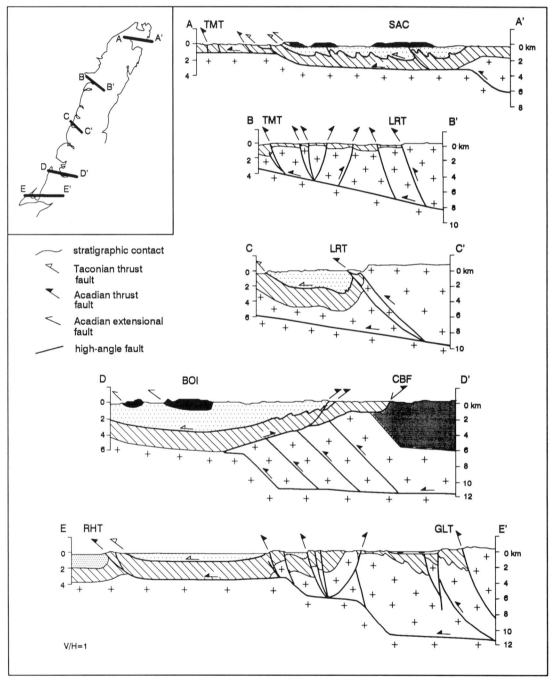

Figure 4. Geologic cross sections of the Newfoundland Humber Zone. Inset shows location of sections. Abbreviations: SAC—St. Anthony ophiolite complex; TMT—Ten Mile Lake thrust; LRT—Long Range thrust; BOI—Bay of Islands ophiolite complex; CBF— Corner Brook Lake fault; RHT—Round Head thrust; GLT—Grand Lake thrust. Patterns as for Figure 1. Horizontal scale = vertical scale.

Head Complex) coincide with along-strike variations in structural style (Cawood and Botsford, 1991). This is also interpreted to reflect the change from west-directed thrusting throughout the crustal section (basement, cover, and allochthon) in the northern and southern regions to the delamination of the west-directed basement duplex from the overlying sequences in the central region.

Along the northern segments of the Humber Arm allochthon, north of Bonne Bay, structural style is similar to the west-directed folds and thrusts in the southern region (Fig. 4, section C–C'). In contrast to the southern region, however, deformation takes place over a relatively narrow zone along the Long Range structural front, which is the zone of deformation associated with the leading (western) edge of the Long Range Inlier of Grenville basement rocks. The inlier is thrust westward over a footwall sequence of Humber Arm allochthon and the structurally under-lying cover sequence consisting largely of shelf carbonates. The Long Range thrust dips to the southwest at around 35° and at its present level of exposure is a high-level, brittle fracture marked by a single, sharp planar detachment (Cawood and Williams, 1986). The inlier has behaved as a relatively rigid block during over-thrusting: The immediate hanging-wall sequence to the Long Range Thrust shows only a narrow zone of brittle fracturing extending less than 1 m into the Grenvillian gneisses (Cawood and Williams, 1986, 1988; Grenier, 1990). Narrow zones of brittle fracturing, postdating Grenvillian metamorphism, occur locally within the inlier (Owen and Erdmer, 1986; Owen, 1986) and probably relate to Acadian deformation. In addition, a broad-spaced fracture cleavage occurs locally along the eastern side of the inlier in association with the more intense zones of Acadian deformation. In Figure 4 (section C–C'), the basal de-tachment associated with overthrusting of the inlier is inferred to propagate offshore into the Gulf of St. Lawrence (Cawood and Williams, 1986). This is the inferred southern extension of the Ten Mile Lake thrust (Fig. 1).

Although the Long Range Inlier lacks a pervasive, post-Grenvillian penetrative fabric, it does show evidence of having undergone broad-scale folding during Acadian overthrusting. The inlier is deformed into a doubly plunging anticline. The anticline is breached along its western side by the Long Range thrust. Basal rock units of the cover sequence unconformably overlying Gren-villian basement occur locally in the hanging wall to the Long Range thrust. The unconformity surface and bedding within these sedimentary rock units is overturned along the southwestern margin of the inlier, dipping at 70 to 80° to the southeast but younging to the northwest. Along the southeast trailing edge of the inlier, in White Bay, the unconformity surface and overlying rock units dip 50 to 60° southeast. Thus in the south the inlier is deformed into an asymmetric hanging-wall anticline. Farther north, bedding in the hanging wall of the thrust is right way up. The footwall sequence to the southern segment of the Long Range thrust lies in an asymmetric syncline with a steep to over-turned eastern limb and a shallow dipping western limb (Fig. 4, section C–C'). A major splay off the Long Range thrust (Parsons

Pond thrust; Williams and others, 1986; Cawood and others, 1987; Grenier and Cawood, 1988) disrupts the footwall sequence and brings the cover sequence up over the Humber Arm alloch-thon. Mesoscopic-scale west-verging folds with an east-dipping axial plane cleavage occur within the cover sequence. Folding in both the hanging wall and footwall of the Long Range thrust decreases to the north, suggesting decreasing foreshortening across the thrust.

Structural style of the more northerly parts of the Long Range structural front is similar to that described above for the region adjacent to the Humber Arm allochthon and is dominated by west-directed thrusting with minor asymmetric folding (Fig. 4, section D–D'; Grenier and Cawood, 1988). However, thrusting, rather than being concentrated along the Long Range thrust as it is in the south, is spread over two additional major thrusts (Ten Mile Lake and St. Margaret Bay thrusts, Fig. 1) and a series of intervening smaller-scale thrusts and high-angle faults. A conse-quence of this broadening of the structural front to the north is that folding of the sedimentary sequence is less pronounced. Rock units within both the hanging wall and footwall to the Long Range, Ten Mile Lake, and St. Margaret Bay thrusts are generally flat lying, and folding, apart from asymmetric synclines in the footwalls to the major thrusts, is limited to broad regional warp-ing. Although thrusting is overall west directed in this region, minor east-directed thrusting occurs immediately to the north of the Humber Arm allochthon (Grenier, 1990). Backthrusting and associated folding are sandwiched between the Long Range and Ten Mile Lake thrusts. At the northern end of the Long Range Inlier, displacement across the Long Range thrust is relatively minor, with most of the regional foreshortening taken up along the outboard Ten Mile Lake thrust. Grenville basement, uncon-formably overlain by the cover sequence, occurs in the immediate hanging wall to this fracture at Castor River (Knight, 1986) and Ten Mile Lake (Bostock and others, 1983). Displacement on the Ten Mile Lake thrust decreases to the north where it merges with the Long Range thrust (Fig. 1). In this region there is a further foreland-directed transfer of the site of maximum displacement to the St. Margaret Bay thrust. There is an overall foreland-directed stepping out of the locus of maximum foreshortening in this zone (Grenier and Cawood, 1988). Total displacement across the northern Long Range is on the order of a few kilometers (Grenier, 1990).

In the Hare Bay region, cover and basement rock units, along with the overlying Hare Bay allochthon, underwent wide-spread west-directed thrusting and folding during Acadian com-pressive deformation (Cawood, 1989; Fig. 4, section A–A'). The intensity of this deformation, like that in regions farther south, increases to the east. The cover sequence west of the allochthon is only mildly deformed. Offset of cover units across thrust faults is relatively minor, generally on the order of a few hundred meters of stratigraphic section or less. The exact level of the sole thrust in this region is unknown, but given its position near the thrust front it is assumed to lie above basement (Fig. 4, section A–A'). Folds are open and lack an axial planar cleavage. Toward the east, thrusting becomes more intense, and folds are tighter and west verging with an axial planar cleavage (Cawood, 1989). Both the offset across thrust faults and the level of the basal detachment increase. On the southern side of Hare Bay, strata of the cover sequence previously thought to be exposed in a window through the allochthon (Whites Arm window; Bostock and others, 1983) are thrust westward over the allochthon (Fig. 1). Farther south in Canada Bay, a sliver of Grenville basement is thrust over clastics at the top of the cover sequence (Williams and Smyth, 1983).

The southern White Bay region provides further evidence for Acadian contraction structures. Silurian and older rock units in this region show multiple penetrative fabrics but with the major fabric axial planar to west-verging folds (Lock, 1969; Smyth and Schillereff, 1982). Thrust faults are widespread and are concentrated along older boundaries marked by major contrasts in competency, such as basement cover and cover allochthon (Smyth and Schillereff, 1981). At Coney Head the igneous upper thrust sheet of the Taconian Southern White Bay allochthon is thrust over the Silurian succession.

The western limit of the Humber Zone in Newfoundland corresponds with the limit of Acadian deformation. On the Port au Port Peninsula, at the southern end of the Humber Arm allochthon, this is also approximately coincident with the limit of Taconian deformation. The Acadian thrust front is generally viewed as emergent and related to west-directed thrusting (Cawood and Williams, 1988; Grenier and Cawood, 1988; Cawood, 1989), although backthrusting on the eastern margin of the Humber Arm allochthon is considered to form the hanging wall of a passive roof duplex (triangle zone) to a locally buried thrust front (Cawood and Williams, 1988). Recently, Stockmal and Waldron (1990) have reinterpreted the character of the Acadian front on the basis of industry seismic lines from offshore of the Humber Arm allochthon. They suggest that these data indicate a buried thrust front between the Port au Port Peninsula and Bonne Bay, in which the Late Ordovician Long Point Group and overlying strata are folded into a northwest-facing monocline and backthrust over the underlying platform succession that lies in a triangle zone. The triangle zone is overprinted by the west-directed Round Head thrust. Both these structural elements affect the latest Silurian Clam Bank Group, indicating that they both constitute part of Acadian orogeny. If correct, this interpretation suggests that the triangle zone lies across strike form, and may be linked to, the back-thrusting associated with the inferred passive roof duplex (triangle zone) along the eastern margin of the central Humber Arm allochthon. In addition, analysis of Acadian thrusting immediately north of the Humber Arm allochthon has recently recognized east-directed thrusting (Grenier, 1990). This structure, termed the Portland Creek Pond thrust by Grenier (1990), may extend south and merge with the back-thrust and triangle zone structures reported by Stockmal and Waldron (1990). Although the seismic data clearly indicate a triangular wedge of a rock below the Long Point Group, Cawood and others (1991) suggest from seismic data west of the Bonne Bay region that it may also represent a primary depositional clastic wedge deposited ahead of, and derived from, the Taconian highlands to the east (cf. Rodgers, 1965).

Extensional Structures

In both the Humber Arm and Hare Bay allochthons, Acadian contractional structures are truncated by subhorizontal brittle faults that are interpreted as extensional faults related to structural collapse of the Acadian mountain front (Cawood, 1989, 1990; Cawood and Williams, 1987).

The Humber Arm allochthon lies in a broad synclinal structure enclosed to the north, south, and east by uplifted blocks of shelf carbonate and Grenville basement. The ophiolitic rocks of the allochthon (Bay of Islands and Little Port complexes) occupy the core of these structural depressions. As outlined above, these depressions are not a consequence of ophiolite obduction and telescoping of the margin during the Taconian orogeny but rather reflect thrusting of Grenvillian basement during the Acadian orogeny. Basal contacts of the ophiolite sheets are subhorizontal detachment surfaces that truncate footwall structures. This is best seen at the southern end of the Lewis Hills massif (Fig. 5), where the ophiolite and lower sedimentary sheets of the Humber Arm allochthon are separated by a thin sliver of platformal carbonate. The carbonate forms part of the cover sequence that originally underlay the allochthon. The basal detachment of the ophiolite also truncates the basement-cored anticlines (Phillips Brook anticline) and associated Acadian thrusts developed immediately to the south, indicating that at least final movement on the basal detachment of the ophiolite sheets postdates Acadian thrusting. The thin sliver of carbonate underlying the southern Lewis Hills is deformed into an anticline that is truncated along both its upper and lower contacts by subhorizontal faults. The anticline may represent the crest of the adjacent Phillips Brook anticline, which was detached and bulldozed to the east during late-stage ophiolite remobilization.

The basal contact of the ophiolitic rocks in the Hare Bay allochthon is also a postmetamorphic, brittle, subhorizontal detachment that truncates Taconian and Acadian structures (Cawood, 1989). The fault truncates the gently dipping Taconian foliation in the metamorphic sole of the overlying ophiolite sheet. In addition, it truncates the moderate east-dipping cleavage and associated asymmetric folds developed in the footwall strata during Acadian compressive deformation (Cawood, 1989, Fig. 2). The hanging-wall sequence lacks a penetrative fabric or metamorphic imprint of Acadian age, both of which are well developed in the footwall sequence. This suggests that the basal detachment of the ophiolite sheet has cut out that part of the section showing a transitional phase of Acadian fabric development and metamorphic grade. These relationships are consistent with down-cutting by the ophiolite during remobilization along an extensional fault. Blocks of sedimentary slices of the Hare Bay allochthon lying immediately west of the ophiolitic rocks are separated from underlying rocks of the cover sequence by a late-stage, subhorizontal fault that truncates both hanging-wall and

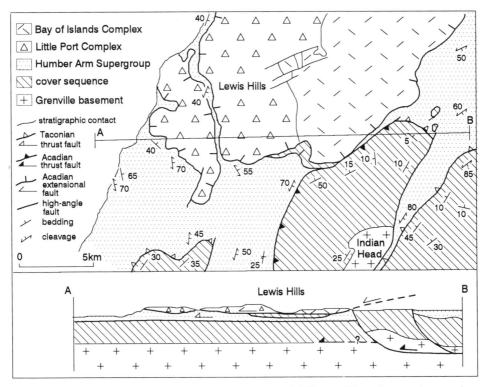

Figure 5. Geologic map and cross section of the southern Lewis Hills massif from the central west region of the Humber Arm allochthon.

footwall structure of inferred Acadian age. These blocks are interpreted to represent material detached from the allochthon and bulldozed ahead of the collapsing, remobilized ophiolite sheet.

Ophiolite complexes in the Humber Arm and Hare Bay allochthons lie at or near the Acadian orogenic (thrust) front (Fig. 6). The front separates rocks unaffected by Acadian orogenic events to the west from the more interior parts of the orogen to the east. The presence of basement-cored, thrust-generated culminations immediately east of the thrust front means that it must have constituted a zone of marked topographic and structural relief (Cawood, 1989, 1990). The potential energy generated through crustal thickening and the unconstrained free western surface of the monoclinal flexure at the mountain front makes this an ideal site for the development of extensional, culmination collapse structures. Relationships outlined for the ophiolite sheets in the Humber Arm and Hare Bay allochthon are consistent with the basal detachment of these sheets being extensional faults formed through gravitational collapse of the Acadian mountain front. Movement of the ophiolite sheets probably occurred during, or soon after, thrusting and development of the mountain front. However, the lack of post-Acadian sediments straddling ophiolite contacts mean there is no upper age constraint on the time of movement.

Recently, Waldron and Milne (1991) have shown that the boundary between the metaclastic sequence around Deer Lake and the allochthonous rocks of the Humber Arm allochthon

around Old Mans Pond is a major ductile shear zone showing normal offset. Although the age of deformation in this region is poorly constrained, they suggest that it may predate the Acadian *sensu strictu* and be related to the widespread Silurian igneous activity in central Newfoundland. O'Brien (1987, 1988) previously proposed that extensional faulting in south central Newfoundland is driven by Silurian intrusive events.

Metamorphism

The distribution of metamorphic facies related to the Acadian orogeny is shown in Figure 7. This diagram is compiled from thermal maturation data collected for the carbonate platform and allochthon sequences from conodont color alteration indices (CAIs) and illite crystallinity studies (Botsford, 1988; Nowlan and Barnes, 1987; F. O'Brien and P. A. Cawood, unpublished data) and for the metaclastic and basement sequences from regional metamorphic (mineral assemblages) facies studies (Hibbard, 1983b; Kennedy, 1982; Owen and Erdmer, 1986). The former only provide a record of thermal metamorphism (under an assumed lithostatic load), whereas the latter provide a record of dynamothermal regional metamorphism. For the purposes of integrating these differing data bases, rocks that have illite crystallinities of greater than 5 mm or that contain conodonts with a CAI of <2 are considered to be unmetamorphosed to weakly metamorphosed. This corresponds to a maximum thermal state

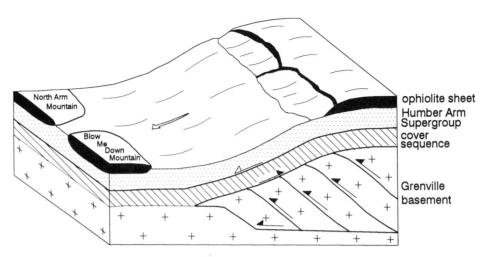

Figure 6. Schematic reconstruction of the Appalachian orogenic front in the vicinity of the Humber Arm allochthon after Acadian thrusting. Diagram depicts the Bay of Islands ophiolite of the Humber Arm allochthon being derived from a collapsing culmination farther east. No attempt is made to show post-Taconian rock units (overburden) at time of collapse.

of less than approximately 100°C (cf. Nowlan and Barnes, 1987; Islam and others, 1982). CAIs of 2 to 5 and illite crystallinities of 5 to 2.4 mm correspond to temperatures of around 100 to 300°C and will approximate subgreenschist facies metamorphic conditions. CAIs of >5 and illite crystallinities of <2.4 mm indicate temperatures of >300°C and are taken to represent greenschist facies conditions. Regions within the Humber Zone shown as being metamorphosed to amphibolite facies are defined solely on metamorphic mineral assemblages. The spatial distribution of boundaries between differing metamorphic facies is poorly constrained, but available data suggest they lie along or near major Acadian structural boundaries (Fig. 7).

Acadian metamorphism was essentially synchronous with contractional deformation. Metamorphic grade increases to the east. Rocks in the west are essentially unmetamorphosed or only weakly metamorphosed with conodont CAI of <2. The onland exposed width of this band narrows to the north of Bonne Bay, corresponding with overthrusting of the Long Range inlier. Rocks of this low grade are absent from the northern tip of the Great Northern Peninsula. This absence appears to be related to the stepping-out of the deformation zone from the Long Range thrust to the Ten Mile Lake and St. Margaret Bay thrusts (Fig. 7). The bulk of the Humber Zone is at subgreenschist facies, including most of the carbonate platform and underlying Grenville basement. However, basement along the eastern side of the Long Range Inlier and overlying rock units exposed along strike to the north and south in Hare Bay and east of Bay of Islands, respectively, reach greenschist grade. Rock units of the metaclastic sequence and Grenville basement exposed immediately west of the Baie Verte Line, along the western side of the Baie Verte Peninsula and in the vicinity of Grand Lake, reach amphibolite grade (Fig. 7). At least on parts of the Baie Verte Peninsula, metamor-

phism is a high temperature/pressure (T/P) overprint of an earlier high P/T eclogite facies event (de Wit and Strong, 1975; Jamieson, 1990).

The timing of amphibolite facies metamorphism has previously been viewed as largely of Taconian age and related to ophiolite obduction and loading of the North American margin (e.g., Jamieson and Vernon, 1987). However, Bursnall and de Wit (1975) suggested that some late-stage deformation and metamorphism may be as young as Acadian. The widespread presence of Siluro-Devonian Ar/Ar cooling ages from metamorphic hornblende, biotite, and muscovite (Dallmeyer, 1977; Dallmeyer and Hibbard, 1984) from the amphibolite facies terrane and the position of this terrane on the eastern side of an overall eastward-increasing Acadian metamorphic grade suggest that metamorphism may be largely mid-Paleozoic (Acadian *sensu lato*) rather than early Paleozoic (Taconian) in age. The $^{40}Ar/^{39}Ar$ ages are cooling ages recording the time at which the temperature of the dated minerals fell below that for Ar diffusion from their mineral lattice. The dates provide an upper age limit on the timing of metamorphism and associated deformation. They will, however, approximate the timing of orogenesis if postmetamorphic exhumation (and cooling) is relatively rapid. Structural relations outlined above indicate that Acadian compressive deformation (and metamorphism) was followed by extensional deformation related to culmination collapse. This sequence implies that crustal thickening associated with deformation and metamorphism along the eastern side of the Humber Zone was followed by rapid exhumation and cooling of this structural pile associated with the extensional structures recognized in the Hare Bay and Humber Arm regions. Such relationships imply that $^{40}Ar/^{39}Ar$ isotopic ages closely approximate the time of metamorphism.

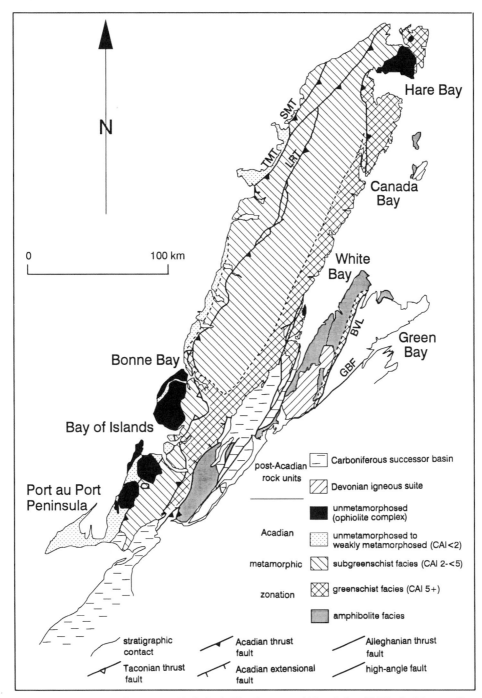

Figure 7. Distribution of metamorphic facies in Humber zone. Abbreviation: SMT—St. Margaret Bay thrust; TMT—Ten Mile Lake; LRT—Long Range thrust; BVL—Baie Verte line; GBF—Green Bay fault; CAI—color alteration index.

Significance

The evolution of the marginal zone of orogenic belts and their adjacent foreland is controlled by, and provides a monitor on, the evolution of the more interior parts of the overthrusting orogen (Jamieson and Beaumont, 1989). The Newfoundland Humber Zone lies within the foreland fold and thrust belt of the Appalachian orogen.

The Acadian orogeny affected all of the Newfoundland tectonostratigraphic zones, but the intensities of its effects are concentrated in the central parts of the orogen in the Gander and Dunnage zones and decrease both westward across the Humber

Zone and eastward across the Avalon Zone (Williams, this volume). The distribution and intensity of Acadian effects appear to be linked to the spatial distribution and interaction of three seismically defined lower crustal blocks: Grenville, Central, and Avalon (Keen and others, 1986; Marillier and others, 1989). Mid-Paleozoic orogenic activity is concentrated above the Central crustal block, diminishing outward in intensity into the zones overlying the Grenville and Avalon blocks. Thus, it appears that mid-Paleozoic accretion and suturing of lower crustal blocks control the character of exposed rock units within the orogen (Williams, this volume). The deformation, metamorphism, and minor igneous activity in the Newfoundland Humber Zone took place in response to, and represent the upper crustal foreland manifestation of, this collisional interaction.

Thrusting and associated metamorphism within the Humber Zone during Acadian compressive deformation have detached it from its root of Grenville crustal block and displaced it toward the foreland.

Discussion

Dunning and others (1990; see also O'Brien and others, 1991) have shown that mid-Paleozoic orogenic events in the southern Newfoundland Appalachians consist of several temporally discrete, but spatially overlapping, pulses. These authors include orogenic events of Silurian age within the Salinic orogeny and propose that only orogenic movements of Devonian age be included in the Acadian orogeny.

Effects of the Salinic orogeny, like those of the Acadian, are concentrated in central Newfoundland. The Humber Zone lay along the margins of the Salinic disturbance. Orogenic effects are minor, and apart from possibly the metaclastic sequences, the zone appears to have been at a high structural level throughout this time frame. Volcanic and sedimentary rocks of the Silurian Sops Arm Group, which are exposed along the eastern side of the Humber Zone, lie on the limit of widespread Silurian igneous activity in central Newfoundland and form part of a high-level manifestation of this orogenic pulse (cf. Bevier and Whalen, 1990). Similarly, the late Silurian terrestrial and marginal marine sedimentary rocks of the Clam Bank Group, which are exposed on the Port au Port Peninsula, probably represent foreland molasse sedimentation to an uplifted Salinic orogenic welt in the hinterland.

Stratigraphic and structural relations from rock units west of the metaclastic sequence indicate that mid-Paleozoic deformation and metamorphism within the Humber Zone are of Devonian (Acadian) age. However, isotopic data that come from the rock units of the metaclastic sequence are equivocal and do provide the possibility for mid-Paleozoic deformation and metamorphism of Silurian (Salinic) age.

K-Ar and ^{40}Ar/^{39}Ar data from the metaclastic sequence along the eastern margin of the Humber Zone show Silurian cooling ages. The metaclastic sequence by Grand Lake has given K-Ar ages on muscovite of 420 ± 14 Ma and 437 ± 14 Ma (Wanless and others, 1973; Fig. 3; Appendix) and a K-Ar age on biotite from granite of 428 ± 20 (Wanless and others, 1965). Ar/Ar ages from the Fleur de Lys Supergroup and East Pond Metamorphic Suite from the central Baie Verte Peninsula have given Ar/Ar ages on hornblende, muscovite, and biotite of 398 to 429 ± 10 Ma, 400 to 421 ± 10 Ma, and 373 to 394 ± 5 Ma, respectively (Dallmeyer, 1977; Dallmeyer and Hibbard, 1984). Although the biotite data from Baie Verte are exclusively of Devonian age, the muscovite and hornblende data from both there and Grand Lake include Silurian ages. It is uncertain if the spread of ages reflects a protracted cooling history, only partial reset of the hornblende and muscovite systems, and/or the competing effects of both Silurian and Devonian metamorphism and cooling. Further work is clearly needed to resolve the age of metamorphism within the metaclastic sequences and its relationship to the remainder of the Humber Zone.

CONCLUSIONS

The Humber Zone in the west Newfoundland Appalachians underwent widespread orogenic deformation and metamorphism in the mid-Paleozoic, following the emplacement of the Ordovician Taconian allochthons but prior to the deposition of relatively undeformed and flat-lying Carboniferous strata. Stratigraphic and structural relations together with radiometric data indicate that the bulk of this orogenic event took place during the latest Silurian or early Devonian at around 400 Ma and forms part of the Acadian orogeny.

West Newfoundland lies along the deformational front of the Acadian orogeny, separating a topographically depressed and structurally undeformed western foreland from the topographically elevated and structurally thickened crust in the interior of the orogen to the east. The Acadian forms an integrated orogenic cycle involving initial contractional deformation and metamorphism, followed by later extensional faulting and exhumation. Contractional deformation is characterized by uplift of Grenville basement associated with west-directed thrusting and associated folding. Thus, the ancient continental margin rocks of the Humber Zone are detached from their lower crustal root and arranged concertina style into an overall west-verging imbricate stack. Deformation and metamorphism are concentrated along the eastern side of the zone, toward the interior of the orogen, and decrease in intensity to the west toward the foreland. High-level contractional structures, developed toward the foreland margin of the orogen, are truncated by postmetamorphic, extensional, low-angle faults. These fractures are interpreted as a gravitationally induced collapse of the orogenic front toward the foreland, resulting in a release of the potential energy built up within the orogen during crustal thickening. Such a collapse leads to rapid exhumation of the orogenic wedge and provides a likely explanation for the latest Silurian to early Devonian Ar/Ar cooling ages developed along the eastern side of the Humber Zone.

ACKNOWLEDGMENTS

Fieldwork in west Newfoundland was supported by the Natural Sciences and Engineering Research Council of Canada and the Department of Energy, Mines, and Resources. I thank Greg Dunning and Hank Williams for numerous discussions on the temporal and spatial character of mid-Paleozoic orogenies in Newfoundland and Jim Hibbard, Brendan Murphy, and Hank Williams for review of the manuscript.

APPENDIX. LIST OF RADIOMETRIC AGE DATA USED IN COMPILATION OF FIGURE 3
(RECALCULATED WITH NEW DECAY CONSTANTS WHERE APPROPRIATE)

Area	Rock Unit	Method	Mineral	Age (Ma)	Ref.*	Area	Rock Unit	Method	Mineral	Age (Ma)	Ref.*
1. Long Range Mtns.-west	Portland Ck. pluton	K/Ar	Biotite	832±11	1	25. Baie Verte	East Pond Metamorphics	Ar/Ar	Biotite	394±5	11
2. Long Range Mtns.-west	Portland Ck. pluton	K/Ar	Biotite	870±11	1	26. Baie Verte	Fleur de Lys Supergroup	Ar/Ar	Hornblende Muscovite	429±10 419±5	11
3. Long Range Mtns.-west	Long Range Complex	K/Ar	Biotite	960±65	2	27. Baie Verte	Fleur de Lys Supergroup	Ar/Ar	Muscovite Biotite	421±15 383±5	11
4. Long Range Mtns.-west	Long Range Complex	K/Ar	Biotite	971±65	2	28. Baie Verte	Fleur de Lys Supergroup	Ar/Ar	Hornblende Muscovite	416±10 400±5	11
5. Long Range Mtns.-east	Long Range Complex	K/Ar	Hornblende	914±37 914±38	3	29. Baie Verte	Fleur de Lys Supergroup	Ar/Ar	Hornblende Muscovite Biotite Biotite	394±5 398±5 375±5 373±5	11
6. Indian Hd. Range	Long Range Complex	K/Ar	Biotite	814±65	4						
7. Indian Hd. Range	Long Range Complex	Ar/Ar	Hornblende	890±15	5	30. Baie Verte	Mings Bight Group	Ar/Ar	Biotite	394±5	11
8. Indian Hd. Range	Long Range Complex	Ar/Ar	Hornblende	895±15	5	31. Baie Verte	Mings Bight Group	Ar/Ar	Biotite	337±5	12
9. Indian Hd. Range	Long Range Complex	Ar/Ar	Biotite	837±15	5	32. Baie Verte	Mings Bight Group	Ar/Ar	Hornblende Hornblende	356±5 357±5	12
10. Indian Hd. Range	Long Range Complex	Ar/Ar	Biotite	845±15	5	33. Baie Verte	Dunamagon Granite	Ar/Ar	Biotite	344±5	12
11. Indian Hd. Range	Long Range Complex	Ar/Ar	Biotite	822±15	5	34. Baie Verte	Burlington Granodiorite	Ar/Ar	Hornblende	417±10	12
12. White Bay	Granite intruding Coney Head Complex	U-Pb	Zircon and titanite	432±2	6	35. Baie Verte	Burlington Granodiorite	Ar/Ar	Hornblende	464±5	12
13. Hare Bay	Dyke	K/Ar	Biotite	416±17	7	36. Baie Verte	Burlington Granodiorite	Ar/Ar	Hornblende Biotite	406±5 409±15	12
14. White Bay	Gull Lake Granite	U-Pb	Zircon	~390	8	37. Baie Verte	Burlington Granodiorite	Ar/Ar	Hornblende	414±5	12
15. White Bay	Devils Room Granite	U-Pb	Zircon	~390	8	38. Baie Verte	Burlington Granodiorite	Ar/Ar	Hornblende	413±5	12
16. Groais Is.	Dyke	K/Ar	Phlogopite	360±16	3	39. Baie Verte	Burlington Granodiorite	Ar/Ar	Hornblende	418±5 414±10	12
17. Long Range (east)	Dyke swarm	Ar/Ar	Whole rock	614±10	9	40. Baie Verte	Burlington Granodiorite	Ar/Ar	Hornblende	414±5 412±10	12
18. Long Range (east)	Dyke swarm	Ar/Ar	Whole rock	396±5	9	41. Baie Verte	Burlington Granodiorite	Ar/Ar	Biotite	343±5	12
19. Long Range (east)	Dyke swarm	Ar/Ar	Whole rock	380±14	9	42. Baie Verte	Burlington Granodiorite	Ar/Ar	Biotite	345±5	12
20. Long Range (east)	Long Range Complex	K/Ar	Biotite	442±18	3	43. Baie Verte	Pacquet Harbour Group	Ar/Ar	Hornblende	356±5	12
21. Long Range (east)	Long Range Complex	K/Ar	Biotite	530±20	3	44. Baie Verte	Pacquet Harbour Group	Ar/Ar	Biotite	347±5	12
22. Steady Brook	Metaclastic terrane	K/Ar	Mustovite	420±14	3	45. Baie Verte	Pacquet Harbour Group	Ar/Ar	Hornblende Hornblende Biotite	355±5 350±5 340±5	12
23. Steady Brook	Metaclastic terrane	K/Ar	Muscovite	437±14	3	46. Baie Verte	Pacquet Harbour Group	Ar/Ar	Hornblende	356±5	12
24. Grand Lake	Metaclastic terrane	K/Ar	Biotite	428±20	10						

*1 = Pringle and others, 1971; 2 = Lowden and others, 1963; 3 = Wanless and others, 1973; 4 = Lowden, 1961; 5 = Dallmeyer, 1978; 6 = Dunning, 1987; 7 = Wanless and others, 1968; 8 = Baadsgaard and others, unpublished data; 9 = Stukas and Reynolds, 1974; 10 = Wanless and others, 1965; 11 = Dallmeyer, 1977; 12 = Dallmeyer and Hibbard, 1984.

REFERENCES CITED

Berry, W.B.N., and Boucot, A. J., 1970, Correlation of the North American Silurian rocks: Geological Society of America Special Paper 102, p. 289.

Bevier, M. L., and Whalen, J. B., 1990, Tectonic significance of Silurian magmatism in the Canadian Appalachians: Geology, v. 18, p. 411–414.

Bostock, H. H., Cumming, L. M., Williams, H., and Smyth, W. R., 1983, Geology of the Strait of Belle Isle area, northwestern insular Newfoundland, southern Labrador and adjacent Quebec: Geological Survey of Canada Memoir, 145 p.

Bosworth, W., 1985, East-directed imbrication and oblique-slip faulting in the Humber Arm Allochthon of western Newfoundland: Structural and tectonic significance: Canadian Journal of Earth Sciences, v. 22, p. 1351–1360.

Botsford, J. W., 1988, Geochemistry and petrology of Lower Paleozoic platform equivalent shales, west Newfoundland: Canada-Newfoundland Mineral Development Agreement 1984–1989, contract No. 23233-7-0223/01ST, unpublished report, 59 p.

Boucot, A. J., 1962, Appalachian Siluro-Devonian, in Coe, K., ed., Some aspects of the Variscan fold belt: Manchester, England, Manchester University Press, p. 155–163.

Boucot, A. J., Field, M. T., Fletcher, R., Forbes, W. H., Naylor, R. S., and Pavlides, L., 1964, Reconnaissance bedrock geology of the Presque Isle quadrangle, Maine: Maine Geological Survey Quadrangle Series 2, 123 p.

Bradley, D. C., 1982, Subsidence in Late Paleozoic basins in the Northern Appalachians: Tectonics, v. 1, p. 107–123.

Bruckner, W. D., 1966, Stratigraphy and structure of West-Central Newfoundland, in Poole, W. H., ed., Guidebook, Geology of part of Atlantic Provinces: Geological Association of Canada and Mineralogical Association of Canada, Guidebook, p. 137–151.

Bursnall, J. T., and de Wit, M. J., 1975, Timing and development of the orthotectonic zone in the Appalachian orogen of northwest Newfoundland: Canadian Journal of Earth Sciences, v. 12, p. 1712–1722.

Cawood, P. A., 1989, Acadian remobilization of a Taconian ophiolite, western Newfoundland: Geology, v. 17, p. 257–260.

———, 1990, Late-stage gravity sliding of ophiolite thrust sheets in Oman and western Newfoundland, in Malpas, J., Moores, E. M., Panayiotou, A., and Xenophonots, C., eds., Ophiolites: Oceanic crustal analogues, Proceedings, Symposium "TROODOS 87": Nicosia, Cyprus, Geological Survey Department, Ministry of Agriculture and Natural Resources, p. 433–445.

Cawood, P. A., and Botsford, J. W., 1991, Reactivation of a continental margin transfer fault during orogenic deformation: the Bonne Bay cross-strike discontinuity, western Newfoundland: American Journal of Science, v. 291, p. 737–759.

Cawood, P. A., and Williams, H., 1986, Northern extremity of the Humber Arm Allochthon in the Portland Creek area, western Newfoundland, and relationships to nearby groups, in Current research, part A: Geological Survey of Canada Paper 86-1A, p. 675–682.

———, 1987, Late stage gravity sliding of ophiolite thrust sheets in Oman and western Newfoundland: Troodos-87, Ophiolites and Oceanic Lithosphere Conference, abstract, p. 88.

———, 1988, Acadian basement thrusting, crustal delamination and structural styles in and around the Humber Arm Allochthon, western Newfoundland: Geology, v. 16, p. 370–373.

Cawood, P. A., Williams, H., and Grenier, R., 1987, Geology of Portland Creek area (12I/4), western Newfoundland: Geological Survey of Canada, Open File 1435, scale 1:50,000.

Cawood, P. A., Williams, H., O'Brien, S. J., and O'Neill, P. P., 1988, A geologic cross-section of the Appalachian Orogen: Geological Association of Canada, Mineralogical Association of Canada, and Canadian Society of Petroleum Geologists Joint Annual Meeting, Field Excursion Guidebook, 160 p.

Cawood, P. A., Shaw, B. R., Etemadi, M., and Stevens, R. K., 1991, Comment on 'Structure of the Appalachian deformation front in western Newfoundland: Implications of multichannel seismic reflection data': Geology, v. 19, p. 951–952.

Chandler, F. W., Loveridge, D., and Currie, K. L., 1987, The age of the Springdale Group, western Newfoundland, and correlative rocks—Evidence for a Llandovery overlap assemblage in the Canadian Appalachians: Transactions of the Royal Society of Edinburgh: Earth Science, v. 78, p. 41–49.

Church, W. R., and Stevens, R. K., 1971, Early Paleozoic ophiolite complexes of the Newfoundland Appalachians as mantle-oceanic crust sequences: Journal of Geophysical Research, v. 76, p. 1460–1466.

Coyle, M., and Strong, D. F., 1987, Geology of the Springdale Group: A newly recognized Silurian epicontinental-type caldera in Newfoundland: Canadian Journal of Earth Sciences, v. 24, p. 1135–1148.

Dallmeyer, R. D., 1977, $^{40}Ar/^{39}Ar$ age spectra of minerals from the Fleur de Lys terrane in northwest Newfoundland: Their beairng on chronology of metamorphism within the Appalachian orthotectonic zone: Journal of Geology, v. 85, p. 89–103.

———, 1978, $^{40}Ar/^{39}Ar$ incremental-release ages of hornblende and biotite from Grenville basement rocks within the Indian Head Range complex, southwest Newfoundland: Their bearing on Late Paleozoic–Early Paleozoic thermal history: Canadian Journal of Earth Sciences, v. 15, p. 1374–1379.

Dallmeyer, R. D., and Hibbard, J., 1984, Geochronology of the Baie Verte Peninsula, Newfoundland: Implications for the tectonic evolution of the Humber and Dunnage Zones of the Appalachian Orogen: Journal of Geology, v. 92, p. 489–512.

Dewey, J. F., 1988, Extensional collapse of orogens: Tectonics, v. 7, p. 1123–1139.

de Wit, M. J., and Strong, D. F., 1975, Eclogite-bearing amphibolites from the Appalachian mobile belt, northwest Newfoundland: Dry versus wet metamorphism: Journal of Geology, v. 83, p. 609–627.

Dunning, G. R., 1987, U-Pb geochronology of the Coney Head Complex, Newfoundland: Canadian Journal of Earth Sciences, v. 24, p. 1072–1075.

Dunning, G. R., and six others, 1990, Silurian Orogeny in the Newfoundland Appalachians: Journal of Geology, v. 98, p. 895–913.

Gillespie, R. T., 1983, Stratigraphic and structural relationships among rock groups at Old Man's Pond, western Newfoundland [M.Sc. thesis]: St. John's, Memorial University of Newfoundland, 198 p.

Grenier, R., 1990, The Appalachian fold and thrust belt, northwestern Newfoundland [M.Sc. thesis]: St. John's Memorial University of Newfoundland, 214 p.

Grenier, R., and Cawood, P. A., 1988, Variations in structural style along the Long Range Front, western Newfoundland, in Current research, part B: Geological Survey of Canada Paper 88-1B, p. 127–133.

Harland, W. B., Cox, A. V., Llewellyn, P. G., Pickton, C.A.G., Smith, A. G., and Walters, R., 1982, A geologic time scale: Cambridge Earth Science Series, Cambridge, Cambridge University Press, 131 p.

Hibbard, J., 1983a, Notes on the metamorphic rocks in the Corner Brook area (12A/13) and regional correlation of the Fleur de Lys Belt, western Newfoundland, in Current research: Newfoundland Department of Mines and Energy Mineral Development Division Report 81-1, p. 41–50.

———, 1983b, Geology of the Baie Verte Peninsula, Newfoundland: Newfoundland Department of Mines and Energy Mineral Development Division Memoir 2, 279 p.

Hyde, R. S., Miller, H. G., Hiscott, R. N., and Wright, J. A., 1988, Basin architecture and thermal maturation in the strike-slip Deer Lake Basin, Carboniferous of Newfoundland: Basin Research, v. 1, p. 85–105.

Islam, S., Hesse, R., and Chagnon, A., 1982, Zonation of diagenesis and lowgrade metamorphism in Cambro-Ordovician flysch of Gaspé Peninsula, Quebec Appalachians: Canadian Mineralogist, v. 20, p. 155–167.

Jamieson, R. A., 1990, Metamorphism of an early Paleozoic continental margin, western Baie Verte Peninsula, Newfoundland: Journal of Metamorphic Petrology, v. 8, p 269–288.

Jamieson, R. A., and Beaumont, C., 1989, Deformation and metamorphism in convergent orogens: A model for uplift and exhumation of metamorphic terrains, in Daly, J. S., Cliff, R. A., and Yardley, B.W.D., eds., Evolution of metamorphic belts: Geological Society Special Publication 43, p. 117–129.

Jamieson, R. A., and Vernon, R. H., 1987, Timing of porphyroblast growth in the Fleur de Lys Supergroup, Newfoundland: Journal of Metamorphic Geology, v. 5, p. 272–288.

Keen, C. E., and 10 others, 1986, Deep seismic reflection profile across the northern Appalachians: Geology, v. 14, p. 141–145.

Kennedy, D.P.S., 1982, Geology of the Corner Brook Lake area, western Newfoundland [M.Sc. thesis]: St. John's, Memorial University of Newfoundland, 370 p.

Knight, I., 1983, Geology of the Carboniferous Bay St. George Subbasin, western Newfoundland: Newfoundland Department of Mines and Energy Mineral Development Division Memoir 1, 358 p.

—— , 1986, Geology of the Port Saunders map sheet (12I/14), western Newfoundland: Newfoundland Department of Mines and Energy, Mineral Development Division, Map 86-59, scale 1:50 000.

Lilly, H. D., 1967, Some notes on stratigraphy and structural style in central west Newfoundland, *in* Neale, E.R.W., and Williams, H., eds., Geology of the Atlantic Region: Geological Association of Canada Special Paper 4, p. 201–211.

Lock, B. E., 1969, The Lower Paleozoic geology of western White Bay, Newfoundland [Ph.D. thesis]: Cambridge, Cambridge University, 343 p.

—— , 1972, Lower Paleozoic history of a critical area; eastern margin of the St. Lawrence Platform in White Bay, Newfoundland, Canada: International Geological Congress, 24th, Montreal,, Section 6, p. 310–324.

Lowden, J. A., 1961, Age determinations by the Geological Survey of Canada: Geological Survey of Canada Paper 61-17, 127 p.

Lowden, J. A., Stockwell, C. H., Tipper, H. W., and Wanless, R. K., 1963, Age determinations and geologic studies (including isotopic ages - report 3): Geological Survey of Canada, Paper 62-17, 117 p.

Marillier, F., and six others, 1989, Crustal structure and surface zonation of the Canadian Appalachians: Implications of deep seismic reflection data: Canadian Journal of Earth Sciences, v. 26, p. 305–321.

McKerrow, W. S., 1988, Wenlock to Givetian deformation in the British Isles and the Canadian Appalachians, *in* Harris, A. L., and Fettes, D. J., eds., The Caledonian-Appalachian Orogen: Geological Society Special Publication 38, p. 437–448.

McKerrow, W. S., Lambert, R.S.J., and Cocks, L.R.M., 1985, The Ordovician, Silurian and Devonian periods, *in* Snelling, N. J., ed., Geochronology and the geological record: Geological Society of London Memoir 10, p. 73–80.

Nowlan, G. S., and Barnes, C. R., 1987, Thermal maturation of Paleozoic strata in eastern Canada from conodont colour alteration index (CAI) data with implications for the burial history, tectonic evolution, hotspot tracts and mineral and hydrocarbon exploration: Geological Survey of Canada Bulletin 367, 47 p.

O'Brien, B. H., 1987, The lithostratigraphy and structure of the Grand Bruit–Cinq Cerf area (parts of 11O/9 and 11O/16), southwestern Newfoundland, *in* Current research: Newfoundland Department of Mines Mineral Development Division Report 87-1, p. 311–334.

—— , 1988, Relationships of phyllite, schist and gneiss in the La Poile Bay–Roti Bay area (parts of 11O/9 and 11O/16), southwestern Newfoundland, *in* Current research: Newfoundland Department of Mines Mineral Development Division Report 88-1, p. 109–125.

O'Brien, B. H., O'Brien, S. J., and Dunning, G. R., 1991, Silurian cover, Late Precambrian-Early Ordovician basement, and the chronology of Silurian orogenesis in the Hermitage flexure (Newfoundland Appalachians): American Journal of Science, v. 291, p. 760–799.

Owen, J. V., 1986, Geology of Silver Mountain area, west Newfoundland, *in* Current research, part A: Geological Survey of Canada Paper 86-1A, p. 515–522.

Owen, J. V., and Erdmer, P., 1986, Precambrian and Paleozoic metamorphism in the Long Range Inlier, western Newfoundland, *in* Current research, part B: Geological Survey of Canada Paper 86-B, p. 29–38.

Palmer, A. R., 1983, The Decade of North American Geology 1983 geologic time scale: Geology, v. 11, p. 503–504.

Pringle, I. R., Miller, J. A., and Warrell, D. M., 1971, Radiometric age determina-

tions from the Long Range Mountains, Newfoundland: Canadian Journal of Earth Sciences, v. 8, p. 1325–1330.

Rodgers, J., 1965, Long Point and Clam Bank formations, western Newfoundland: Geological Association of Canada Proceedings, v. 16, p. 83–94.

Smyth, W. R., and Schillereff, H. S., 1981, 1:25,000 geology field maps of Jackson's Arm northwest (Map 81-109), Jackson's Arm southwest (Map 81-110), Hampden northwest (Map 81-111), and Hampden southwest (Map 81-112): Newfoundland Department of Mines and Energy Mineral Development Division, Open File maps.

—— , 1982, The Pre-Carboniferous geology of southwest White Bay, *in* Current research: Newfoundland Department of Mines and Energy Mineral Development Division Report 82-1, p. 78–98.

Stevens, R. K., 1970, Cambro-Ordovician flysch sedimentation and tectonics in west Newfoundland and their possible bearing on a Proto-Atlantic Ocean, *in* Lajoie, J., ed., Flysch sedimentology in North America: Geological Association of Canada Special Paper 7, p. 165–177.

Stockmal, G. S., and Waldron, J.W.F., 1990, Structure of the Appalachian deformation front in western Newfoundland: Implications of industrial seismic reflection data: Geology, v. 18, p. 765–768.

Stukas, V., and Reynolds, P. H., 1974, ^{40}Ar/^{39}Ar dating of the Long Range dikes Newfoundland: Earth and Planetary Science Letters, v. 27, p. 256–266.

Suppe, J., 1983, Geometry and kinematics of fault-bend folding: American Journal of Science, v. 283, p. 684–721.

Tucker, R. D., Krogh, T. E., Ross, R. J., and Williams, S. H., 1990, Time-scale calibration by high-precision U-Pb zircon dating of interstratified volcanic ashes in the Ordovician and Lower Silurian stratotypes of Britain: Earth and Planetary Science Letters, v. 100, p. 51–58.

Waldron, J.W.F., 1985, Structural history of continental margin sediments beneath the Bay of Islands Ophiolite, Newfoundland: Canadian Journal of Earth Sciences, v. 22, p. 1618–1632.

Waldron, J.W.F., and Milne, J. V., 1991, Tectonic history of the central Humber Zone, western Newfoundland Appalachians: Post-Taconian deformation in the Old Man's Pond area: Canadian Journal of Earth Sciences, v. 28, p. 398–410.

Wanless, R. K., Stevens, R. D., Lachance, G. R., and Rimsaite, J.Y.H., 1965, Age determinations and geological studies, K-Ar isotopic ages, report 5: Geological Survey of Canada Paper 64-17, pt. 1, 136 p.

Wanless, R. K., Stevens, R. D., Lachance, G. R., and Edmonds, C. M., 1968, Age determinations and geological studies, K-Ar isotopic ages, report 8: Geological Survey of Canada Paper 67-2, pt. A, 141 p.

Wanless, R. K., Stevens, R. D., Lachance, G. R., and Delabio, R. N., 1973, Age determinations and geological studies, K-Ar isotopic ages, report 11: Geological Survey of Canada Paper 73-2, 139 p.

Whalen, J. B., Currie, K. L., and van Breeman, O., 1987, Episodic Ordovician-Silurian plutonism in the Topsails igneous terrane, western Newfoundland: Transactions of the Royal Society of Edinburgh: Earth Sciences, v. 78, p. 17–28.

Williams, H., 1977, The Coney Head Complex: Another Taconic allochthon in West Newfoundland: American Journal of Science, v. 277, p. 1279–1295.

—— , 1978, Tectonic lithofacies map of the Appalachian Orogen: Memorial University of Newfoundland Map 1, scale 1:1,000,000.

—— , 1979, Appalachian Orogen in Canada: Canadian Journal of Earth Sciences, Tuzo Wilson volume, v. 16, p. 792–807.

—— , 1984, Miogeoclines and suspect terranes of the Caledonian-Appalachian Orogen: Tectonic patterns in the North Atlantic region: Canadian Journal of Earth Sciences, v. 21, p. 887–901.

Williams, Harold, 1992, Acadian orogeny and its controls in the Newfoundland Appalachians, this volume.

Williams, H., and Bursnall, J., 1982, Newfoundland, *in* Keppie, J. D., coordinator, Structural map of the Appalachian Orogen in Canada: Memorial University of Newfoundland Map 4, scale 1:2,000,000.

Williams, H., and Cawood, P. A., 1986, Relationships along the eastern margin of the Humber Arm Allochthon between Georges Lake and Corner Brook, western Newfoundland, *in* Current research, part A: Geological Survey of

Canada Paper 86-1A, p. 759–765.

—— , 1989, Geology, Humber Arm Allochthon, Newfoundland: Geological Survey of Canada Map 1678A, scale 1:250,000.

Williams, H., and Hiscott, R. N., 1987, Definition of the Iapetus rift-drift transition in western Newfoundland: Geology, v. 15, p. 1044–1047.

Williams, H., and Smyth, W. R., 1983, Geology of the Hare Bay Allochthon, *in* Geology of the Strait of Belle Isle area, Northwestern Insular Newfoundland, Southern Labrador, and adjacent Quebec: Geological Survey of Canada Memoir 400, p. 109–141.

Williams, H., and Stevens, R. K., 1974, The ancient continental margin of eastern North America, *in* Burk, C. A., and Drake, C. L., eds., The geology of continental margins: New York, Springer-Verlag, p. 781–796.

Williams, H., and St. Julien, P., 1982, The Baie Verte–Brompton Line: Continent-ocean interface in the northern Appalachians, *in* St. Julien, P., and Bernard, J., eds., Major structural zones and faults of the northern Appalachians: Geological Association of Canada Special Paper 24, p. 177–207.

Williams, H., Cawood, P. A., James, N. P., and Botsford, J., 1986, Geology of the St. Pauls Inlet area, 12H/13, western Newfoundland: Geological Survey of Canada, Open File 1238, scale 1:50 000.

Williams, H. S., 1893, The elements of the geological time scale: Journal of Geology, v. 1, p. 283–295.

MANUSCRIPT ACCEPTED BY THE SOCIETY JUNE 8, 1992

Geological Society of America
Special Paper 275
1993

Stratigraphic effects of the Acadian orogeny in the autochthonous Appalachian basin

Terence Hamilton-Smith
Kentucky Geological Survey, 228 Mining and Mineral Resources Building, University of Kentucky, Lexington, Kentucky 40506-0107

ABSTRACT

The autochthonous Appalachian basin is located in western West Virginia, southern Ohio, eastern Tennessee, and eastern Kentucky. It is bounded on the east and the south by detachment fronts formed in the Alleghenian orogeny and to the west by the uplifts of the Cincinnati arch and Waverly arch. The Wallbridge unconformity of Early Devonian age corresponds to the boundary between the Tippecanoe and Kaskaskia sequences. It resulted primarily from a major eustatic drawdown but also marks the initial effects of the Acadian orogeny. Subsequent development of a foreland basin included deposits of the Needmore, Marcellus, and Mahantango shales. In the Middle Devonian, growth of this initial foreland basin was terminated by the regional Taghanic unconformity. The Taghanic unconformity was developed during a period of eustatic sea-level rise and is tectonic in origin, resulting from peripheral bulge reactivation during the Acadian orogeny. Development of a successor foreland basin followed, including deposition of the Genesee through the Ohio shales. Maximum Catskill delta progradation in the latest Devonian was followed by a period of postorogenic delta destruction, accompanied by glacio-eustatic sea-level drawdown.

Recognition of two distinct foreland basins suggests that the Acadian orogeny outboard of the autochthonous Appalachian basin resulted from two distinct impacts on the continental margin. The first occurred in the Early Devonian and was relatively small. The second occurred in the Middle Devonian and was large. This interpretation requires a three-plate tectonic model. One possible interpretation would attribute the first impact to obduction of an Avalonian microcontinent and the second to collision with Gondwanaland.

INTRODUCTION

Two major tectonic events in the Acadian orogeny are documented in the Devonian shale stratigraphy of the autochthonous Appalachian basin. These events were the formation of two distinct foreland basins, each preceded by a major unconformity. The Wallbridge unconformity was initially developed in the Early Devonian, at a time of major eustatic sea-level drawdown. This interregional unconformity marks the boundary between the Tippecanoe and Kaskaskia sequences of Sloss (1963) but was also influenced by initial structural movements of the Acadian orogeny. Foreland basin development began immediately follow-

ing the Wallbridge unconformity, marked by continuous deposition of the Needmore through the Marcellus and Mahantango shales. There is no evidence in this sequence for the distinction between two initial tectophases suggested by Ettensohn (1985).

Deposition in the initial foreland basin was terminated by the Taghanic unconformity, initially developed in the late Middle Devonian. The regional Taghanic unconformity was developed by peripheral bulge reactivation of the Cincinnati arch system at a time of rising eustatic sea level and is entirely of tectonic origin. Uplift of the Cincinnati arch system occurred by widespread reactivation of basement faults and was constrained by the Precambrian roots of the arch system in the Grenville front. It is very

Hamilton-Smith, T., 1993, Stratigraphic effects of the Acadian orogeny in the autochthonous Appalachian basin, *in* Roy, D. C., and Skehan, J. W., eds., The Acadian Orogeny: Recent Studies in New England, Maritime Canada, and the Autochthonous Foreland: Boulder, Colorado, Geological Society of America Special Paper 275.

unlikely that growth of the peripheral bulge proceeded simply by elastic wave propagation as suggested by Quinlan and Beaumont (1984).

The Taghanic unconformity was followed by the growth of a second, successor foreland basin, which accommodated the continuous deposition of the Genesee, Sonyea, West Falls, Java, and Ohio shales. Deposition in this second foreland basin was succeeded by a postorogenic phase of reworking of the Catskill delta, controlled by a fluctuating and generally falling glacio-eustatic sea level. There is no evidence for the view of Ettensohn (1985) that Acadian orogenesis continued into a fourth tecto-phase in the Mississippian.

Two foreland basins preceded by two unconformities suggests two episodes of thrust fault obduction on the continental margin outboard of the autochthonous Appalachian basin. Max-imum accumulation rates for the two basins are consistent with the inferred continental margin loads of Beaumont and others (1988) and suggest that the first tectonic episode was relatively small compared to the second. These two episodes of obduction can be best accommodated with the three-plate model of McKer-row and Ziegler (1972), as modified by subsequent workers. Collision of a microcontinent in the Early Devonian would have been followed by the subsequent collision of Laurentia with the African part of Gondwanaland in the Middle Devonian. A signif-icant improvement in plate tectonic models may be obtained by combining this three-plate interaction with current models of oblique convergence.

This thesis is based on a comprehensive review of the Devo-nian stratigraphy of the autochthonous Appalachian basin, much of which is documented in relatively inaccessible publications of regional conferences, state surveys, the U.S. Department of Energy, and the Gas Research Institute. This substantial body of work is concerned almost entirely with subsurface rather than outcrop relationships and is frequently motivated by economic concerns. As such, this set of publications has not previously been used to a significant degree in regional synthesis. This chapter is an attempt to redress that situation, which inevitably requires a more detailed discussion than would have been the case if the relevant literature were better known.

AUTOCHTHONOUS APPALACHIAN BASIN

In this section I would like to set the geological context for the subsequent discussion of the effects of the Acadian orogeny. In the autochthonous Appalachian basin, the relationship be-tween Devonian stratigraphy and deep structural movements is preserved. Detailed consideration of the structural geology is beyond the scope of this chapter, and will be documented else-where. In general, formation of both foreland basins in the Aca-dian orogeny was influenced by widespread reactivation of basement faults inherited from the late Proterozoic and Early Cambrian. Finally, consideration of the region immediately be-fore the Acadian orogeny indicates a period of tectonic quies-cence. Shallow shelf sedimentation controlled by eustacy and

characterized by internal drainage has no suggestion of the prior existence of a foreland basin.

The autochthonous Appalachian basin (Fig. 1) is located in eastern Tennessee, eastern Kentucky, western West Virginia, and southern Ohio and makes up part of the Alleghenian Plateau province (Price and Hatcher, 1983) and the adjacent craton. To the southeast, the Alleghenian front is defined by the edge of thrust sheets such as the Pine Mountain and St. Clair faults. To the east of the autochthonous Appalachian basin is another al-lochthonous terrane, the boundary of which is marked by the Mann Mountain and the Burning Springs anticlines and by the Cambridge arch (Rodgers, 1963). Although these structures are anticlines at the surface, at the Devonian level they consist of stacked imbricate thrust sheets, transported during Alleghenian orogenesis from the east and southeast and detached from the basement in the Salina salts or the Marcellus shale of the middle Paleozoic (Calvert, 1983; Perry, 1978; Gray, 1982).

To the west, the autochthonous Appalachian basin is bounded by a complex of arches and uplifts that includes the Nashville dome, the Cumberland saddle, the Jessamine dome, the Cincinnati arch, the Waverly arch, and the Findlay arch (Wood-ward, 1961; Shearrow, 1966; Collinson and others, 1988; see Fig. 1). This arch complex has episodically been a positive structural element in the region since at least the Ordovician (McFarlan, 1939; Wilson and Stearns, 1963; Stearns and Reesman, 1986; Collinson and others, 1988). The Grenville front underlies the Paleozoic arch complex within the Precambrian basement (Lidiak and others, 1985; Bickford and others, 1986; Lucius and Von Frese, 1988; Pratt and others, 1989; Culotta and others, 1990), which suggests that episodic development of the various arches in the Paleozoic is related to reactivation of structures of the Grenville front. The basement of the autochthonous Appa-lachian basin is characterized by numerous faults inherited from the Grenville orogeny, trending approximately north-south, paral-lel to the Grenville front (Hamilton-Smith, 1988).

In the Late Proterozoic and the Early Cambrian, some 200 m.y. after the Grenville orogeny, an extensional crustal regime developed in the Appalachian region, producing widespread crustal rupture (Rankin, 1975; Thomas, 1977). The Rome trough of the autochthonous Appalachian basin was a product of this extensional regime (Thomas, 1991) and is characterized by a set of east-west– to northeast-southwest–trending faults of generally dip-slip displacement. Faults associated with the Rome trough, together with those inherited from the Grenville terrane, define a system of mobile intersecting basement faults, which were repeat-edly reactivated during the Paleozoic and influenced thickness and facies in Devonian rocks (Shumaker, 1986, 1987; Hamilton-Smith, 1988).

Before the initiation of the Acadian orogeny, the Appala-chian depositional basin was a shallow stable shelf, with internal depocenters in north-central Pennsylvania and northeast West Virginia (Colton, 1970, Fig. 20; Faill, 1985, Fig. 4). This basin configuration was initiated in the late Middle Silurian during the deposition of the McKenzie Formation and culminated with the

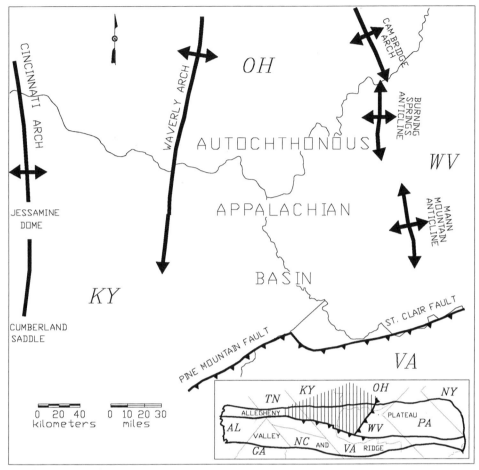

Figure 1. Map of the autochthonous Appalachian basin. Modified from Hamilton-Smith (1988).

accumulation of the Salina Formation in a closed evaporite basin (Smosna and Patchen, 1978). Subsequent deposition of the Helderberg Group was dominated by eustatic variations, grading upward into the shallow marine beds of the Oriskany Sandstone (Head, 1974; Dennison and Head, 1975). The depositional axis of the Helderberg basin remained internal to the shelf, with relatively shallow water facies deposited along the eastern margin (Head, 1974, Fig. 10).

The stratigraphy of the Devonian section in the autochthonous Appalachian basin is outlined in Figure 2. The contents of this figure are drawn from various sources, including Sevon and Woodrow (1985), Patchen and others (1985a, b), Baranoski and Riley (1987), Baranoski and others (1988), Filer (1985), Frankie and others (1986a, b), Moody and others (1987a, b), Sweeney (1986), and Neal and Price (1986). Age dates assigned to stage boundaries are taken from Harland and others (1990).

WALLBRIDGE UNCONFORMITY

The Wallbridge unconformity is one of the two major erosional surfaces in the autochthonous Appalachian basin associated with the Acadian orogeny. The Wallbridge unconformity is mainly of eustatic origin and forms the boundary between the Tippecanoe and the Kaskaskia stratigraphic sequences. At the same time, it has some characteristics specifically associated with the Acadian orogeny, including structural uplift along the Cincinnati arch system.

The earliest known stratigraphic effect of the Acadian orogeny in the autochthonous Appalachian basin was marked by the Wallbridge unconformity (Wheeler, 1963; Dennison, 1985; see Fig. 2). This interregional unconformity is widely developed in the continental interior as the bounding hiatus between the Tippecanoe and the Kaskaskia stratigraphic sequences (Sloss, 1963). The exact placement of the Wallbridge unconformity remains a matter of some controversy. Sloss (1963), followed by Johnson and others (1985), placed it below the Oriskany Sandstone. In contrast, Wheeler (1963), followed by Dennison (1985), maintained that the unconformity should be placed at the top of the Oriskany, below the Onondaga Limestone. Local work suggests that the Oriskany Sandstone is locally bounded by unconformities above and below but that the upper hiatus is more significant (Dennison, 1961; Head, 1974; Heyman, 1977; Diecchio, 1985; Sweeney, 1986; Neal and Price, 1986). In this chapter I have followed the usage of Dennison (1985).

The age of initiation of the Wallbridge unconformity, based on the age of the contact of the Oriskany Sandstone with the overlying Onondaga Limestone, is latest Siegennian stage (Patchen and others, 1985a). In chronological terms, this corresponds to approximately 390 Ma (Harland and others, 1990), a time for which Dennison and Head (1975) had suggested a eustatic sea-level lowstand more pronounced than at any other time during the Silurian and the Devonian. This eustatic interpretation was generally confirmed by Hallam (1984) and by Johnson and others (1985).

INITIAL FORELAND BASIN

In this section I describe the stratigraphy and the configuration of the initial foreland basin of the Acadian orogeny, between the Wallbridge and the Taghanic unconformities. Some attention to stratigraphic detail is necessary to establish the fact that the fill of this basin constitutes a continuous, conformable sequence that suggests only one episode of tectonic activity.

The main sedimentary product of the Acadian orogeny in the central Appalachians was the Catskill delta. The first distal deposits of the Catskill delta consist of the Needmore and Esopus shales (Ettensohn, 1985). In the autochthonous Appalachian basin, deposition of the Needmore shales began in the earliest Emsian, immediately following the erosional hiatus of the Wall-

bridge unconformity (Patchen and others, 1985a, b; Fig. 2). In West Virginia, the Needmore passes laterally into the Huntersville Chert, which is the lateral equivalent of the Onondaga Formation (Sevon and Woodrow, 1985; Kissling and Ehrets, 1985; Patchen and others, 1985a). The top of the Onondaga is marked by the Tioga ash fall, a series of volcanic ash layers derived from the vicinity of the present Virginia Piedmont and spread throughout the Appalachian, Illinois, and Michigan basins by east-southeast prevailing winds (Dennison and Textoris, 1987). The Tioga ash fall is a direct indicator of early Acadian tectonic activity to the east. In West Virginia, the facies assemblage of Onondaga limestone, Huntersville chert, and Needmore shale is conformably succeeded by the shales of the Hamilton Group. Basal deposition in the Hamilton Group is represented by the black Marcellus shale, first deposited in the middle Eifelian (Sevon and Woodrow, 1985). The black shales of the Marcellus are conformably succeeded by the green-gray shales of the Mahantango Formation.

In eastern Kentucky, the shales of the Hamilton Group are truncated by the regionally significant Taghanic unconformity, which has removed most evidence of equivalent facies in the vicinity of the Cincinnati arch (Fig. 2). Coeval deposits of the eastern Illinois basin are all carbonates (Shaver, 1985; Collinson and others, 1988). Biofacies of the Marcellus and the Mahantango shales indicate contributions of terrestrial organic material

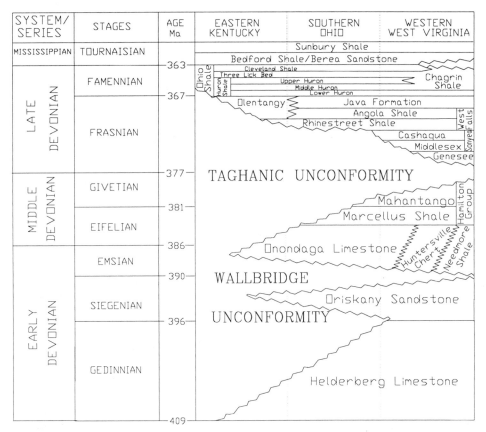

Figure 2. Stratigraphy of the Devonian section of the autochthonous Appalachian basin.

only from the southeast (Zielinski and McIver, 1982), suggesting that the Cincinnati arch was not significantly emergent during the Middle Devonian. A facies change from shales to shelf carbonates in Kentucky, analogous to that developed earlier in central West Virginia between the Needmore shale and the Onondaga limestone, was probably removed by subsequent erosion.

The initial configuration of the depositional basin in the Early Devonian consisted of two distinct subbasins (Faill, 1985, Fig. 5; Ettensohn, 1985) separated by an arch extending perpendicular to the Acadian orogenic front from Maryland to northwestern Pennsylvania. I suggest that the irregular configuration of the Needmore-Esopus basin may indicate uneven initial loading at the continental margin. The trend of minimum accumulation rates in central Ohio suggests an effective hinge line between basin and cratonic platform in the vicinity of the Waverly arch. Basin subsidence in the Middle Devonian substantially changed from that of the Early Devonian. Maximum subsidence rates increased from 14 to 72 m per million years, according to Faill (1985, Figs. 5, 6), recalculated with the age dates of Harland and others (1990). The irregularities of the depositional subbasins of the Esopus and the Needmore shales were eliminated, and a simpler configuration was developed. A uniform basin axis trending northeast-southwest appeared, with increasing subsidence to the southeast toward the continental margin, suggesting a simple foreland basin. Minimal subsidence in the western portion of the basin remained unchanged, indicating a continued hinge line in the vicinity of the Waverly arch.

Deposition in the initial foreland basin of the Acadian orogeny persisted uninterrupted into the middle Givetian, a period of approximately 11 m.y. (Harland and others, 1990). During this time interval, eustatic sea level was characterized by a consistent and general rise (Hallam, 1984; Johnson and others, 1985). The resulting sequence of shales from the base of the Needmore to the top of the Mahantango is completely conformable, with no break in deposition that can be attributed to tectonic activity. The fact that the Needmore shales pass upward conformably through the shales of the Hamilton Group suggests that there is no stratigraphic basis for Ettensohn's (1985) distinction between what he considered a first and second tectophase.

TAGHANIC UNCONFORMITY

Here I describe the Taghanic unconformity in some detail, as it is essential to a good understanding of the Acadian orogeny in the region. I will demonstrate regionally consistent and substantial truncation and onlap at the unconformity, indicating a low-magnitude angular relationship suggestive of structural movement of some magnitude. By mapping the hiatus associated with the angular unconformity, it is possible to see the character of the Taghanic unconformity as essentially tectonic. Finally, through a review of the eustatic evidence, I suggest that the Taghanic unconformity was developed during a period of continual sea-level rise.

The Devonian shale sequence of the autochthonous Appalachian basin is divided in two by the Taghanic unconformity, which was initially developed in the late Givetian (Fig. 2). This unconformity has been widely recognized by various workers from New York through Pennsylvania into West Virginia (e.g., Brett and Baird, 1982; Wright, 1973; Filer, 1985; Neal and Price, 1986), but neither its regional extent nor its tectonic significance have previously been emphasized. Figure 3 is a subsurface gamma-log correlation from central West Virginia to eastern Kentucky that shows the character of the unconformity. The Taghanic unconformity is characterized by a pattern of truncation and onlap that indicates a low-magnitude angular relationship between beds below and above the unconformity. The orientation of this angular break suggests significant uplift and erosion, increasing toward the Cincinnati and Waverly arches.

Truncation under the Taghanic unconformity increases from central West Virginia into eastern Kentucky. The Mahantango and then the Marcellus shales are increasingly truncated toward the Cincinnati arch, so that in eastern Kentucky they are absent, and the Upper Devonian shale sequence rests unconformably on the Onondaga Limestone (Sweeney, 1986; Neal and Price, 1986; Filer, 1987; Frankie and others, 1986a, b). Closer to the crest of the Cincinnati arch, the degree of truncation is more substantial. The Taghanic unconformity merges with the Wallbridge unconformity, and possibly older erosional surfaces, so that Upper Devonian shale successively overlies the Helderberg Limestone, the Salina Formation, and the Lockport Dolomite (Currie, 1981; Shaver, 1985).

Above the Taghanic unconformity, units of the Upper Devonian shale sequence are progressively lost by onlap westward toward the Cincinnati Arch (Fig. 3). The Genesee Formation pinches out in central West Virginia. Proceeding farther west, the Cashaqua Member of the Sonyea Formation and then the Rhinestreet Shale form the basal shales of the Upper Devonian section (Filer, 1987). In eastern Kentucky, the Rhinestreet section has been reduced to less than 15 m (Frankie and others, 1986a). From east to west, the gray-green shales of the Angola Member of the West Falls Formation and the Java Formation merge to form the Olentangy Formation of Ohio and Kentucky (Sevon and Woodrow, 1985; Frankie and others, 1986c). Farther toward the Cincinnati arch the Olentangy disappears by onlap, so that the Ohio Shale unconformably overlies older units (Kepferle and Roen, 1981). The angular relationship indicated by Figure 3 is on the order of 0.06 degrees, consistent with a tectonic interpretation of the unconformity but not necessarily decisive.

The regional character of the Taghanic unconformity is shown in Figure 4. The contour values represent the magnitude of the hiatus at the Taghanic unconformity, measured in millions of years, and include the effects of both truncation and onlap. The base stratigraphic data were taken from Correlation of Stratigraphic Units of North America (COSUNA) correlations for the region (Patchen and others, 1985a, b; Shaver, 1985). At the crest of the Cincinnati arch the maximum hiatus that can be attributed specifically to the Taghanic unconformity is between 16 and 25

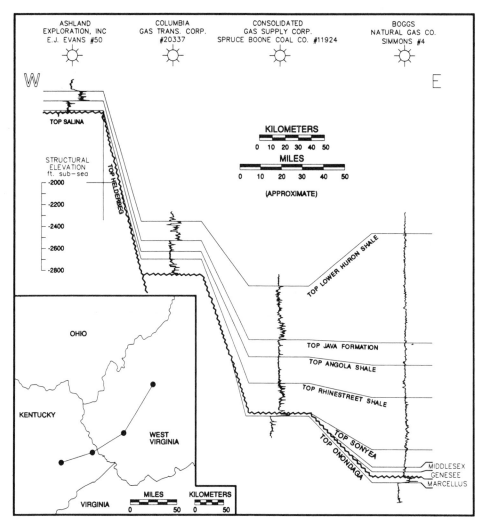

Figure 3. Taghanic unconformity, subsurface structural cross section across autochthonous Appalachian basin.

m.y., constrained by the range of possible age assignments to truncated Onondaga Limestone and to onlapping Huron Shale (see Fig. 2).

The unconformity is most prominently developed along the Waverly arch and the eastern margin of the Cincinnati arch, with a secondary extension along the trend of the Kankakee arch in northern Indiana (Fig. 4). This character suggests the unconformity is tectonic in origin, related to a reactivation of the arch system. The hiatus at the Taghanic unconformity diminishes from the Cincinnati arch westward and virtually disappears into the continental interior. This feature does not correspond to the character of a stratigraphic sequence boundary resulting from eustatic drawdown; that is, most intensely developed in the cratonic interior and disappearing at the cratonic margin (Sloss, 1963).

Hallam (1984) and Johnson and others (1985) both indicated a minor eustatic sea-level fall in the Middle Devonian approximately coincident with the initial time of development of the Taghanic unconformity. The Taghanic onlap (Johnson, 1970) has long been recognized as a significant eustatic sea-level rise in the Middle Devonian, but the existence of a preceding minor sea-level fall is very questionable. Of the two European and the three North American sections considered by Johnson and others (1985), only the two sections in New York and in western Canada show an associated unconformity. The hiatus in New York is the Taghanic unconformity itself, which I suggest is entirely of tectonic character, resulting from the Acadian orogeny. In western Canada, the evidence cited by Johnson and others (1985) for a mid-Givetian eustatic sea-level fall consists of the Watt Mountain unconformity. In the Great Slave Lake area of western Canada, Givetian reef carbonates stood as much as 90 m above sea level before burial in marine muds at the beginning of the Late Devonian (Fuller and Pollack, 1972). However, Lenz (1982) showed that the Watt Mountain unconformity is regional in character, restricted to northern Alberta and British Columbia and southern Yukon and Northwest Territories. Braun and Mathison (1986) suggested that the unconformity was the result of

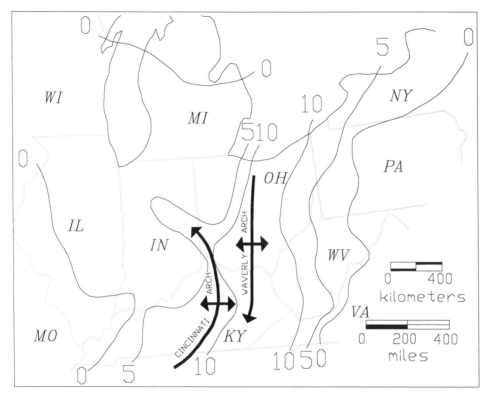

Figure 4. Taghanic unconformity chronostratigraphic map, Appalachian basin and eastern midcontinent. Contour values are millions of years of hiatus at the unconformity.

tectonic movements of the Peace River arch and associated structures in the vicinity of Great Slave Lake. In his review of several sections in western Canada from the Williston basin to the Arctic islands, Lenz (1982) found positive evidence for continuous sea-level rise during the Middle Devonian in every case.

SUCCESSOR FORELAND BASIN

In this section I describe the stratigraphy and configuration of the second successor basin of the Acadian orogeny, particularly in the context of eustacy and maximum Catskill delta progradation. A major eustatic highstand occurred in the early Famennian, with the deposition of the lower Huron Shale, followed by the maximum progradation of the Catskill delta in the late Famennian, marked by the deposition of the Three Lick Bed. Acadian tectonism ended with the Devonian, bringing the foreland basin development to a close.

Following the Taghanic unconformity, deposition resumed in the autochthonous Appalachian basin with the shales of the Genesee Formation (Fig. 2). The Genesee Formation was conformably succeeded by Sonyea and West Falls shales. At least part of the Cincinnati arch was emergent and vegetated during this period, as indicated by the influx of terrigenous organic material from the west into the biofacies assemblage of the Rhinestreet, Angola, and Java shales (Zielinski and McIver, 1982). By

the middle Late Devonian, gray pro-delta Olentangy shales had reached the Cincinnati arch in Ohio and Kentucky. The overlying lower Huron black shales extended over the Cincinnati arch into the midcontinent, passing laterally into equivalent beds of the New Albany Shale of the Illinois basin (Ettensohn and Geller, 1987; Collinson and others, 1988). Lower Huron black shales were succeeded by the gray pro-delta middle Huron shales, laterally equivalent to the prograding coarser clastic wedge of the Catskill delta (Ettensohn, 1985). The middle Huron is recognized in the subsurface of eastern Kentucky (Negus-de Wys, 1979) but does not extend as far as the outcrop belt farther west (Kepferle and Roen, 1981). Middle Huron shales were succeeded by upper Huron black shales.

Deposition of the Lower Huron also marked a major eustatic highstand in the earliest Famennian (Hallam, 1984; Johnson and others, 1985) and was followed by a predominantly regressive period. This interpretation was put forward by Boswell (1988), who suggested that the "shifting of Famennian lithofacies of the Catskill delta complex indicates that eustacy was a dominant depositional control" (p. 9). He suggested a major eustatic change from that of the Frasnian stage, in which "shoreline progradation during a period of major sea-level rise clearly indicates that tectonism was the dominant depositional control. This change is attributed to the waning of Acadian deformation in this area and increased sea-level fluctuation resulting from Famennian glaciation" (p. 9). Boswell's interpretation is supported by the fact that in the Devonian prior to the Famennian, glaciation was not effective because of the location of the paleo-pole in the Paleo-Pacific ocean (Caputo and Crowell, 1985). Glaciation began in the early Famennian and continued into the Mississippian. The resulting Famennian regression has been specifically characterized by Veevers and Powell (1987) as glacio-eustatic.

Maximum Catskill delta progradation occurred in the late Famennian (Dennison, 1985). This maximum progradation was marked in Pennsylvania and eastern West Virginia by the deposition of the Gordon sandstone of the Hampshire Formation. The Gordon was deposited in a wave-dominated delta complex along the eastern boundary of the autochthonous Appalachian basin (Boswell, 1985). Maximum deltaic progradation is also marked by the "Gordon" turbidite siltstones in central West Virginia (Filer, 1985, p. 14) and by the Three Lick Bed of eastern Kentucky (Provo and others, 1978; Fig. 2). A subsequent marine transgression drowned the Gordon delta, producing broad estuarine to bay environments more susceptible to tidal than to wave reworking and producing the marine sandstones of the Oswayo Member of the Price Formation (Boswell, 1985). The offshore product of this brief marine transgression was the black Cleveland Shale (Beuthin and Dennison, 1985; Boswell, 1988; Fig. 2). Cleveland Shale deposition was the final stage of development of the Kaskaskia I subsequence (Sloss, 1982) in the autochthonous Appalachian basin.

Ettensohn (1985) considered the sequence from the Sunbury Shale up to the Middle Mississippian carbonates in the Appalachian basin to be a product of his suggested fourth tectophase of the Acadian orogeny. Although I agree with Ettensohn (1985) that the Middle Mississippian carbonates are indicative of tectonic quiescence, I see no evidence that the subjacent siliciclastic beds make up the fill of a foreland basin. Lower Kaskaskia II deposition in the autochthonous Appalachian basin was apparently posttectonic, controlled predominantly by eustatic events. Maximum progradation of the Catskill delta in the late Famennian indicates a peak of Acadian uplift and erosion at that time or earlier.

Bond and Kominz (1991) compared subsidence curves from the central Appalachians to one for the Iowa platform, for the time period from the Frasnian through the Visean stages. With the reasonable assumption that the Iowa curve could be used as a "baseline" representing eustatic variation, the major relative increase in subsidence in the central Appalachians during the Frasnian and the Famennian can be interpreted as the product of Acadian tectonism. Similarly, the parallelism of the Appalachian and the Iowa curves during the Tournasian and the Visean suggests the absence of any significant tectonic effects on the stratigraphy in either Iowa or the Appalachians during the early Carboniferous. Their work indicated that the lower Kaskaskia II subsequence in the Appalachians postdates the Acadian orogeny.

In contrast, Kominz and Bond (1991) claimed that "basin modeling documents a nearly synchronous episode of rapid subsidence between Middle Devonian and earliest Mississippian time

in almost all pre-existing margins and interior basins of North America" (p. 56). This statement is clearly in error as applied to the earliest Mississippian in the Appalachians. The R1 curves of Kominz and Bond (1991, Fig. 3) are apparently intended to be identical to those of Bond and Kominz (1991, Figs. 6 through 10). The R1 curve for Allegheny County, Maryland, is the same in both publications and shows no incremental subsidence in the Tournaisian and Visean beyond that indicated for the Iowa baseline. The R1 curve for West Virginia is *not* the same in both publications. The R1 curve for West Virginia of Kominz and Bond (1991, Fig. 3) would appear to show significant incremental subsidence during the Tournaisian. However, the more detailed presentation of this curve in Bond and Kominz (1991, Fig. 9) shows that this subsidence episode was marked by the deposition of the Ohio Shale and occurred in the Famennian and that no incremental subsidence compared to the Iowa baseline occurred in the Tournaisian and Visean. Detailed comparison of the West Virginia R1 curve as presented in the two publications shows that the entire curve in Kominz and Bond (1991, Fig. 3, curve 5) has been erroneously shifted to the right by some 5 Ma (M. A. Kominz, 1992, personal communication), giving the misleading impression of Tournaisian subsidence.

The beginning of repeated episodes of glacio-eustatic regression in the early Famennian provides a mechanism for resedimentation of detritus eroded from the top of the Catskill delta. The Kaskaskia II subsequence began with the development of a minor interregional unconformity indicative of eustatic drawdown resulting in a "brief hiatus and minimal stripping of older strata" (Sloss, 1982, p. 34; see also Johnson and others, 1985). Simultaneously, deposition on the Catskill coastal alluvial plain was replaced by erosion (Sevon, 1985). The base Kaskaskia II unconformity is only locally developed in the autochthonous Appalachian basin (Patchen, 1985a, b), at the base of the Berea/Bedford sequence in central West Virginia (Fig. 2). The subsequence boundary unconformity was followed by progradation of the Berea and Bedford delta systems southward from the craton of Ontario and northern Ohio and westward from the eroding Catskill delta of western Virginia and Pennsylvania (Sloss, 1982). In the distal portion of the delta in eastern Kentucky, black shale of the Bedford cannot be distinguished lithologically from Cleveland Shale equivalents below and Sunbury Shale equivalents above (Ettensohn and Elam, 1985), and the sequence appears to be conformable.

In the successor foreland basin of the Acadian orogeny, subsidence had generally the same pattern as in the initial foreland basin, but the subsidence rate at the eastern margin was increased from 72 to over 200 m per million years (Faill, 1985, Figs. 6, 7), recalculated according to the age dates of Harland and others (1990). If the chronostratigraphy of Cowie and Bassett (1989) is used, the Late Devonian maximum subsidence rate was only 140 m per million years, which still represented a significant increase over the subsidence rate of the middle Devonian. The relative increase of sediment input probably was even greater, as suggested by the progradation of subaerial depositional environments to the west throughout the period (Faill, 1985). This pattern suggests a substantially increased loading of the continental margin in the Late Devonian as compared to the Middle Devonian.

REGIONAL TECTONIC INTERPRETATION

This review of the Devonian stratigraphy of the autochthonous Appalachian basin has shown the existence of two distinct foreland basins, separated by a major regional unconformity. This interpretation has been developed to this stage entirely from basic stratigraphic evidence, without the assumption of any particular tectonic model. At this point I will assume that the development of a foreland basin is due to an episode of loading of the continental margin with stacked thrust sheets (Jordan, 1981; Beaumont, 1981). Two distinct foreland basins then logically suggest two distinct episodes of loading on the continental margin. The relatively small accumulation rate of the initial foreland basin as compared to the successor basin further suggests that the initial impact was of relatively small magnitude compared to the second.

Initial loading of the continental margin immediately outboard of the autochthonous Appalachians began in the late Siegennian. This event was coincident with a major eustatic sea-level fall, with the result that the character of the Wallbridge unconformity, although fundamentally eustatic, has some distinct tectonic features in this region. This episode of tectonic loading resulted in the development of a single foreland basin containing the continuous sequence of the Needmore, Marcellus, and Mahantango shales. Structural development of the foreland basin was facilitated by the reactivation of basement faults inherited from Grenville suturing and Iapetean rifting.

The Taghanic unconformity that separates the two foreland basins of the Acadian orogeny can be readily interpreted as entirely tectonic in character. It is not associated with any major eustatic sea-level fall and is regional rather than interregional in extent, disappearing into the continental interior. The fact that its maximum hiatus occurs along the Cincinnati and Waverly arches not only indicates its tectonic affinities but suggests a specific tectonic origin in the growth of the peripheral bulge associated with foreland basin development (Quinlan and Beaumont, 1984).

The location of the axis of maximum uplift of the peripheral bulge could not have been uniquely controlled by loading and flexure because of the influence of the major heterogeneity represented by the Grenville crustal suture, which underlies the Cincinnati and Waverly arches in the Precambrian basement. Jordan (1981) pointed out a similar constraint on the location of the Moxa arch of Wyoming. Preexisting major heterogeneities in the crust may act as significant constraints to crustal behavior. In the case of the autochthonous Appalachian basin, the presence of a major crustal suture at the Grenville front would be expected to act as a significant constraint on peripheral bulge migration and to act as a major contributing factor to the Paleozoic evolution of the Cincinnati and Waverly arches. I consequently do not agree

with Quinlan and Beaumont (1984, p. 973) when they attribute the origin of the Cincinnati and Waverly arches entirely to flexural interactions between the Appalachian basin and the Michigan and Illinois basins.

A second episode of loading of the continental margin outboard of the autochthonous Appalachian basin began in the late Givetian. This event was marked by the development of the Taghanic unconformity and the growth of a second, successor foreland basin. This single basin contained the continuous sequence of the Genesee, Sonyea, West Falls, Java, and Ohio shales. Accumulation rates in the successor foreland basin two to three times as large as those in the initial foreland basin suggest a comparable increase in the magnitude of the loading of the continental margin. This interpretation is supported by the increase in margin load thickness from two to ten times between the Middle Devonian and the Late Devonian timesteps in the model of Beaumont and others (1988). Structural development of the foreland basin was again accommodated by widespread reactivation of older basement faults.

In the autochthonous Appalachian basin, there is no evidence of effects of the Acadian orogeny following the Devonian. Progradation of the Catskill delta ended in the late Famennian and was followed by a prolonged delta-destructive phase in a variable but generally regressive eustatic environment. A local unconformity of small magnitude was developed in central West Virginia in response to the same eustatic sea-level fall that produced the boundary between the Kaskaskia I and Kaskaskia II subsequences of Sloss (1982).

THREE-PLATE TECTONIC MODEL

This tectonic interpretation of the effects of the Acadian orogeny in the autochthonous Appalachian basin offers some useful constraints to the development of plate-tectonic models for the Acadian orogeny in general. In particular, any plate-tectonic model viable in the autochthonous Appalachian basin must result in two distinct episodes of loading at the continental margin, the first—of relatively small magnitude—beginning in the Early Devonian and the second—of relatively large magnitude—beginning in the late Middle Devonian. This sequence of events suggests that any viable plate-tectonic model must involve at least three plates.

McKerrow and Ziegler (1972) originally suggested that the Acadian orogeny resulted from the collision of Gondwanaland and Laurentia, with a relatively small peninsula of the Baltic Shield in between. They further suggested that Gondwanaland first collided with the Baltic fragment (Armorica) and subsequently pushed it against Laurentia. Scotese and others (1979) reversed the order of the interaction by using paleomagnetic evidence that the Iapetus Ocean closed prior to the Theic Ocean, a conclusion later supported by other geophysical evidence presented by Lefort (1983). They visualized an initial plate collision between Laurentia and a relatively small microcontinent, followed by a second collision with a relatively large portion of Gondwanaland. In Van der Voo's recent review (1988) of the relevant paleomagnetic data, he suggested a further modification of the model of McKerrow and Ziegler, indicating that Africa must have been the specific part of Gondwanaland involved in the Acadian collision. Such a model is generally consistent with the tectonic interpretation presented in this chapter of the stratigraphy of the autochthonous Appalachian basin.

The Devonian stratigraphy of the autochthonous Appalachian suggests that the Acadian orogeny resulted from the successive impacts of two continental plates on the adjacent continental margin. The smaller of the two plates, possibly an Avalonian microcontinent, impacted first in the Early Devonian. The second of the two plates, probably the African sector of Gondwanaland, impacted in the Middle Devonian. Including Laurentia, this interpretation constitutes a three-plate tectonic model.

The evidence cited from the autochthonous Appalachian basin permits only a discussion of the margin-normal, as distinct from the margin-parallel, component of the collisions. Nothing in what I have proposed here in any way precludes the development of a plate tectonic model that includes margin-parallel displacement, or oblique convergence. Recent models of oblique convergence attempt to explain the Acadian orogeny as the result of the collision of only two plates, although additional plates are not precluded. Ettensohn (1987) favored "oblique convergence and/or strike-slip movement between Laurussia and a single plate rather than a series of independently acting continental fragments, although a 'regular' collision of independent fragments cannot be precluded" (p. 577). Ferrill and Thomas (1988) proposed "dextral motion between two irregularly shaped plates, the North American plate and another plate" but commented that "neither the identity of the other plate nor a single continuous other plate is essential to the model" (p. 607). The Devonian stratigraphy of the autochthonous Appalachian basin offers positive evidence that three plates were involved in Acadian tectonic events in the central Appalachians. I suggest that a more viable tectonic model for the Acadian orogeny in the Appalachians in general could result from the explicit combination of oblique convergence with the movement of three or more plates.

ACKNOWLEDGMENTS

This chapter resulted from work originally sponsored by the Gas Research Institute and conducted at K&A Energy Consultants, concerning the gas potential of the Devonian shales of the Appalachian basin. More recent work has been supported by the Kentucky Geological Survey. An earlier version of this chapter was reviewed for the Geological Society of America by an anonymous reviewer and by Roger Faill of the Pennsylvania Geological Survey. Reviews were also provided by Jim Drahovzal and Dan Walker of the Kentucky Geological Survey. I am also grateful for the helpful comments of Nick Rast and Pete Goodman of the Department of Geological Sciences of the University of Kentucky, M. C. Noger of the Kentucky Geological Survey, Pat Lowry of K&A Energy Consultants, and Dick Scheper of the Gas Research Institute.

REFERENCES CITED

Baranoski, M., and Riley, R., 1987, Analysis of stratigraphic and production relationships of Devonian shale gas reservoirs in Ohio—Annual report, October 1, 1985 to September 30, 1986: Chicago, Gas Research Institute, GRI-86/0287, 61 p.

Baranoski, M. T., Riley, R. A., and Gray, J. D., 1988, Analysis of stratigraphic and production relationships of Devonian shale gas reservoirs in Ohio—Annual report, October 1, 1986 to September 30, 1987: Chicago, Gas Research Institute, GRI-88/0015, 55 p.

Beaumont, C., 1981, Foreland basins: Geophysical Journal of the Royal Astronomical Society, v. 65, p. 291–329.

Beaumont, C., Quinlan, G., and Hamilton, J., 1988, Orogeny and stratigraphy: Numerical models of the Paleozoic in the eastern interior of North America: Tectonics, v. 7, p. 389–416.

Beuthin, J. D., and Dennison, J. M., 1985, The Late Devonian–Early Mississippian transgressive event in western Maryland and immediately adjacent areas: Appalachian Basin Industrial Associates, v. 8, p. 77–104.

Bickford, M. E., Van Schmus, W. R., and Zietz, I., 1986, Proterozoic history of the midcontinent region of North America: Geology, v. 14, p. 492–496.

Bond, G. C., and Kominz, M. A., 1991, Paleozoic sea level and tectonic events in cratonic margins and cratonic basins of North America: Journal of Geophysical Research, v. 96, p. 6619–6639.

Boswell, R. M., 1985, Nature of the Upper Devonian Catskill shoreline of northern West Virginia: Appalachian Basin Industrial Associates, v. 9, p. 19–56.

—— , 1988, Eustacy, Acadian orogeny, and Famennian shorelines of the central Appalachian basin: Proceedings, 19th Annual Appalachian Petroleum Geology Symposium: Morgantown, West Virginia, West Virginia Geological and Economic Survey and West Virginia University, p. 8–10.

Braun, W. K., and Mathison, J. E., 1986, Mid-Givetian events in western Canada: The Dawson Bay–Watt Mountain–"Slave Point" interlude: Bulletin of Canadian Petroleum Geology, v. 34, p. 426–451.

Brett, C. E., and Baird, G. C., 1982, Upper Moscow–Genesee stratigraphic relationships in western New York: Evidence for regional erosive beveling in the late Middle Devonian, in Beucher, E. J., and Calkin, P. E., eds., Geology of the northern Appalachian basin, western New York: New York State Geological Association, 54th Annual Meeting, Field Trips Guidebook, p. 19–52.

Calvert, G., 1983, Subsurface structure of the Burning Springs anticline, West Virginia, evidence for a two-stage structural development: Appalachian Basin Industrial Associates, v. 4, p. 147–164.

Caputo, M. V., and Crowell, J. C., 1985, Migration of glacial centers across Gondwana during Paleozoic era: Geological Society of America Bulletin, v. 96, p. 1020–1036.

Collinson, C., Sargent, M. L., and Jennings, J. R., 1988, Illinois Basin region, in Sloss, L. L., ed., Sedimentary cover–North American craton (The geology of North America, v. D-2): Boulder, Colorado, Geological Society of America, p. 383–426.

Colton, G. W., 1970, The Appalachian basin—Its depositional sequences and their geological relationships, in Fisher, G. W., Pettyjohn, F. J., Reed, J. C., Jr., and Weaver, K. N., eds., Studies of Appalachian geology: Central and southern: New York, Wiley, p. 5–47.

Cowie, J. W., and Bassett, M. G., 1989, International Union of Geological Sciences 1989 global stratigraphic chart: Episodes, v. 12, supplement.

Culotta, R. C., Pratt, T., and Oliver, J., 1990, A tale of two sutures: COCORP's deep seismic surveys of the Grenville province in the eastern U.S. midcontinent: Geology, v. 18, p. 646–649.

Currie, M. T., 1981, Subsurface stratigraphy and depositional environments of the "Corniferous" (Silurian-Devonian) of eastern Kentucky [M.S. thesis]: Lexington, University of Kentucky, 107 p.

Dennison, J. M., 1961, Stratigraphy of Onesquethaw stage of Devonian in West Virginia and bordering states: West Virginia Geological Survey Bulletin 22, 49 p.

—— , 1985, Catskill delta shallow marine strata, in Woodrow, D. L., and Sevon, W. D., eds., The Catskill delta: Geological Society of America Special Paper

201, p. 91–106.

Dennison, J. M., and Head, J. W., 1975, Sea-level variations interpreted from the Appalachian basin Silurian and Devonian: American Journal of Science, v. 275, p. 1089–1120.

Dennison, J. M., and Textoris, D. A., 1987, Paleowind and depositional tectonics interpreted from Tioga ash bed: Appalachian Basin Industrial Associates, v. 12, p. 107–132.

Diecchio, R. J., 1985, Regional controls of gas accumulation in Oriskany Sandstone, central Appalachian basin: American Association of Petroleum Geologists Bulletin, v. 69, p. 722–732.

Ettensohn, F. R., 1985, The Catskill delta complex and the Acadian orogeny: A model, in Woodrow, D. L., and Sevon, W. D., eds., The Catskill delta: Geological Society of America Special Paper 201, p. 39–49.

—— , 1987, Rates of relative plate motion during the Acadian orogeny based on the spatial distribution of black shales: Journal of Geology, v. 95, p. 572–582.

Ettensohn, F. R., and Elam, T. D., 1985, Defining the nature and location of a Late Devonian–Early Mississippian pycnocline in eastern Kentucky: Geological Society of America Bulletin, v. 96, p. 1313–1321.

Ettensohn, F. R., and Geller, K. L., 1987, Tectonic control of interbasinal relationships between Devonian black shales in the Illinois and Appalachian basins of Kentucky: in Proceedings, 1986 Eastern Oil Shale Symposium, Lexington: Lexington, Kentucky Energy Cabinet Laboratory, p. 303–314.

Faill, R. T., 1985, The Acadian orogeny and the Catskill delta, in Woodrow, D. L., and Sevon, W. D., eds., The Catskill delta: Geological Society of America Special Paper 201, p. 15–37.

Ferrill, B. A., and Thomas, W. A., 1988, Acadian dextral transpression and synorogenic sedimentary successions in the Appalachians: Geology, v. 16, p. 604–608.

Filer, J. K., 1985, Oil and gas report and map of Pleasants, Wood, and Ritchie counties, West Virginia: West Virginia Geological and Economic Survey Bulletin B-11A, 87 p.

—— , 1987, Geology of Devonian shale oil and gas in Pleasants, Wood, and Ritchie counties, West Virginia: Formation Evaluation (Society of Petroleum Engineers), December 1987, p. 420–427.

Frankie, W. T., Moody, J. R., and Kemper, J. R., 1986a, Geologic and hydrocarbon report of Letcher County: Lexington, Kentucky Geological Survey for Gas Research Institute, 59 p.

Frankie, W. T., Moody, J. R., Kemper, J. R., and Johnson, I. M., 1986b, Geologic and hydrocarbon report of Floyd County: Lexington, Kentucky Geological Survey for Gas Research Institute, 52 p.

Frankie, W. T., Moody, J. R., Kemper, J. R., and Johnson, I. M., 1986c, Geologic and hydrocarbon report of Knott County: Lexington, Kentucky Geological Survey for Gas Research Institute, 94 p.

Fuller, J.G.C.M., and Pollack, C. A., 1972, Early exposure of Middle Devonian reefs, southern Northwest Territories, Canada: 24th International Geological Congress, Section 6, Montreal, p. 144–155.

Gray, J. D., 1982, Subsurface structure mapping of eastern Ohio, in Gray, J. D., and eight others, eds., An integrated study of the Devonian-age black shales in eastern Ohio: Columbus, Ohio Department of Natural Resources, p. 3.1–3.13.

Hallam, A., 1984, Pre-Quaternary sea-level changes: Annual Review of Earth and Planetary Sciences, v. 12, p. 205–243.

Hamilton-Smith, T., 1988, Significance of geology to gas production from Devonian shales of the central Appalachian basin: An undeveloped asset: Chicago, Gas Research Institute, GRI-88/0178, 117 p.

Harland, W. B., Armstrong, R. L., Cox, A. V., Craig, L. E., Smith, A. G., and Smith, D. G., 1990, A geological time scale 1989: New York, Cambridge University Press, 263 p.

Head, J. W., 1974, Correlation and paleogeography of upper part of Helderberg Group (Lower Devonian) of central Appalachians: American Association of Petroleum Geologists Bulletin, v. 58, p. 247–259.

Heyman, L., 1977, Tully (Middle Devonian) to Queenston (Upper Ordovician) correlations in the subsurface of western Pennsylvania: Commonwealth of Pennsylvania, Department of Environmental Resources, Topographic and Geologic Survey, Mineral Resource Report 73, 16 p.

Johnson, J. G., 1970, Taghanic onlap and the end of North American Devonian provinciality: Geological Society of America Bulletin, v. 81, p. 2077–2106.

Johnson, J. G., Klapper, G., and Sandberg, C. A., 1985, Devonian eustatic fluctuations in Euramerica: Geological Society of America Bulletin, v. 96, p. 567–587.

Jordan, T. E., 1981, Thrust loads and foreland basin evolution, Cretaceous, western United States: American Association of Petroleum Geologists Bulletin, v. 65, p. 2506–2520.

Kepferle, R. C., and Roen, J. B., 1981, Chattanooga and Ohio Shales of the southern Appalachian basin, in Roberts, T. G., ed., GSA Cincinnati '81 field trip guidebooks: Falls Church, Virginia, American Geological Institute, p. 259–361.

Kissling, D. L., and Ehrets, J. R., 1985, Regional facies of the Huntersville chert: Proceedings, 16th Annual Appalachian Petroleum Geology Symposium: Morgantown, West Virginia, West Virginia Geological and Economic Survey and West Virginia University, p. 10–12.

Kominz, M. A., and Bond, G. C., 1991, Unusually large subsidence and sea-level events during middle Paleozoic time: New evidence supporting mantle convection models for supercontinent assembly: Geology, v. 19, p. 56–60.

Lefort, J. P., 1983, A new geophysical criterion to correlate the Acadian and Hercynian orogenies of western Europe and eastern America, in Hatcher, R. D., Jr., Williams, H., and Zietz, I., eds., Contributions to the tectonics and geophysics of mountain chains: Geological Society of America Memoir 158, p. 3–18.

Lenz, A. C., 1982, Ordovician to Devonian sea-level changes in western and northern Canada: Canadian Journal of Earth Sciences, v. 19, p. 1919–1932.

Lidiak, E. G., Hinze, W. J., Keller, G. R., Reed, J. E., Braile, L. W., and Johnson, R. W., 1985, Geologic significance of regional gravity and magnetic anomalies in the east-central midcontinent, in Hinze, W. J., ed., The utility of regional gravity and magnetic anomaly maps: Tulsa, Oklahoma, Society of Exploration Geophysicists, p. 287–307.

Lucius, J. E., and Von Frese, R.R.B., 1988, Aeromagnetic and gravity anomaly constraints on the crustal geology of Ohio: Geological Society of America Bulletin, v. 100, p. 104–116.

McFarlan, A. C., 1939, Cincinnati arch and features of its development: American Association of Petroleum Geologists Bulletin, v. 23, p. 1847–1852.

McKerrow, W. S., and Ziegler, A. M., 1972, Paleozoic oceans: Nature, v. 240, p. 92–94.

Moody, J. R., Kemper, J. R., Johnston, I. M., and Elkin, R. R., 1987a, Geologic and hydrocarbon report of Pike County: Lexington, Kentucky Geological Survey for Gas Research Institute, 51 p.

Moody, J. R., Kemper, J. R., Johnston, I. M., Frankie, W. T., and Elkin, R. R., 1987b, Geologic and hydrocarbon report of Martin County: Lexington, Kentucky Geological Survey for Gas Research Institute, 50 p.

Neal, D. W., and Price, B. K., 1986, Oil and gas report and maps of Lincoln, Logan, and Mingo counties, West Virginia: West Virginia Geological and Economic Survey Bulletin B-41, 68 p.

Negus-de Wys, J., 1979, The eastern Kentucky gas field—A geological study of the relationships of oil shale gas occurrences to structure, stratigraphy, lithology, and inorganic geochemical parameters [Ph.D. thesis]: Morgantown, West Virginia University, 199 p.

Patchen, D. G., Avary, K. L., and Erwin, R. B., 1985a, Northern Appalachian region, in Lindberg, F. A., ed., Correlation of stratigraphic units of North America (COSUNA) project: Tulsa, Oklahoma, American Association of Petroleum Geologists, chart.

—— , 1985b, Southern Appalachian region, in Lindberg, F. A., ed., Correlation of stratigraphic units of North America (COSUNA) project: Tulsa, Oklahoma, American Association of Petroleum Geologists, chart.

Perry, W. J., Jr., 1978, Mann Mountain anticline: Western limit of detachment in south-central West Virginia, in Wheeler, R. L., and Dean, C. S., eds., Proceedings, Western Limits of Detachment and Related Structures in the Appalachian Foreland: Morgantown, West Virginia, U.S. Department of

Energy, p. 82–99.

Pratt, T., and six others, 1989, Major Proterozoic basement features of the eastern midcontinent of North America revealed by recent COCORP profiling: Geology, v. 17, p. 505–509.

Price, R. A., and Hatcher, R. D., Jr., 1983, Tectonic significance of similarities in the evolution of the Alabama-Pennsylvania Appalachians and the Alberta–British Columbia Canadian Cordillera, in Hatcher, R. D., Jr., Williams, H., and Zietz, I., eds., Contributions to the tectonics and geophysics of mountain chains: Geological Society of America Memoir 158, p. 149–106.

Provo, L. J., Kepferle, R. C., and Potter, P. E., 1978, Division of black Ohio Shale in eastern Kentucky: American Association of Petroleum Geologists Bulletin, v. 62, p. 1703–1713.

Quinlan, G. M., and Beaumont, C., 1984, Appalachian thrusting, lithospheric flexure, and the Paleozoic stratigraphy of the eastern interior of North America: Canadian Journal of Earth Science, v. 21, p. 973–996.

Rankin, D. W., 1975, The continental margin of eastern North America in the southern Appalachians: The opening and closing of the proto-Atlantic ocean: American Journal of Science, v. 275-A, p. 298–336.

Rodgers, J., 1963, Mechanics of Appalachian foreland folding in Pennsylvania and West Virginia: American Association of Petroleum Geologists Bulletin, v. 47, p. 1527–1536.

Scotese, C. R., Bambach, R. K., Barton, C., Van der Voo, R., and Zeigler, A., 1979, Paleozoic base maps: Journal of Geology, v. 87, p. 217–277.

Sevon, W. D., 1985, Nonmarine facies of the Middle and Late Devonian Catskill coastal alluvial plain, in Woodrow, D. L., and Sevon, W. D., eds., The Catskill delta: Geological Society of America Special Paper 201, p. 79–90.

Sevon, W. D., and Woodrow, D. L., 1985, Middle and Upper Devonian stratigraphy within the Appalachian basin, in Woodrow, D. L., and Sevon, W. D., eds., The Catskill delta: Geological Society of America Special Paper 201, p. 1–7.

Shaver, R. H., 1985, Midwestern basin and arches region, in Lindberg, F. A., ed., Correlation of stratigraphic units of North America (COSUNA) project: Tulsa, Oklahoma, American Association of Petroleum Geologists, chart.

Shearrow, G. G., 1966, Paleozoic rocks in Ohio: American Association of Petroleum Geologists Cross Section Publication 4, p. 41–45.

Shumaker, R. C., 1986, Structural development of Paleozoic continental basins of eastern North America, in Proceedings, International Conference of Basement Tectonics, 6th, Santa Fe, New Mexico: Salt Lake City, International Basement Tectonics Association, p. 82–95.

——, 1987, Structural parameters that affect Devonian shale gas production in West Virginia and eastern Kentucky: Appalachian Basin Industrial Associates, v. 12, p. 133–201.

Sloss, L. L., 1963, Sequences in the cratonic interior of North America: Geological Society of America Bulletin, v. 74, p. 93–113.

——, 1982, The mid-continent province, United States, in Palmer, A. R., ed., Perspectives in regional geological synthesis: Geological Society of America Decade of North America Geology Special Publication 1, p. 27–39.

Smosna, R., and Patchen, D., 1978, Silurian evolution of central Appalachian basin: American Association of Petroleum Geologists Bulletin, v. 62, p. 2308–2328.

Stearns, R. G., and Reesman, A. L., 1986, Cambrian to Holocene structural and burial history of Nashville dome: American Association of Petroleum Geologists Bulletin, v. 70, p. 143–154.

Sweeney, J., 1986, Oil and gas report and maps of Wirt, Roane, and Calhoun counties, West Virginia: West Virginia Geological and Economic Survey Bulletin B-40, 102 p.

Thomas, W. A., 1977, Evolution of Appalachian-Ouachita salients and recesses from reentrants and promontories in the continental margin: American Journal of Science, v. 277, p. 1233–1278.

——, 1991, The Appalachian-Ouachita rifted margin of southeastern North America: Geological Society of America Bulletin, v. 103, p. 415–431.

Van der Voo, R., 1988, Paleozoic paleogeography of North America, Gondwana, and intervening displaced terranes: Comparisons of paleomagnetism with paleoclimatology and biogeographical patterns: Geological Society of America Bulletin, v. 100, p. 311–324.

Veevers, J. J., and Powell, C. McA., 1987, Late Paleozoic glacial episodes in Gondwanaland reflected in transgressive-regressive depositional sequences in Euramerica: Geological Society of America Bulletin, v. 98, p. 475–487.

Wheeler, H. E., 1963, Post-Sauk and pre-Absaroka Paleozoic stratigraphic patterns in North America: American Association of Petroleum Geologists Bulletin, v. 47, p. 1497–1526.

Wilson, C. W., Jr., and Stearns, R. G., 1963, Quantitative analysis of Ordovician and younger structural development of Nashville dome, Tennessee: American Association of Petroleum Geologists Bulletin, v. 47, p. 823–832.

Woodward, H. P., 1961, Preliminary subsurface study of southeastern Appalachian interior plateau: American Association of Petroleum Geologists Bulletin, v. 45, p. 1634–1655.

Wright, N. A., 1973, Subsurface Tully Limestone, New York and northern Pennsylvania: New York State Museum and Science Service Map and Chart Series 14, 6 p.

Zielinski, R. E., and McIver, R. D., 1982, Resource and exploration assessment of the oil and gas potential in the Devonian gas shales of the Appalachian basin: Miamisburg, Ohio, Mound Facility for U.S. Department of Energy, DOE/DP/0053-1125, 326 p.

MANUSCRIPT ACCEPTED BY THE SOCIETY JUNE 8, 1992

Index

[Italic page numbers indicate major references]

accretion, 20, 28, 67, 68, 69, 85
  Cambrian, *72*
  Caradocian, *72*
  timing, *33*
Ackley batholith, 129
alcids, 42
Alleghenian Plateau province, 154
Allegheny County, Maryland, 160
allochthons, 136, 148
Allsbury Formation, 75
Alto allochthon, 14
Alton quadrangle, 56, 58, 60
American Cordillera, 67
Amity quadrangle, 75
amphibolites, 9, 106, 129
andalusites, 63, 64, 129
Angers-Dugal Outcrop Belt, 110
Angola member, 157
Annapolis Valley region, 48
Annieopsquotch Complex, 36
Anse Cascon Formation, 109
Anseà Pierre-Loiselle Formation, 109
anticlines, 141
anticlinoria, 8, 51
Antigonish County, 44, 45
Appalachian basin, 154
  autochthonous, *153*
Appalachian-Caledonian orogen, 28, 135
Appalachian Mountains
  central, 2
  northern, 2, 5, 12, 15, 16, 17, *27*, 43, 47, 56, 67, 78, 80, 102
  south-central, *13*
  southern, 2, 19
Appalachian orogen, 13
Appledore Diorite, 10
arches, 154
arcs
  island, *31*, 68
  volcanic, 35, 96
Armorica, 17
Armorican massif, 19
Armorican–Nova Scotian Acadian–Ligerian orogenic belt, 19
Armorican Peninsula, 13, 18
Aroostook-Matapedia basin, 80
Aroostook-Matapedia belt, 8, *75*, 77
  southern lobe, *73*
Aroostook-Percé anticlinorium, 102, *106*, 110, 111, 114, 115, 116, 118
Arsenault area, 118
Arsenault Formation, 106, *107*
Ascot Complex, 85, *87*, 89, 94, 97
Ascot volcanic belt, 89
Ascot volcanics, 96
ash deposits, 109
ash layers, volcanic, 156
ashflows, 76
Ashgillian, *73*
Aspy terrane, 130

Assémetquagan River area, 111
Atlantic Realm, 45
Atrypa reticularis, 48
Avalon, 35
Avalon belt, 8, 11, 77
Avalon block, 19, *78*, 80, 126, 130
Avalon Composite Terrane, 44
Avalon crust, 11
Avalon island arc, 17
Avalon superterrane, 9, 11, 12, 13, 14
Avalon terrane, 20, *28*
Avalon Zone, 15, 102, 126, 128, 129, 130, 148
Awantjish Formation, 110
Aylmer Pluton, 95

backfolds, 64
Badger Bay area, 32
Baie Verte, 130
Baie Verte–Brompton Line, 28, 86, 87, *106*, 117, 136
Baie Verte Flexure, 128
Baie Verte Peninsula, 128, 129, 139, 146
Bailiella, 44
Balmoral Group, 118
Baltic Realm, 44, 47
  faunas, 45
Baltica, 17
basalt, 31, 71, 80
  pillow, 31, 69, 71
basement
  Cambro-Ordovician, *106*
  crystalline, 136
basins
  back-arc, *31*, 37, 96
  depositional, 80
  filling, *112*
  forearc, 86, 106
  foreland, 136, 153, 154, *156*, *159*, 161
  intraplate, 77
  marine, 113
  ocean, 35, 37, 68, 77, 80
  sedimentary, 72
  See also specific basins
Baskahegan Lake Formation, *69*, 74
Bassin Nord-Ouest fault, 115
batholiths, 129
Battery Point Formation, 111
Bay du Nord Group, 129
Bay of Islands complex, 141, 144
Bay St. George basin, 136
Bear Pond Rhyolite, 138
Beauce area, 85, *88*, 93, 95, 96, 97
Beauceville Formation, 89, 91, 95
bedding, *58*
Belchertown pluton, 64
Belle Isle fault, 8
Belleoram Granite, 124
belts, *7*, 17, 20, 68
  supracrustal, *69*

Benner Hill belt, 8, 9, 17
bentonite beds, 111
Berwick Formation, 11, 52
Berwick quadrangle, 60
Bethlehem pluton, 15
Big Berry Mountains syncline, 116
Bigelow Brook fault complex, 10
Billings Fold, 56
biogeography
  analysis, *43*
  Cambrian, *41*, *44*, *47*
  Carboniferous, *41*, *44*, *47*
bioherms, 109
biostromes, 109
biotite, 58, 129, 139, 146, 148
Black-Cape Volcanics, 109
blocks, 20, 35, 67, *78*, 130, 131
  basement, *78*
  continental, *32*
Blue Hills Nappe, 60
Blue Ridge, 14
Bonaventure Formation, 109
Bonnie Bay region, 144
Boston area, 44
Boston Avalon block, 11
Boston Basin, 28
Boston terrane, 19
Botwood Group, 32, 34, *35*, 37, 128
boudins, 58
Boundary Mountain anticlinorium, 6
Boundary Mountain antiform, 75
Boundary Mountain block, *78*, 80
Boundary Mountain terrane, 6
Bowers Mountain Formation, 69, 74
brachiopods, 107, 109
Bras d'Or terrane, 130
breccia, 31, 76, 87, 114
Brevard fault zone, 14
British Caledonides, *12*
British Isles, 13, 17
Brompton-Cameron Terrane, 52
Bronson Hill anticlinorium, 6, 52, 60
Bronson Hill belt, 6, 8, 10
Bronson Hill island arc, 6
Burgeo pluton, 129
Burgeo type, 130
Burnt Jam Brook Formation, 110, 111
Byrne Cove Mélange, 33

Cabot Fault System, 136
Calais Formation, 69, 71
calcarenites, 75, 109, 111, 112
calcilutites, 107, 109, 110
calc-mylonite, 94
calc-silicate, 58
calderas, 129
Caldwell Group, 87, 96
Caledonian, late, 2
Caledonian Belt, 56
Caledonian orogeny, 13
  defined, 2, 17

Cambrian, *41*, *44*, *69*, *72*
Cambro-Ordovician, *106*
Campbell Hill–Nonesuch River fault
    zone, 52, 60
Canadian Appalachians, 16, 17, 102
Canterbury Basin, 47
Cape Breton Island, 11, 44
Cape Elizabeth Formation, 9
Cape George area, 45
Cape Ray, 124, 128, 129
Cape St. John Group, 128, 129
Caradocian, *69*, 72
carbonates, 56, 136, 138, 141, 144, 156,
    160
    platform, 145
    sedimentation, 48
Carboniferous, *41*, *46*
Cardigan Pluton, 63
Carolina block, 19
Carolina Slate Belt, 44
Carolina terrane, 19
Carrabassett Formation, 75
Carys Mills Formation, *75*
Casco Bay area, 9
Casco Bay belt, 8, *9*
Casco Bay Group, 9
Cashaqua Member, 157
Castine Formation, 76
Castor River, 143
Catamaran fault, 117
Catskill Delta, 48, 156, 159, 160
Celtic unit, 45
Central block, *78*, 80, 126, 129, 130
Central Maine basin, 80
Central Maine Boundary Fault, 75
Central Maine Terrane, 52, 56, 60, 62
Central Mobile Belt, 11, 15, 17, *28*, 35,
    37
Central New Hampshire anticlinorium,
    52, 56, 60, 62, 63
Chain Lakes Massif, 6, 72, 78, 80, 96
Chalceurs Group package, 108
Chaleur Bay, 8, 119
Chaleur Bay succession, 8
Chaleurs Bay synclinorium, 102, *108*,
    111, 113, 114, 116, 117, 118
Chaleurs Group, 102, 107, 108, *109*,
    111, 113, 118
Chanceport Group, 31, 32, 33
Change Islands, 33, 34, 35
Charlotte belt, 14
Chase Brook Formation, 69
Chaudière River area, 93
cherts, 71
Chinese Realm, 44
chlorite, 75
Cincinnati arch system, 153
Cinq Isles Formation, 124, 130
Clam Bank Group, 123, 128, 130, 138,
    144, 148
clastics, 56, 85
clasts, 58, 73, 74, 75, 107, 110
claystone, 107, 110
cleavage, 2, 5, 8, 10, 12, 16, 33, 62, 85,
    87, 89, 90, 91, 95, 96, 110, 116,
    119, 141, 143
Clemville Formation, 109
Cleveland Shale, 160

Climacograptus spiniferus Zone, 108,
    118
climatic gradients, global, *46*, 48
Clinton-Newbury fault, 11
Cloridorme Formation, 118
coal swamps, 47
Coastal Acadia, *43*, 46, 47
Coastal Maine zone, 11
Coastal Volcanic belt, *76*, 80, 81
Cobequid-Chedabucto fault system,
    117
collision, 13, 17, 18, 19, 20, 118, 119
    continent-continent, 20, 28
    continent-volcanic arc, 96
    island arc, 85
Concord, New Hampshire quadrangle,
    56, 60
Concord plutonic suite, 14
    early, 63
Concord Tectonic Zone, 56
Coney Head, 144
conglomerate, 34, 58, 69, 73, 74, 75, 76,
    93, 107, 109, 110, 111, 113, 130
    assemblage, 111
Connecticut Valley–Gaspé belt, 5, 6,
    75, 80
Connecticut Valley–Gaspé–Notre
    Dame belt, 6
Connecticut Valley–Gaspé
    Synclinorium, 5, 86, 102, 106, *110*,
    113, 114, 115, 116
Connecticut Valley belt, 6
Connecticut Valley trough, 5
conodonts, 107, 109
continental crust, 119
continental drift, 13
continental margin, 76, 130, 136, 161
convergence, 80, 118, 162
Cookson Group, 69, 71
Cookson Island, 76
corals, 34, 46, 74
Cornish nappe sequence, 60
Costistrophonella punctulifera, 48
Cottrell's Cove Group, 31
cover, subaerial, *32*, 35
Cranbourne Formation, 96
crust
    continental, 119
    lower, 11
    oceanic, 19, 31, 36, 67, 78, 80, 86,
        136
    rupture, 154
    thickening, 145
    upper, 78
Cumberland fault, 89, 91, 95
currents, turbidity, 8
Cushing Formation, 9

Daggett Ridge Formation, 74, 75
Dalejina sp., 48
DalHousie Formation, 109
Danforth quadrangle, 75
Dashwoods subzone, 126, 129
Deblois granite pluton, 8
debris, 81
décollement, 60
Deer Lake, 145
Deer Lake Basin, 138

deformation, 8, 10, 11, 52, 72, 116, 119,
    124, 125, *127*, 130, 141, 143, 148
    Acadian, *51*, *85*, 95
    Alleghanian, *136*
    episodes, 87
    Late Ordovician, *114*
    mechanisms, 56
    mid-Devonian, *114*
    periods, *60*, 64
    regional, 28, *85*
    Taconian, *136*
    timing, *63*, *101*
Delmar deformation, 19
detritus, 73, 75, 81, 130, 138, 160
Devils Room granite, 129, 138
Devonian, *46*
    Early, *56*, *73*
    mid-, *101*, *114*
Dicoelosia sp., 48
Digdeguash Formation, *76*
disconformity, 47
discontinuities, 60, 78
Dog River fault zone, 6
domains
    external nappe, 106
    internal nappe, 106
    parautochthon, 106
    tectonic, 106
    tectonostratigraphic, 110
Dome-Crescent-Mushroom pattern, 63
Dover Fault, 28, 36, 128
Dunn Point Formation, 33
Dunnage belt, 11
Dunnage Zone, 5, 17, 85, 87, 96, 97,
    102, *106*, 109, 117, 118, 126, 128,
    129, 130, 136, 147
dykes, felsic, 114

East Pond suite, 139, 148
Eastern Americas Realm, 46, 47
Eastern Townships, 106
Eastport, Maine, 76
Eastport Formation, 47, 48
elements, light rare earth (LREE), 80
Eliot Formation, 52
Ellen Wood Ridge Formation, 73, 74,
    75
Ellsworth belt, 8
Emsian, *56*
erosion, 13, 111, 160
Escuminac Formation, 48, 102, 108,
    110, 113
Esopus shale, 156
European Province, 46
European Realm, 44, 45
Eustis domain, *87*
evolution
    Ordovician, *27*
    Silurian, *27*
    tectonic, 68
Exploits-Meelpaeg subzone boundary,
    128
Exploits Subzone, 35, 126, 128, 130
extension, 80

facies
    amphibolite, *128*, 138, 141, 146

carbonate, 102
greenschist, *128*, 138, 139, 146
siliciclastic, 102, 110
subgreenschist, 146
Famine Formation, 93, 94
faults, *114*
bedding-parallel, 33
extensional, 144, 145
normal, 73, 114, 119
oblique-slip, 115
premetamorphic, 8
reverse, 87, 88, 97, 114, 115, 127
sinistral, 18, 19, 20
strike-slip, 11, 73, 93, 95, 97, 114,
117, 119
thrust, 52, 60, 73, 75, 89, 93, 115,
141, 144
transcurrent, 7, 13, 14, 17, 20, 67, 68,
119
See also specific faults
fauna, 108
Celtic, 29, 32
Silurian, 43
Tremadocian, 45
Findlay arch, 154
Fleur de Lys Supergroup, 129, 139, 148
Fleurant Formation, 102, 108, *110*, 113
floatstone, 93
Flume Ridge Formation, 76
flysch, 47, 110
folding, 33
folds, 8, 10, 16, 35, *52*, 60, 85, 87, 90, 95,
116, 118, 141, 142
map-scale, *63*
foliation, 87, 141
Forillon Formation, 111
Fortin Group, 102, 110, 111
Fortune Bay, 124, 130
Fortune Harbour Peninsula, 32
fossils, 32, 34, 43, 87
marine, 44, 47, 109
Silurian, 14
France, 2
Fredericton belt, 8, 72, 73, *76*, 78, 82
Fredericton–Central Main belt, 8, 10
Fredericton Trough, 47, 72, 73, 77, 81
Frenchville Formation, 75
Frontière Formation, 95

gabbros, 8
Gander-Avalon zone boundary, 128
Gander belt, 11
Gander Group, 11
Gander Lake Subzone, 126, 128
Gander River Ultramafic Belt, 36
Gander Terrane, 60
Gander zone, 17, 35, 102, 126, 128,
129, 130, 147
Garin Formation, 107, 108, 118
garnet, 58
Gascons Formation, 109, 110
Gaspé Belt, *102*, 114, 116, 118, 119
basement, *117*
sedimentary history, *111*
stratigraphy, *106*
Gaspé Peninsula, 5, 93, 95, 97, 103, 110,
114, 115, 118, 119
regions, 114

Gaspé Peninsuladian orogeny, *101*
Gaspé Sandstones, 111
Gaspé Sandstones Group, 102, 108,
112, 113
Gaspé segment, 102
Gaspésie, 96
Gastonguay anticline, 110, 116
Gastonguay fault, 115
Gedinnian, 56
Genesee Formation, 157, 159
Genesee shale, 154, 161
geology, structural, *58*
geometry, *63*
Gilmanton quadrangle, 63
glaciation, 160
gneiss, 10, 15, 19, 129, 139
gneiss domes, 56, 63
Goldson Formation, 128
Gondwana, 17
Gondwana Realm, 45
Gondwanaland, 161
Gordon Sandstone, 160
Gordon siltstones, 160
graben, continental, 8
gradients
climatic, 42
latitudinal diversity, 42
Grampian terrane, 13
Grand Banks, 128, 130
Grand Lake, 148
Grand Lake thrust, 128, 141
Grand Pabos fault, 106, 107, 114, *115*
Grand Pitch Formation, 69, 75
Grande Rivière fault, 106, 115, 116
granites, 12, 14, *15*, 123, 124, 138, 148
Acadian, *15*
granitoids, 8, 14, 15
granofels, 58
graptolites, 45, 69, 107, 109, 113, 118
gravels, 113
graywackes, 8, 34, 35
Great Bay de l'Eau Formation, 124
Great Northern Peninsula, 146
Great Slave Lake area, 158
Greater Acadia, *4*, 43, 44
Grenville block, *78*, 80, 126
Grenville-Munsungun-Weeksboro-
Miramichi terrane, 73
Grenville orogeny, 154
Griffon Cove River Member, 111
grit, 58
Grog Brook Group, 112, 118
Gulf of St. Lawrence, 11, 143
Gull Lake Intrusive Suite, 138
Gypidula coeymanensis, 48

Hamilton Group, 48, 156
Hampshire Formation, 160
Hare Bay allochthon, 28, 36, 136, 143,
144
Hare Bay region, 138, 143
Hartin Formation, 47, 48
Hebron Gneiss, 10
Hedeina (Macropleura) macropleura,
48
Helderberg basin, 155
Helderberg Group, 155
hematite, 34

Hermitage Flexure, 128
Hirnantia, 47
history, accretionary, *27*
Honorat Formation, 5
Honorat Group, 102, *107*, 112, 116, 118
Honorat-Matapédia Group package,
108
Hope Valley block, 19
hornblende, 139, 146, 148
Horton Group, 48
hot spots, 15, 56, 63
Houlton area, 75
Humber Arm Allochthon, 28, 36, 128,
136, *141*, *143*
Humber Arm region, 138
Humber Zone, 15, *28*, 85, 96, 102, *106*,
110, 117, 118, 126, 127, 128, 129,
130, *135*, 144, 147, 148
Humboldt Current, 43
Hungry Grove, 129
Huntersville Chert, 156
Huron Shale, 158, 159
Hurricane Mountain belt, 7
Hurricane Mountain line, 7
Hurricane Mountain mélange, 71
hydrothermal alteration, 114

Iapetus I, 36, 37
Iapetus II, 36, 37, 118
Iapetus basin, 36
Iapetus Ocean, 13, *17*, 28, 29, 36, 72,
102, 161
Iapetus suture, 13, 17
Iberian peninsula, 2
Illinois basin, 159
eastern, 156
illite crystallinity, 145
Indian Cove Formation, 111
Indian Head Complex, 139, 143
Indian Point Formation, 109, 110
Inner Piedmont, 14
intrusion, 124
intrusive bodies, 15
Iowa platform, 160
Ireland, 47
iron ores, 46
island arc, *31*, 68
island blocks, 17
Isleboro-Rockport belt, 8
isograds, 15

Java Formation, 157
Java shales, 143, 159, 161
Jessamine dome, 154
Joey's Cove mélange, 33

Kankakee arch, 158
Kaskaskia II, 160
Kearsage–Central Maine belt, *75*, 77
Kearsage–Central Maine synclinorium,
8, 10, 51, 52, 60
Kellyland quadrangle, 69
Kendall Mountain Formation, 69, 71
kinematics, *93*
King George IV Lake area, 32
Kinsman pluton, 15
Kinsman suite, 63
Kittery Formation, 52

Knoydart Formation, 48
Kootenia, 44
Kossuth Group, *69, 73*
kyanite, 19, 129

La Garde Formation, 102, *110*, 113
La Guadeloupe fault, 85, *87*, 89, 91, 97,
    115
    age constraints, *93*
    kinematics, *93*
La Poile Group, 128
La Vieille Formation, 109
Lac à la Raquette structural domain,
    90
Lac Aylmer Formation, 95
Lac Fortin structural domain, *89*
Lac Lambton Formation, 95
Lac Lanigan structural domain, *90*
Lac McKay Volcanics, 111
Laforce Formation, 111
Lake Char fault, 11
Lake District, 47
Lake Matapédia syncline, 110
Lake Temiscouta area, 46
Laurentia, 17, 32, 35, 161
Laurentian craton, 17
Laurentian Realm, 44, 45
lava flows, 76, 109
Lawler Ridge Formation, 76
Lebanon Antiformal Syncline, 52, 62
Leptaena rhomboidalis, 48
leucogranites, 129
Levenea subarinata, 48
Lewis Hills, 144
Lewis Hills massif, 144
Ligerian belt, 2
Ligerian orogeny, 18, 19
limestones, 44, 58, 75, 93, 107, 109, 110,
    111, 112, 113
linkages, sedimentary, 77
lithosphere, 136
Little Harbour Formation, 31
Little Port complex, 144
Littleton Formation, 52, *56*, 58
Llandoverian, *58*
Lobster Cove–Chanceport Fault, 31, 35
Lobster Mountain anticlinorium, 6
Logan's Line, 106, 127
Long Point Group, 141, 144
Long Range Complex, 139
Long Range dykes, 139
Long Range Front, 127, 143
Long Range inlier, 129, 142, *143*, 146
Long Range thrust, 128, *143*, 146
Long Reach region, 44
Lower Kaskaskia II, 160
Lower Rangeley Formation, 58, 60
Lubec–Belle Isle fault, 8
Lucerne intrusive, 15
Ludlovian, *58*

Madawaska County, New Brunswick,
    48
Madeleine River area, 110
Madrid Formation, *58*, 75, 76, 82
Magdalen basin, 11
magmatism, 16

magnetite, 31
magnetization, 31, 32
Magog Group, 85, 86, 87, 88, 95, 96, 97
Mahantango shale, 153, 156, 157, 161
Maine
    eastern, *67, 79*
    western, 8
Maine Appalachians, 17
Maine belts, 11
Maine flysch belt, 8
Malbare Formation, 111
Mann Formation, 109
Mann Mountain anticline, 154
mantle, 11, 86, 136
Mapleton Formation, 48
Maquereau Group, 106, 109, 117
Maquereau-Mictaw inlier, 106, 108,
    116, 117, 118
marble, 9
Marcellas shale, 153, 154, 156, 157, 161
Marcil Nord fault, 115
margin, continental, 76, 130, 136, 161
Maritime Canada, 2
Martic line, 13
Mascarene belt, 15, 17
Mascarene Group, 11
Mascarene peninsula, 76
Massabesic Gneiss, 10
Massabesic Gneiss Complex, 52
Massabesic-Merrimack sequence, 52,
    63, 64
Matapédia area, 115, 116
Matapédia Group, 102, 106, *107*, 110,
    112, 113, 118
Matapédia syncline, 110
Mattawamkeag Lakes quadrangle, 75
McCrea, mélange, 106, 107, 108
McKenzie Formation, 154
Meelpaeg division, 129
Meelpaeg subzone, 126, 129
megafossils, 46
Megakozlowskiella sp., 48
mélange, 19, 33, 71, 80, 128, 136
    See also specific mélanges
Meristella sp., 47, 48
Merrimack Belt, 51
Merrimack Group, 10, 52
Merrimack synclinorium, 51
Merrimack trough, 8, 11, 17, 18
Merrimack trough sequence, *10*
metabasalts, 87
metaclastics, 128
metagraywackes, 8
metalimestones, 58
metamorphism, 7, 11, 12, 14, *15*, 19, 56,
    63, 71, 114, *128*, 130, 136, *145*, 148
    timing, *63*
metapelite, 56
metasandstones, 9
metaturbidite, 56
Mexa arch, 161
Mic Mac Lake Group, 128
mica, 46, 58
microplates, 20
Mictaw Group, 96, 106, 107, 109, 118
Middle Ridge pluton, 129
Middle Ridge type, 129, 130

Midland Valley terrane, 13
migmatites, 60
Mill Priviledge Brook Formation, *73,
    75*
Millstream fault, 117
Mira Terrane, 130
Miramichi belt, 8, *69*, 72, 77, 78, 80, 81,
    82
Miramichi terrane, 47
models
    Acadian events, 68
    pre-Acadian events, 68
    structural, *52*
    tectonic, 69, *161*
molasse, 130
monazite, 64
Monroe fault, 52
Monroe line, 6
Mont Albert Complex, 106
Mont Albert nappe, 106
Mont Serpentine inlier, 106
Montreal region, 46
Moreton's Harbour Group, 31, 33
Morisset structural domain, *89*
Morocco, 2
Mount Alexander-Pellegrin area, 110
Mount Cormack division, 129
Mount Cormack subzone, 126, 129
Mt. Katahdin intrusive, 15
Mt. Kearsage quadrangle, 60
Mt. Kearsage summit, 63
Mt. Monadnock, 56
Mount Peyton, 129
Mount Peyton pluton, 129
Mountain House Wharf Formation, 93
muds, 113
mudshales, 107
mudstone, 76, 107, 109, 110, 111
Munsungun-Winterville belt, 69, 72, 80
Murphy Creek Formation, 107
Murphy Creek inlier, 106, 117
muscovite, 139, 146, 148
mylonite, 14, 94, 128
mylonitization, 10, 124

Nadeau ophiolitic mélange, 106
nappes, 52, 56, 60, 64, 96
    domain, 107, 109
    emplacement, 118
Nashville dome, 154
Natlins Cove Formation, 138
Neckwick Formation, 107, 108
Needmore-Esopus basin, 157
Needmore shale, 153, 156, 161
Neseuretus, 45
New Albany Shale, 159
New Brunswick, 5, 8, 15
New Brunswick Group, 112
New Hampshire, 8
    central, *51*
New Hampshire anticlinorium, central,
    64
New Mills Conglomerate, 111
New Mills Formation, 109
New Richmond area, 109
New World Island, 31, 33, 128
New York, 5
Newburyport Quartz-diorite, 10

Newfield quadrangle, 58
Newfoundland, 5, *11*, *123*
 central, 138
 southern, 138
 west, *135*
Newfoundland Appalachians, 17, 28, 35, 37, 123, 148
Newfoundland Reentrant, 130
Nonesuch River fault, 8, 10
North American block, 80
North American Platform, 47
North Atlantic Region, *1*, *2*, 45
North Brook anticline, 141
North-Central Maine belt, 7, 8
North Silurian Realm, 45
Northern Outcrop Belt, 110, 111, 113
Norumbega Fault zone, 8, 69, 71
Notre Dame Bay area, 29, 32, 33, 128
Notre Dame subzone, 35, 106, 126, 128, 129
Nouvelle River area, 110, 111
Nova Scotia, 5, 11, *12*
Nucleospira sp., 48

Oak Bay Formation, 76
Obturamentella sp., 48
Ocmulgee fault, 19
Ohio shale, 154, 160, 161
Old Mans Pond, 145
Old Red Sandstone, 13, 16
Old World Realm, 46, 47, 48
Oldhamia-bearing turbidites, 69
Olentangy Formation, 157
Onondaga Formation, 156
Onondaga Group, 48
Onondaga Limestone, 155, 157, 158
ophiolite, 28, 36, 85, 86, 96, 136, 144
Ordovician, *27*, *44*
 evolution hypothesis, *27*
 Late, *101*, *114*
Oriskany Sandstone, 155, 156
orogenesis, mid-Paleozoic, *1*
orogeny
 defined, 2, 135
 Gaspé Peninsuladian, *101*
 salinic, 138, 148
orthogenesis, 138
orthoquartzites, 71
Oswayo Member, 160
Owl Capes Member, 111

Pabos Formation, 107
Pacific Realm, 45
Pacquet Harbour Group, 129
paleogeography, *27*, *29*
paleolatitudes, 32, 33, 36
paleomagnetic data, *29*
paleopoles, 32
Paleozoic, mid-, *1*
Parsons Pond thrust, 143
Patagonian Andes, 80
Patch Mountain member, 58
Paxton Formation, 10
Peace River arch, 159
pegmatites, 10
pelites, 8, 69, 71, 75
penguins, Galapagos, 42
Penobscot Bay area, 76

Penobscot orogeny, 71, 72, 80
Penobscotian Disturbance, 45
Percé area, 106, 107, 110
peridotite, 106
Permian, *46*
Perry Formation, 48
Perry Mountain Formation, *58*
Phillips Brook anticline, 141, 144
phyllites, 87
Pine Mountain fault, 154
Pipestone Pond Complex, 36
Pirate Cove Formation, 102, 110, 113
Piscataquis volcanic belt, 8, 80
Placentia Bay, 129
plagioclase, 58
plagiogranite, 87
plate convergence, 93
plate tectonics, *17*
Platyorthis sp., 48
Plicoplasia sp., 48
plutonism, *16*, 28, 97, *129*
 timing, *63*
plutons, 14, 15, 18, 47, 63, 77, 78, 79, 81, *129*
Pocomoonshine Lake Formation, 76
Pokiok Batholith, 47
Ponacook quadrangle, 60, 62
Pools Cove Formation, 124, 130
Popelogan anticline, 118
porphyroblasts, 62, 64
Port Albert Peninsula, 34
Port au Port Peninsula, 123, 127, 138, 144
Port-Daniel fault, 106
Portland Creek Pond thrust, 144
Prentiss Group, *73*
Prescott pluton, 64
Presque Isle, Maine, 16
Price Formation, 160
Pridolian, 58
prism, accretionary, 86, 96
progradation, 160
provincialism, 47
pulses, 2, 37, 113
pumice, 109
Putnam-Nashoba belt, 11
Putnam-Nashoba zone, 18
Pyrenees, 62
pyroclastics, 31

quartz, 107
quartz arenites, 34, 111
quartzites, 9, 56, 58
quartz-wacke, 107
Quebec, 5
Québec Appalachians, southwestern, *85*
Québec Eastern Townships, 115
Québec Reentrant, 111, 117

Randell's Head, 35
Rangeley, Maine, 56
Rangeley Formation, 56, *58*
Raudin volcanics, 114
reconstruction, palinspastic, 118
red beds, 32, 46, 109, 111, 113, 128, 130
Red Island, 129
Red Rock Cove, 35

reef carbonates, 158
reefs, 75
regression, glacio-eustatic, 160
Restigouche fault, 106, 110, *115*
Restigouche syncline, 108, *109*, 111, 113, 115, 116, 118
Restigouche Volcanics, 109
Rhenish-Bohemian Region, 47, 48
Rhenorensselaeria, 47
Rhinestreet, Angola, 159
Rhinestreet section, 157
Rhinestreet Shale, 157
rhyolite, 31, 138
Richmond belt, 9
rifting, 8
Rivière fault, 106
Rivière Garin fault, 115
Rivière Noire fault, 91
Rivière Port-Daniel Mélange, 106, 107
Robert's Arm Group, 32, 33, 37
rocks
 Ashgillian, *73*
 calc-silicate, 9
 clastic, 12, 73
 country, 15
 cover, 7, 15, 56, 77, 80
 Early Devonian, *73*
 folded, 52
 granitic, 78
 igneous, 78, 136, 138
 intrusive, 110
 mafic, 31, 87
 marine, 128
 metamorphic, 17, 138
 metasedimentary, 52, 56, 58, 60, 87, 118, 128
 ophiolitic, 144
 pelitic, 69, 73, 75
 plutonic, 68, 69, 103
 pyroclastic, 109
 sedimentary, 47, 68, 69, 109, 128, 136, 138, 148
 siliclastic, 43, 110, 111
 supracrustal, 69, 82
 volcanic, 9, 11, 31, 37, 47, 68, 69, 73, 76, 80, 87, 103, 109, 110, 118, 119, 128, 130, 138, 148
 volcanogenic, 9
Rocky Brook fault, 117
Rome trough, 154
Roncelles Formation, 110
roof duplex, 144
rotations, 20, 91, 93, 95, 116
Round Head thrust, 127, 138, 141, 144
Rye Formation, 10

Saint-Benjamin fault, 91, 95
St. Clair fault, 154
St. Croix belt, 8, 9, *69*, 76, 77, 82
St. Croix–Ellsworth belt, 8
Saint-Daniel argillites, 96
Saint-Daniel Mélange, 85, 86, 87, 96
Saint-Francis Group, 87, 88, 95, 97
St. George area, 46
St. George batholith, 15
Saint-George-de-Beauce, 93
St. George fault, 9
St. Helen's Island, 46, 48

St. Jean River anticline, 110, 111, 116
Saint John, New Brunswick region, 44, 45
Saint John Group, 71
St. Julian Island, 138
St. Lawrence Promontory, 111, 117, 119, 130
St. Lawrence River estuary, 11
Saint-Leon Formation, 111
St. Margaret Bay thrust, 143, 146
Saint-Victor Formation, 89, 93
Saint-Victor Synclinorium, 86, 87, 90
Sainte-Florence fault, 110, 115
Salina Formation, 155
Salina salts, 154
Salinic disturbance, 17, 113
Salinic unconformity, 116, 119
salt diapirs, 63
Sam Rowe Ridge Formation, 73, 74
Samson Formation, 128
sand, 113
sandstone, 34, 74, 75, 76, 87, 107, 109, 110, 111, 112, 160
Sangerville Formation, 58, 76
Sayabec Formation, 110
Scandian orogeny, 17
Scandinavia, 2
schist, 8, 9, 56, 58, 64, 87, 94, 107, 129
schistosity, 60, 62, 87
scoria, 109
sea level, fall, 158
seamounts, 29, 32, 33
Sebago block, *78*, 80, 82
Sebago pluton, 15
Seboomook Formation, 8
Seboomook Group, 75
sedimentation, 8, 72, 76
    cyclic, *113*
sediments
    carbonate, 8
    marine, 6
Sellarsville fault zone, 115
Sellarsville slice, 109
shales, 69, 71, 73, 75, 76, 136, 153, 157, 160
Sherbrooke area, 85, *87*, 89, 93, 95, 96
Sherbrooke domain, *87*
Shickshock Group, 106
Shickshock-Sud fault, 106, 110, 115
Shiphead Formation, 111
Siberian Realm, 44
siliciclastics, 46, 48, 109
sillimanite, 129
sillimanite zone, 63
sills, 107
siltstone, 76, 107, 110, 111
Silurian, *27, 45*
    Early, *58*
    evolution hypothesis, *27*
    Late, *58*
Silurian-Devonian succession, 5
Skinner Cove volcanics, 36
Skitchewaug Fall Mountain sequence, 60
slate, 12, 75
slate belt, 110, 116
Smalls Falls Formation, 8, *58*, 63
Smyrna Mills quadrangle, 75

soapstones, 56
Sonyea Formation, 157
Sonyea shale, 154, 159, 161
Sops Arm Group, 129, 136, 138, 148
Sources Formation, 110
South-Central Maine belt, 9
Southeastern Province, 46
Southern Irumbide Belt, 56
Southern Uplands, 13
Southern White Bay allochthon, 136
Spaulding suite, 63
Springdale Group, 34, 128, 138
Springfield quadrangle, 75
Squameofavosites, 75
Stapeley Formation, 33
Stetson Mountain Formation, 69, 74
Stoke domain, *87*
Stony Lake volcanics, 128
Strait of Belle Isle, 130
stratigraphy, *56, 153*
    Gaspé Belt, *106*
structures
    contractional, *141*
    extensional, *144*
subduction, 37, 82
subsidence, 97, 157, 160
Summerford Gruop, 32, 33
Sunbury Shale, 160
Suncook quadrangle, 60
superdomain, structural, *89*
supracrust belts, *77*
suture, 13, 68
syncline, 63, 143
synclinorium, 51
synthesis, *32*

Table Mountain anticline, 141
Taconian, *95*
Taconic orogeny, 28, 72, 78, 80, 85, 93, 96, 97, 102, 103, 106, 113
Taconic unconformity, 103, 109, 113, 118
Taghanic onlap, 158
Taghanic unconformity, 153, 156, *157*, 161
Talladega belt, 15
Taylors Pond Conglomerate, 138
tectonic domains, 106
tectonic evolution, 68
tectonic hinge, 8
tectonic quiescence, 154
tectonic regime, defined, 2
tectonic transport, 56
tectonics, 19, *27, 35*, 85, 97, 116
    pop up, 56
    regional, *161*
tectonism, 12, 68, 118
    Acadian, *77*
tectonostratigraphic data, *33*
tectonostratigraphy, *33*
Témiscouata area, 93, 95
Témiscouta Formation, 48
Ten Mile Lake thrust, 143, 146
terrane
    displaced, 28
    exotic, 52, 67, 68
    suspect, 9
Tetagouche Group, 8

Thetford Mines ophiolites, 87
tholeiites, 80
Three Lick Bed, 159, 160
thrust fault obduction, 154
thrusts, low-angle, 64
tillites, 47
Tioga ash fall, 156
Tobique volcanic belt, 8
Topsails plutons, 129
Tornquist ocean, 19
Touladi Formation, 93
trilobites, 43, 44, 45
triple junction, 17
Troisième Lac fault, 115
Tryplasma, 75
tuffs, 9, 87, 107
turbidites, 8, 34, 56, 69, 71, 76, 112, 113
Turtle Head fault, 11

ultramafics, 31
ultramylonite, 94
unconformity, 2, 5, 16, 69, 102, 103, 109, 111, 112, *113*, 118, 128, 153, 161
uplift, 13, 48, 78, 153, 154, 160, 161
Upper Gaspé Limestones, 110
Upper Gaspé Limestones Group, 102, *111*, 113

Val-Brillant Formation, 110
Valley and Ridge, 14
Variscan Belt, 56
Vassalboro Formation, 8
Vermont Spear, 6
Virginia Piedmont, 14, 156
volcanic arc, 35, 96
volcanic suites, 80
volcaniclastics, 31
volcanics, 8, 9, 31, 32, 34, 56, 87, 109
    intra-oceanic, *31*
volcanism, 71, 72, 80, 114, 129

wackes, 69, 73, 74, 75, 76, 107, 111
Waite quadrangle, 69
Wallbridge unconformity, 153, *155*, 161
warping, 96
Watt Mountain unconformity, 158
Waverly arch, 154, 157, 158, 161
Waweig Formation, *76*
weathering, 32, 34
wedge
    accretionary-tectonic, 96
    basement, 11
    clastic, 14, 19
Weedon area, 95
Weeksboro–Lunksoos Lake belt, 69, 72, 75, 80
Wenlockian, *58*
West Falls Formation, 157
West Falls shale, 154, 159, 161
West Point Formation, 111
Western Head Formation, 31, 107
White Bay, 129, 138, 144
White Bay allochthon, 138, 144
White Rock Formation, 47
Whites Arm window, 144

Wild Goose Grits, 58
Windsor Group, 47
Windsor Point Group, 124, 128, 129, 130
Winnipesaukee quartz diorite, 63
Wold Cove Pond Intrusive Suite, 138

Wolfeboro quadrangle, 58
Woodland Formation, 69
Wytopitlock quadrangle, 75

York Lake Formation, 111

zones
  detrital, *56*, 60, 64
  structural, 102
  tectonostratigraphic, 102
Zoophycos-rich assemblage, 109

Typeset by WESType Publishing Services, Inc., Boulder, Colorado
Printed in U.S.A. by Malloy Lithographing, Inc., Ann Arbor, Michigan